RADIATION T[illegible]OGY FOR ADVANCED MATERIALS

Radiation Technology for Advanced Materials

From Basic to Modern Applications

Edited by

GUOZHONG WU

Shanghai Institute of Applied Physics, Chinese Academy of Sciences, Shanghai, P.R. China

MAOLIN ZHAI

College of Chemistry and Molecular Engineering, Peking University, Beijing, P.R. China

MOZHEN WANG

Department of Polymer Science and Engineering, University of Science and Technology of China, Hefei, P.R. China

ELSEVIER

ACADEMIC PRESS

An imprint of Elsevier

SHANGHAI JIAO TONG UNIVERSITY

Academic Press is an imprint of Elsevier
125 London Wall, London EC2Y 5AS, United Kingdom
525 B Street, Suite 1650, San Diego, CA 92101, United States
50 Hampshire Street, 5th Floor, Cambridge, MA 02139, United States
The Boulevard, Langford Lane, Kidlington, Oxford OX5 1GB, United Kingdom

British Library Cataloguing-in-Publication Data
A catalogue record for this book is available from the British Library

Library of Congress Cataloging-in-Publication Data
A catalog record for this book is available from the Library of Congress

ISBN: 978-0-12-814017-8

For Information on all Academic Press publications
visit our website at https://www.elsevier.com/books-and-journals

Publisher: Matthew Deans
Acquisition Editor: Glyn Jones
Editorial Project Manager: Naomi Robertson
Production Project Manager: Debasish Ghosh
Cover Designer: Matthew Limbert

Typeset by MPS Limited, Chennai, India

CONTENTS

LIST OF CONTRIBUTORS

Yinyong Ao
College of Chemistry and Molecular Engineering, Peking University, Beijing, P.R. China

Pei Dai
College of Chemistry and Molecular Engineering, Peking University, Beijing, P.R. China

Xuewu Ge
Department of Polymer Science and Engineering, University of Science and Technology of China, Hefei, P.R. China

Yan Hao
College of Chemistry and Molecular Engineering, Peking University, Beijing, P.R. China

Jiangtao Hu
Shanghai Institute of Applied Physics, Chinese Academy of Sciences, Shanghai, P.R. China

Jingye Li
Shanghai Institute of Applied Physics, Chinese Academy of Sciences, Shanghai, P.R. China

Rong Li
Shanghai Institute of Applied Physics, Chinese Academy of Sciences, Shanghai, P.R. China

Weihua Liu
Shanghai Institute of Applied Physics, Chinese Academy of Sciences, Shanghai, P.R. China

Jun Ma
College of Chemistry and Molecular Engineering, Peking University, Beijing, P.R. China

Jing Peng
College of Chemistry and Molecular Engineering, Peking University, Beijing, P.R. China

Mouhua Wang
Shanghai Institute of Applied Physics, Chinese Academy of Sciences, Shanghai, P.R. China

Mozhen Wang
Department of Polymer Science and Engineering, University of Science and Technology of China, Hefei, P.R. China

Shuojue Wang
College of Chemistry and Molecular Engineering, Peking University, Beijing, P.R. China

Yuhai Wang
EVE Rubber Institute Co., Ltd., Qingdao, P.R. China; Qingdao Key Laboratory for Advanced Tire Materials, Qingdao, P.R. China

Guozhong Wu
Shanghai Institute of Applied Physics, Chinese Academy of Sciences, Shanghai, P.R. China

Zhe Xing
Shanghai Institute of Applied Physics, Chinese Academy of Sciences, Shanghai, P.R. China

Lu Xu
Shanghai Institute of Applied Physics, Chinese Academy of Sciences, Shanghai, P.R. China

Chenguang Yang
Shanghai Institute of Applied Physics, Chinese Academy of Sciences, Shanghai, P.R. China

Ming Yu
Shanghai Institute of Applied Physics, Chinese Academy of Sciences, Shanghai, P.R. China

Xinmiao Zeng
College of Chemistry and Molecular Engineering, Peking University, Beijing, P.R. China

Maolin Zhai
College of Chemistry and Molecular Engineering, Peking University, Beijing, P.R. China

Cong Zhang
Shanghai Institute of Applied Physics, Chinese Academy of Sciences, Shanghai, P.R. China

Lin Zhang
EVE Rubber Institute Co., Ltd., Qingdao, P.R. China; Qingdao Key Laboratory for Advanced Tire Materials, Qingdao, P.R. China

Maojiang Zhang
Shanghai Institute of Applied Physics, Chinese Academy of Sciences, Shanghai, P.R. China

Wenli Zhang
Shanghai Institute of Applied Physics, Chinese Academy of Sciences, Shanghai, P.R. China

Zhicheng Zhang
Department of Polymer Science and Engineering, University of Science and Technology of China, Hefei, P.R. China

PREFACE

The interaction of a high energy ionizing beam with a substance leads to a variety of chemical reactions that may be followed by time-resolved measurements such as the pulse radiolysis technique, which is especially useful for understanding the reactions that occur in the liquid state. Radiation-induced chemical reactions result in the splitting or recombination of molecules for many substances, including water, gas, organic molecules, and polymers. The utilization of such radiation-induced reactions may lead to an improvement or better control of certain properties of the target material. Radiation technology has emerged as a valuable tool for the application of radiation reactions to a variety of disciplines. It has found a wide range of industrial applications, including environmental protection, food preservation, health care, and materials processing. Food irradiation and cable/wire cross-linking by radiation are widely applied in China and around the world. The application of radiation in polymer processing is very attractive and efficient, since radiation-induced radical reactions can be initiated in any type of polymer, independent of temperature and pressure. The high-power EB accelerator has assured the commercial utilization of radiation processing at a reasonable cost.

Although the possible applications of radiation technology are wide-ranging, it is not our aim to describe well-established radiation applications such as food irradiation in this book. Radiation technology can play a very important role in materials processing. In addition to cross-linking and degradation, radiation grafting will generate many new functional materials, because desirable functional groups can be anchored to the polymer material surface by this technology. We aim to provide a comprehensive survey of possible radiation applications in advanced materials, such as SiC fiber, carbon fiber, textiles, fiber adsorbent, polymer micropowder, polymer micro-spheres, polymer foam, emulsion adhesives, biomaterials, metal nanoparticles, graphene, and so on. The contributors are all from China and have extensive knowledge of radiation technology, and some have wide experience in radiation processing. More importantly, many of the studies described in this book provide clear guidelines for readers to make better use of the knowledge and master the skills. We believe that the knowledge provided in this book is valuable for scientists and engineers to understand the merit of radiation technology in advanced materials from fundamental radiation research to real industrial applications.

CHAPTER ONE

Radiation Sources and Radiation Processing

Guozhong Wu
Shanghai Institute of Applied Physics, Chinese Academy of Sciences, Shanghai, P.R. China

Ionizing radiation is radiation with sufficient energy that can move tightly bound electrons of atoms or molecules. When matter is subjected to ionizing radiation, energy is deposited on molecules and induces excitation, ionization, and final scission of chemical bonds. The matter can be in the form of gas, liquid, or solid. This radiation-induced chemical reaction is independent of temperature. Industrial radiation processing is usually performed at ambient temperature, since it is unnecessary to raise or lower the temperature. Radiation-induced chemical reactions in liquid can be followed using the time-resolved pulse radiolysis technique. This technique enables the observation of the formation and decay of radiation-induced solvated electrons, radical ions, or neutral radicals. Water is an exceptional liquid because it is present everywhere, including in the human body; and high-temperature water is used as a coolant in pressurized and boiling water nuclear reactors. Radiation chemical reactions in room-temperature water or high-temperature water have been investigated over different timescales for a very long time and a number of papers have been published. Solvated electrons, generated by water radiolysis, are strong reductive agents and many metal ions can be reduced to atoms. Utilizing this approach, many metal nanoparticles have been prepared by radiation-induced reduction in water. Irradiation of other liquid materials such as alcohols and paraffins also generates reductive solvated electrons and intermediate ion species, but these systems are normally investigated in basic research, since organic liquid is rarely used to produce high added-value products by radiation treatment. Radiation

Radiation Technology for Advanced Materials.
DOI: https://doi.org/10.1016/B978-0-12-814017-8.00001-9

effects in gas molecules (e.g., O_2, N_2, CO_2, Ar, He) are similar to those in liquid media, but the density difference is huge.

Since polymer material is normally composed of macromolecules containing C, H, O, and N atoms, ionization-induced scission of C–C, C–H, C–O, and C–N bonds leads to the formation of gaseous products (H_2, CH_4, CO, CO_2, etc.) and macromolecular radicals or radical ions. The gaseous products can be measured by gas chromatography. The radiation chemical yields (G values) of gaseous products are dependent on the type of polymer and the gas atmosphere. Radical species formed in polymer by irradiation can be determined by electron spin-resonance spectroscopy, which provides clear evidence of radiation effects. The lifetime of polymeric radicals is dependent on the backbone structure of the polymer as well as the temperature of storage. The characterization of radical species and their decay behavior is important for the understanding of radiation effects in polymer. Compared to other polymers, radicals in polytetrafluoroethylene (PTFE) are very stable over a long period at room temperature in air, and its lifetime may be longer than dozens of years at room temperature. Moreover, the radicals in PTFE are stable even at an elevated temperature, and it decays at temperatures higher than 200°C. Polyethylene (PE) is one of the most common plastics, and radiation cross-linking of PE has been investigated for a long time. Furthermore, this technology has been commercialized for many years. Radical species formed in PE by irradiation have relative short lifetimes when stored at room temperature in air. However, radicals in irradiated ultrahigh molecular weight PE fiber at room temperature have comparatively long lifetimes, due to their high crystallinity and high degree of orientation. To prolong the lifetime of macromolecular radicals after irradiation, polymer material is usually stored at low temperature. In contrast, to eliminate radicals formed in polymer by radiation and avoid postoxidation, polymer material is normally heated to a high temperature (annealing) because the decay of radicals is enhanced by raising the temperature.

When one attempts to apply radiation technology to the processing of polymers in an effort to manufacture high-performance material with desirable properties, at the beginning it is crucial to investigate the overall radiation effects, in terms of cross-linking or degradation. With the exception of some specific applications, radiation cross-linking is mostly used for upgrading polymer properties, for example, higher mechanical properties, better thermal resistance, and so on. Multifunctional agents are normally added to enhance the cross-linking, for example, the radiation cross-linking of PE and polypropylene. The addition of a cross-linking agent can significantly reduce the absorbed dose for cross-linking to the same gel content, thus reducing the cost of radiation processing. There are many cases of radiation cross-linking of polymer in industry, including heat-shrinkable tubes and films, electric wires and cables, automobile tires, and some polymeric foams. In comparison, the industrial application of radiation degradation is much more limited, and mainly focuses on the degradation of PTFE and natural polymers. The indispensable merit of radiation degradation is that the process can be completed at room temperature to avoid the formation of side products.

For the application of radiation technology to the fabrication of advanced materials, the choice of a suitable radiation source or machine is important when considering the shape and dimensions of the material, the penetration depth required, and the production capacity. The key issue is to fabricate the target materials by radiation processing at an economically effective cost. Herein we provide a brief introduction to the gamma (γ) ray facility and electron beam (EB) accelerator with different beam energies, in combination with their most suitable applications in material processing.

1.1 BASIC PRINCIPLES OF RADIATION APPLICATION FOR ADVANCED MATERIAL DEVELOPMENT

The radiation effect is the influence of radiation on the irradiated materials, as the name suggests. As shown in Fig. 1.1, the

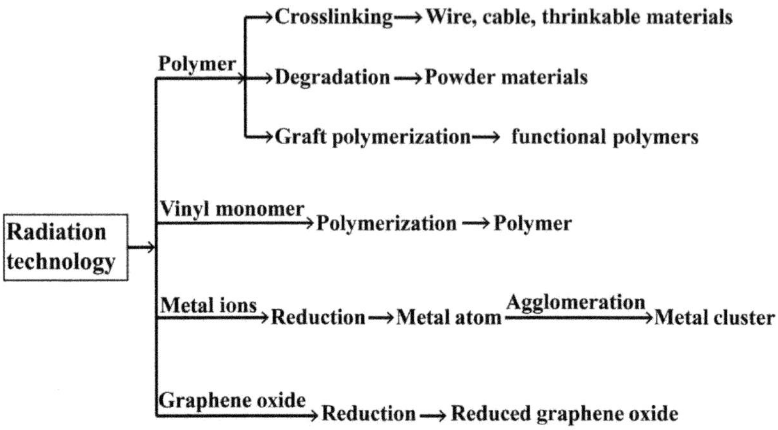

Figure 1.1 Radiation effects used for preparation of various types of materials.

radiation effect for different materials can be utilized for the fabrication of various functional materials. For example, for polymer material, radiation technology can be applied to produce (1) crosslinked polymeric materials such as wires, cables, shrinkable tubes, and films; (2) degraded polymeric materials such as PTFE micropowder; and (3) functional polymeric materials via grafting with vinyl monomers. For metal ions and graphene oxide, chemical reactions can be induced in the solid dispersed in water without the use of chemical initiators, since radiolysis of water generates reductive species such as hydrated electrons, which can serve as efficient redox agents to produce metal nanoparticles and reduced graphene oxide.

1.2 GAMMA RAY

There are three types of natural radioactivity: alpha, beta, and γ radiation. γ-Ray is electromagnetic radiation, like X-ray, while alpha and beta radiation are in the form of particles. The emission of γ-ray from a nucleus does not alter the number of protons or neutrons in the nucleus, but has the effect of moving the nucleus from a higher to a lower energy state. For radiation processing, γ-ray (1.17 and 1.33 MeV) emitted from cobalt-60 (^{60}Co) is widely applied. ^{60}Co is a synthetic radioactive isotope of cobalt with a half-life of 5.27 years. It is produced artificially from isotope ^{59}Co by absorbing neutrons in nuclear reactors. Cesium-137 can also be used as a γ-ray radiation source.

The penetration depth of γ-ray is very high, and γ-ray radiation is convenient for processing material of any shape and size. γ-Ray radiation is commonly used for sterilization of disposable medical equipment, such as syringes, needles, cannulas, IV sets, and food. Compared to other sterilization methods such as steam, dry heat, and ethylene oxide, γ-ray radiation is safe, clean, and easy to handle. The absorbed dose for the killing of bacteria is commonly 3–5 kGy, and may be 8–10 kGy or even higher, depending on the target to be sterilized. There are a number of γ-ray facilities around the world for food preservation, sterilization, and R&D. In China, more than 140 γ-ray facilities are currently in operation.

γ-Ray is rarely utilized for industrial applications of radiation processing of material. However, γ-ray radiation is especially suitable for the processing of material in the liquid state compared to electron accelerator irradiation. Radiation-induced emulsion polymerization of acrylate monomers has been commercialized in China for more than 20 years. Several plants are still in operation with competitive costs and product quality through the use of γ-ray sources. Another example is the radiation treatment of rubber latex by γ-ray in Malaysia. The plant is operated on an industrial scale, and rubber latex is irradiated to decompose the protein or to achieve a low degree of cross-linking of rubber in latex.

1.3 ELECTRON BEAM ACCELERATOR

There are several different methods for accelerating electrons to form a beam, and there are different types of accelerators. EB accelerators are multipurpose machines that deliver EB energies

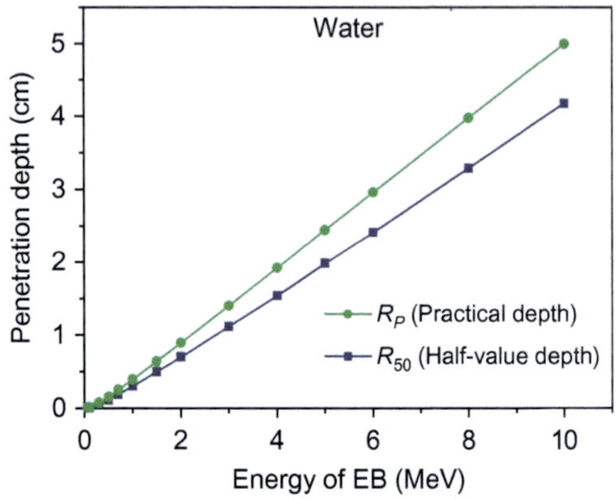

Figure 1.2 Penetration depth as a function of EB energy.[1] *EB*, Electron beam.

spanning a few orders of magnitude, and are used in many applications, ranging from fundamental research on particle interactions to medical use and polymer processing. Fig. 1.2 shows the relationship between the penetration depth in water and the energy of EB.[1] Each type of accelerator has its own characterization and specific design for certain applications. Here we only consider the application of EB accelerators in sterilization and material processing, which are the most common applications of EB accelerators in industry. These applications have been commercialized in many countries.

1.3.1 High energy (10 MeV) electron beam accelerator

EB accelerators with energy of higher than 10 MeV are usually used for radiation therapy and the production of isotopes for nuclear medicine (e.g., F-18 isotope), since energy higher than at least 10 MeV is needed to induce radioactivity in a metal material. High energy EB accelerators can also be applied in the irradiation of gemstones. For radiation processing, the beam energy of the accelerator is normally not higher than 10 MeV. Many EB machines are currently operating in China and other countries. EB processing is also commonly used for sterilization and food irradiation. EB uses an on—off technology and provides a much higher dose rate than γ-ray. Due to the much higher dose rate, less exposure time is needed and potential degradation is therefore reduced. Compared with γ-ray facilities, there is no problem in decommissioning and no radioactive waste to dispose of. Although the penetration depth of EB accelerators is much less than that of γ-ray, when two sides of an item are irradiated by 10 MeV EB accelerators, the effective penetration depth is ~9 cm when the density is assigned to be 1 g cm^{-3}. Since the bulk density of most items is less than 1 g cm^{-3}, this penetration depth is sufficient for sterilization and food irradiation. The processing capacity is greatly determined by the beam current of the EB accelerator, or in other words, by the power. In China, the beam current of 10 MeV accelerators is usually ~2 mA, which means a power of ~20 kW (Fig. 1.3). The Rhodotron accelerator of the IBA Company has beam energy of ~10 MeV and an adjustable beam current. Its maximum beam current could be ~6 mA or even higher. Therefore, the Rhodotron accelerator is very powerful for radiation sterilization and has many benefits in operation. The processing capacity of Rhodotron for food irradiation and sterilization is ~3 times that of 10 MeV EB accelerators presently being operated in China. An EB accelerator (10 MeV) is difficult to use for material processing because the beam current is relatively low; however, it may be useful for radiation degradation of polymer, and the cost is at least competitive with γ-ray irradiation.

1.3.2 Middle energy (~5 MeV) electron beam accelerator

For radiation processing of polymer material, if the penetration depth of an EB accelerator can meet the requirement, the higher the beam current (or power), the larger the productive capacity. This leads to a significant cost reduction of radiation processing. The beam energy of 4.5–5 MeV is relatively large, and it can be applied for radiation cross-linking of tubes, cables, and sheets of a large thickness that cannot be penetrated by a lower energy beam. This is useful only for the radiation cross-linking of items with a large thickness or some other specific materials. However, since the

Figure 1.3 Main structure of a linear accelerator (10 MeV, 20 kW).

beam current is large (up to 30 mA or even higher) and the penetration depth is relatively high, this type of machine can be effectively applied for the radiation degradation of PTFE and other polymers, in which a high absorbed dose is normally required for significant degradation. For instance, the absorbed dose for radiation degradation of recycled PTFE may exceed 1000 kGy to produce PTFE micropowder.

An EB accelerator with beam energy of 5 MeV or higher is suitable for the generation of X-ray. When a high energy EB bombards an X-ray target, fractions of electron energy are converted into photon energy in the form of X-ray beams, while most electron energy is absorbed by the X-ray target, resulting in a temperature rise of the target. Such a high energy X-ray beam has high penetration depth and can be utilized for sterilization, very similar to sterilization by γ-ray. The efficiency of the conversion of an EB to an X-ray beam is dependent on the material and thickness of the target. This efficiency is low at present, and a high output power (>150 kW) of the EB accelerator is required in order to make X-ray generation economically feasible. If the efficiency can be significantly improved in the future, this kind of facility would be very competitive with γ-ray facilities and 10 MeV EB accelerators for sterilization and food irradiation.

1.3.3 Middle energy (1–3 MeV) electron beam accelerator

EB accelerators with the beam energy of 1–3 MeV are mostly used in China, and ~400 installations are in operation. This range of beam energy is suitable for cross-linking of tubes, cables, wires, and shrinkable materials. The beam energy is selected according to the requirement of penetration depth and production capacity. Beam currents of 3 MeV (see Fig. 1.4) EB accelerators of Dynamitron type can be 40–50 and 80–100 mA, respectively. The EB accelerator with middle energy is very attractive for the industry in terms of machine cost, reliability, and processing capacity.

1.3.4 Low-energy (100–500 keV) electron beam accelerator

For irradiation of fiber, film, sheet, and cable with a small thickness, a low-energy accelerator is normally used because of the very low penetration depth required. The beam current can be 100–200 mA or even higher. For continuous operation in industry, the under-beam transportation system moves at high speed, depending on the absorbed dose of treatment. Ebeam Technologies, a brand of the *COMET Mechanical Equipment Co., Ltd.*, has produced a series of low-energy EB accelerators with an energy range from 80 kV up to 300 kV, widths ranging from 150 mm to 2.74 m, and beam power up to 600 kW (see Fig. 1.5) in order to meet the market demand. The application of low-energy EB accelerators is essential for the

Figure 1.4 Dynamitron type accelerator (3.0 MeV, 90 kW).

manufacturing of multilayer polyolefin shrinkable films for packaging. After cross-linking, the thermal resistance and mechanical strength of films are enhanced; more importantly, the heat-sealing strength can be significantly improved. Low-energy EB irradiation can also be used for cross-linking or surface modification of polymer fiber. Radiation-generated free radicals on the surface of the fiber and membrane can initiate grafting polymerization of functional groups. This type of modified material can be used in composites to enhance the bonding between polymer fiber and the matrix, or used as an independent functional material such as heavy metal ion adsorption and wastewater purification.

In addition to radiation processing of polymer material, low-energy EB can be used for curing of coatings, which is more

Figure 1.5 Low-energy eBeam system of COMET Mechanical Equipment Co., Ltd.

environmentally friendly than UV curing, since a photo-initiator is added in UV curing. It can also be utilized for the sterilization of grains by killing the surface bacteria. The beam energy can be lower than 100 keV for surface curing and sterilization.

REFERENCE

1. Grosswendt B. Determination of electron depth-dose curves for water, ICRU tissue, and PMMA and their application to radiation protection dosimetry. *Radiat Protect Dosim* 1994;**75**(2):85–97.

CHAPTER TWO

Radiation Technology Application in High-Performance Fibers and Functional Textiles

Weihua Liu, Mouhua Wang, Lu Xu, Wenli Zhang, Zhe Xing, Jiangtao Hu, Ming Yu, Jingye Li and Guozhong Wu
Shanghai Institute of Applied Physics, Chinese Academy of Sciences, Shanghai, P.R. China

2.1 RADIATION APPLICATION IN CARBON FIBERS AND SILICON CARBIDE FIBERS

Carbon fibers (CFs) and silicon carbide (SiC) fibers are two representative high-performance fibers. CFs have high tensile strength and modulus, and their composites are widely used in many industries. As an inorganic ceramic material, SiC fiber exhibits excellent thermal resistance and can even withstand temperatures as high as above 1500°C. It is used in aero engines owing to its advantages over other materials for its long-term use at high temperatures in air.

The fabrication of CFs and SiC fibers involves spinning, curing, low-temperature sintering, high-temperature sintering, and surface sizing. Interestingly, radiation technology has been found to play an important role in the production of CFs and SiC fibers and can improve the tensile strength and surface performance via modification to CFs.

2.1.1 Carbon fibers

CFs, as excellent reinforcement for composites, have gained a wide range of applications in aerospace, national defense, and the military due to their high strength and modulus. With the development of

Radiation Technology for Advanced Materials.
DOI: https://doi.org/10.1016/B978-0-12-814017-8.00002-0

production technology and reduced production costs, CFs have been also gradually used in sports goods, medical equipment, and other civilian industries. They have attractive application prospects in the automotive industry, the new energy industry, and many other areas. Therefore, the market demand for CFs is growing. The production of high-quality CF and reducing the cost of CF to expand its application in civilian industry are urgent tasks for China's CF manufacturing industry. Meanwhile, the tensile strength of the highest level of CF, T1000, is only ~5% of its theoretical value. Therefore, there is great potential for the improvement of CF performance, which has inspired scientists around the world to use a variety of ways to improve the tensile strength of CFs.

The precursors for preparing CFs include fibers of viscose, pitch, polyacrylonitrile (PAN), polyimide, and polybenzothiazole. However, of these only pitch-based CF, viscose-based CF, and PAN-based CF have reached industrial production scale. Although CFs can be made from pitch precursor, processing and purifying it to the fiber form is more expensive than PAN-based fibers. As we all know, most of the CFs used today are made from PAN precursor. PAN-based fibers have also been found to be the most suitable precursors for producing high-performance CFs compared to pitch, rayon, etc., because of their higher melting point, greater carbon yield (>50% of the original precursor mass), and fast rate of pyrolysis without changing their basic structure.

2.1.1.1 Chemical structure of polyacrylonitrile

PAN, which is a polymer with a chain of carbon atoms connected to one another, is a hard, horny, relatively insoluble, and high-melting point material. The dominant characteristic of the PAN molecule is the presence of strongly polar nitrile groups due to the greater electronegativity of the nitrogen atom compared to the carbon atom. Adjacent nitrile groups of the same macromolecule repel each other. This intramolecular repulsion compels the individual macromolecules into a somewhat irregular helical conformation. The structure of the model is shown in Fig. 2.1.[1] The twisted, kinked macromolecule may be thought of as a more or less rigid

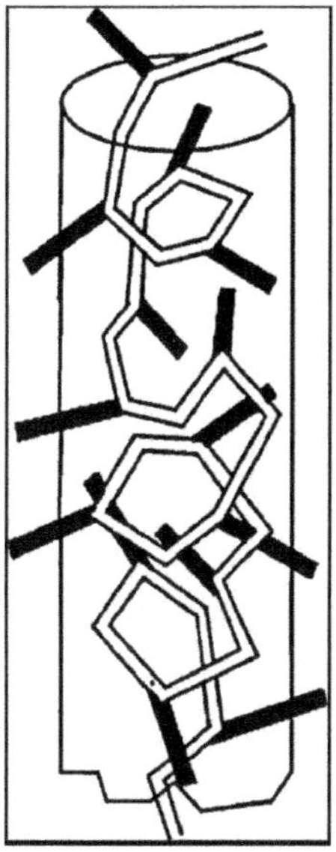

Figure 2.1 Rigid helical conformation of highly orientated acrylic fibers.[1]

structure, fitting within a cylinder of ~6 Å in diameter. Some of the nitrile (CN) groups extend beyond the confines of the cylinder. PAN is a two-phase quasicrystal structure. The PAN molecular chain is rigid due to the strong dipole–dipole interaction between CN groups and a cylinder of ~0.6 nm diameter is formed. Dozens of cylinders are arranged in parallel to form the ordered quasicrystal region, and the disordered amorphous region is composed of the molecular segments that do not form a regular conformation. The structure model is shown in Fig. 2.2.[2] It shows two orders of fine structure within a typical PAN fibril. The major order region consists of rods ~8 Å long aligned parallel to the fibril and the fiber axis. The diameter of the rods is ~6 Å. The disordered region consisting of ring chain, cross-linking chain, defects and monomer, is ~4 Å in length.

2.1.1.2 Radiation effect on polyacrylonitrile

It has been previously found that PAN is irradiated to produce a predominant cross-linking structure in vacuum.[3] The G value of cross-linking $G(x)$ and chain scission $G(s)$ is 0.59 and 0,

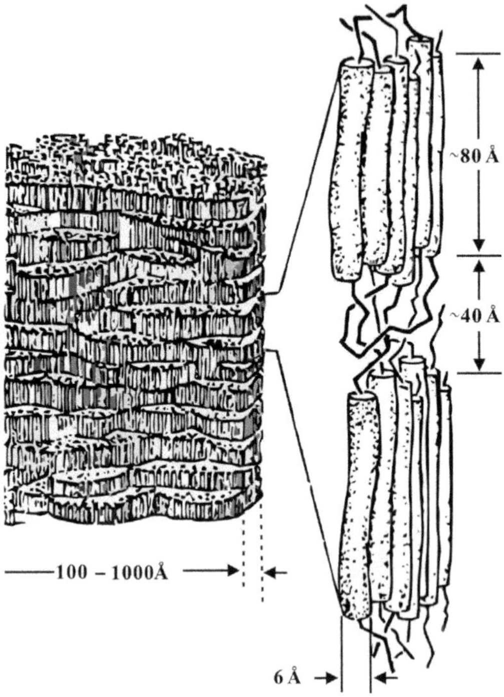

Figure 2.2 Schematic of molecular structure of polyacrylonitrile molecule.[2]

respectively, when irradiated in vacuum at room temperature. However, the molecular weight of PAN is gradually decreased and the oxidation degradation is evident when irradiation is carried out in the presence of oxygen (O_2).

The precursor for producing CFs is PAN copolymers, which contain a second monomer or even a third monomer. Compared with pure PAN, PAN fiber has a special molecular structure and morphology. After irradiation in vacuum, PAN fibers are cross-linked to form a three-dimensional structure and become insoluble in dimethyl sulfoxide (DMSO). As shown in Fig. 2.3, compared with the original fibers completely dissolved in DMSO during 4 h, a yellow gel was formed for the PAN fibers irradiated with an absorbed dose of 200 kGy.[4] Moreover, shrinkage was clearly observed for the irradiated sample due to the disorientation of the PAN fibers after swelling by the solvent.

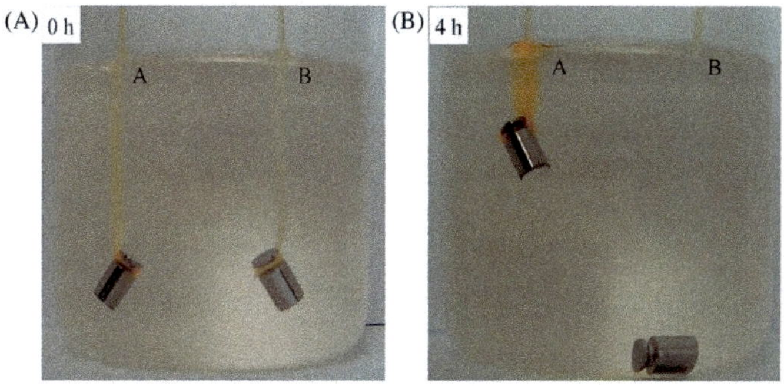

Figure 2.3 PAN fibers before (A) and after (B) immersion in DMSO (A: 200 kGy; B: original fiber).[4] *PAN*, Polyacrylonitrile; *DMSO*, dimethyl sulfoxide.

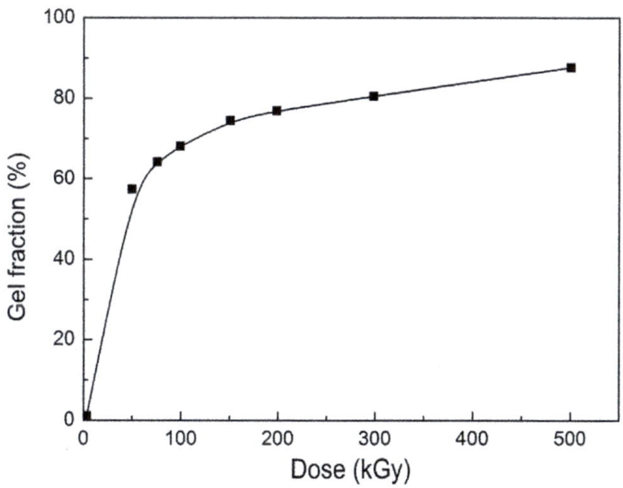

Figure 2.4 Gel fraction of PAN fibers irradiated at various doses in vacuum.[4] *PAN*, Polyacrylonitrile.

Gel fraction is an important parameter to characterize the extent of cross-linking. The gel fraction increases rapidly with the absorbed dose and it is ~60% at an absorbed dose of 50 kGy and then increases gradually with increasing the dose.[4] A gel fraction of 90% is achieved at 500 kGy, as seen in Fig. 2.4.

According to the Charlesby–Pinner formula, a linear relationship exists between $s + s^{1/2}$ (s: soluble fraction) and the reciprocal of dose. Fig. 2.5 shows the Charlesby–Pinner plot of PAN fibers

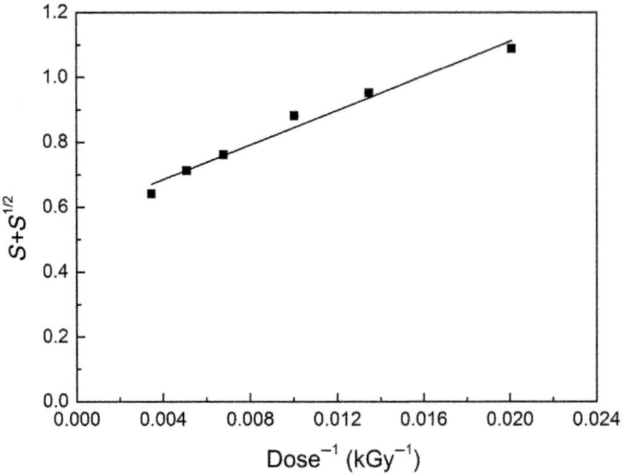

Figure 2.5 Charlesby–Pinner plot for PAN fibers irradiated in vacuum at room temperature (*s*: soluble fraction).[4] *PAN*, Polyacrylonitrile.

irradiated with various doses. The G values of cross-linking and scission were calculated to be $G(x) = 0.28$ and $G(s) = 0.16$ (per 100 eV), respectively.

Therefore, radiation cross-linking predominantly occurs in PAN fibers in vacuum, although chain scission occurs to some extent. The critical absorbed dose at the gel point for PAN copolymers is ~19 kGy. Under the same irradiation conditions, the G value for cross-linking and scission is closely related to the physical state of the polymer. Therefore, the $G(x)$ or $G(s)$ may be different for different PAN materials. The $G(x)$ and $G(s)$ for PAN powder are reported to be 0.59 and 0, respectively.[3]

PAN is a predominantly cross-linking type of polymer upon irradiation. The cross-linking may result from the recombination of the backbone radicals or the addition of backbone radicals to the nitrile groups on the adjacent chains, resulting in cyclization of the nitrile group, as seen in Fig. 2.6.[4]

The cross-linking predominantly occurs in the amorphous region and preferentially near the surface of crystalline regions for general polymers due to the higher mobility in these regions. It may also result in structural changes in the fibers.

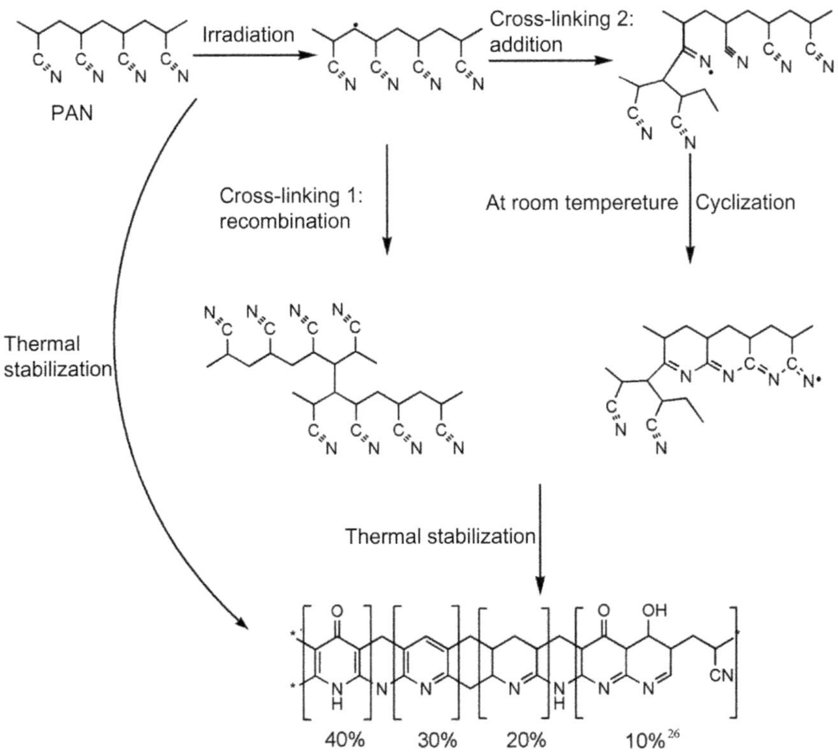

Figure 2.6 Hypothesized mechanism of radiation cross-linking of PAN fibers.[4] *PAN*, Polyacrylonitrile.

By irradiation, free radicals are produced in both the crystalline and amorphous parts of PAN. Due to the higher mobility of molecular chains in amorphous regions, the radicals are more active in inducing cross-linking during irradiation. The radicals trapped in the crystalline region cannot induce cross-linking immediately, but they can migrate in crystals across the molecular chains to the surface of the crystal region to induce cross-linking. Therefore, the cross-linking induced by irradiation mainly occurs in the amorphous region and the surface of crystalline regions.

The radicals induced by irradiation may react with O_2 penetrating into the amorphous region and result in radiation oxidation when irradiation is performed in the presence of O_2. At the same time, the radicals trapped in both the crystalline and amorphous parts react preferentially with O_2 and finally form oxidative products when the samples are stored in air after irradiation.

The relationship between the absorbed dose and the gel fraction of PAN fibers irradiated in three different atmospheres is shown in Fig. 2.7.[5] In vacuum, the critical absorbed dose at the gel point is ~15 kGy. The gel fraction increases sharply with absorbed dose until 50 kGy; then the gel fraction increases gradually up to 100%. This behavior is typical for radiation-induced cross-linkable polymers. In air and O_2, the shape of the dose–gel fraction curves is similar, but the gel point dose shifts to a higher absorbed dose; moreover, the saturated gel fraction decreases with the increase of O_2 pressure.[5]

Generally, the primary radicals produced by radiation react with O_2 to form secondary radicals that amplify the propagation process and lead to new oxidation products. Radiation oxidation is an O_2 diffusion-controlled reaction. In PAN fibers, oxidation reactions should be limited to the skin area because it is difficult for O_2 to penetrate into the fiber. Therefore, cross-linking occurs in the inner part of PAN fibers. The radiation oxidation extent is closely related to the O_2 pressure and absorbed dose. The oxidation area can be calculated according to the saturated gel fraction of PAN

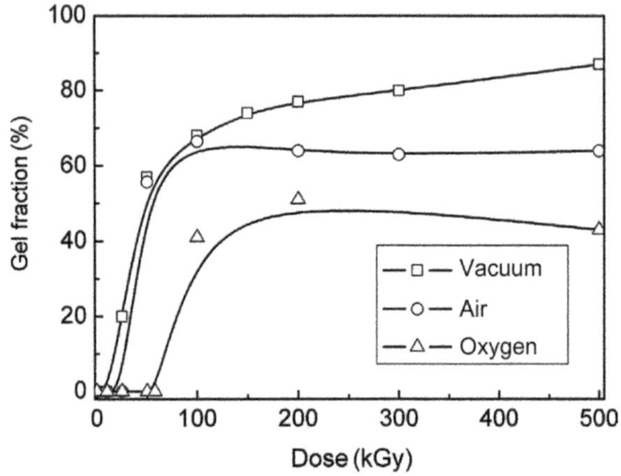

Figure 2.7 Absorbed dose dependence on gel fraction for PAN fibers irradiated under three different atmospheres (O_2: 0.3 MPa).[5] *PAN*, Polyacrylonitrile.

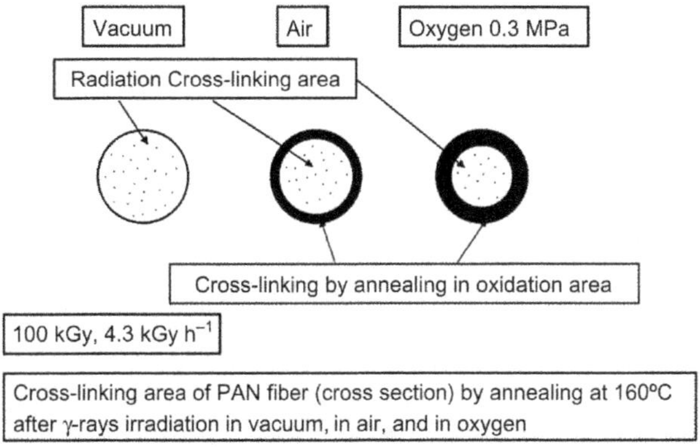

Figure 2.8 Model of PAN fiber curing by radiation oxidation (the dots stand for the cross-linking point by irradiation and the annealing process). *PAN*, Polyacrylonitrile.

fibers in air and O_2. In air, the gel fraction is saturated at 100 kGy. Approximately 35% of the fiber cross section is oxidized and 65% is cross-linked; in O_2, the saturated absorbed dose is also 100 kGy and 50% of the fiber cross section is oxidized. The oxidation and cross-linking processes in air are illustrated in Fig. 2.8.

2.1.1.3 Radiation effect on the preoxidation process of polyacrylonitrile fibers

The manufacture of CFs from the PAN precursor usually involves three processes: preoxidation, carbonization, and graphitization. The structure of PAN fibers changes from linear to a ladder structure in the preoxidation process. Preoxidation plays a critical role in the performance and structure of the resulting CFs. Various reactions are involved in this process, including dehydrogenation, cyclization, oxidation, and cross-linking (as seen in Fig. 2.9).[6] Since these reactions are considered to be exothermic, the preoxidation process is intensely exothermic.

If the exothermal behavior is not well controlled, the heat cannot be removed in time, leading to heat storage in the fibers; the local temperature may rise, resulting in excessive oxidation or even burning of the PAN fibers. It is necessary to control the exothermic behavior in the time-consuming preoxidation process of PAN fibers.

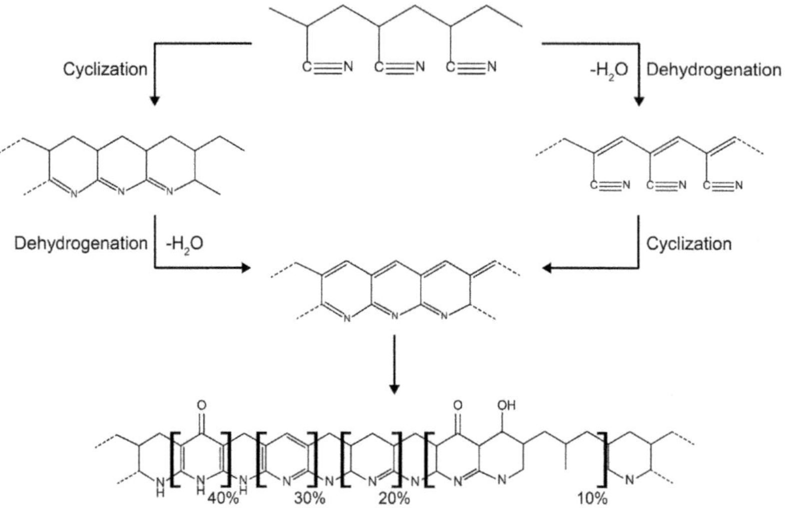

Figure 2.9 Proposed chemistry of PAN stabilization.[6] *PAN*, Polyacrylonitrile.

Shortening the preoxidation time and avoiding the skin–core structure formation is conducive to reducing energy consumption and cutting the production cost of CFs. Radiation modification of PAN fibers can significantly alleviate the exothermic behavior of the preoxidation process and improve the preoxidation process, which affect the performance of the final CFs. Irradiation treatment can also accelerate the preoxidation process, and shorten the preoxidation time, which is especially attractive for the cost reduction of CF manufacturing.

The radiation effect of X-ray on the structure and properties of PAN has been studied using ultraviolet (UV) spectroscopy, infrared spectroscopy, X-ray diffraction, and electron spin resonance (ESR).[7] The free radicals formed by hydrogen abstraction induced by irradiation react with the main chain to initiate cross-linking and the intramolecular cyclizing of the nitrile groups, and then the polyalkene structures are formed. The differences in the preoxidation process between irradiated and original PAN fibers were also investigated. The results of differential scanning calorimetry (DSC) and dynamic mechanical spectroscopy indicate that the molecular debris formed by γ-ray irradiation increases the plasticity of PAN fiber.[8] At the same time, irradiation treatment can lower the weight loss during the preoxidation process and accelerate the preoxidation process. By contrasting the thermomechanical curves of PAN fibers before and after irradiation, it is found that the alpha transformation temperature decreases after irradiation. This indicates that the thawing temperature on the transverse vibration of the PAN macromolecular chain decreased, which is beneficial in reducing the cyclization reaction temperature.

The thermal properties of PAN fibers irradiated at different absorbed doses were investigated by DSC. It was found that the cross-linking structure of PAN fibers induced by radiation reduced the initial temperature of the cyclization reaction and alleviated the exothermic behavior. This effect can be controlled by adjusting the absorbed dose (as shown in Fig. 2.10 and Table 2.1).

The irradiated PAN fibers show a higher density under the same preoxidation process (shown in Fig. 2.11), indicating that irradiation promotes the preoxidation process. The γ-irradiation pretreatment of PAN fiber can effectively reduce the reaction temperature and alleviate the exothermic reaction, resulting in a significant reduction in weight loss during the preoxidation process; this effect can be controlled by adjusting the absorbed dose. The free radicals

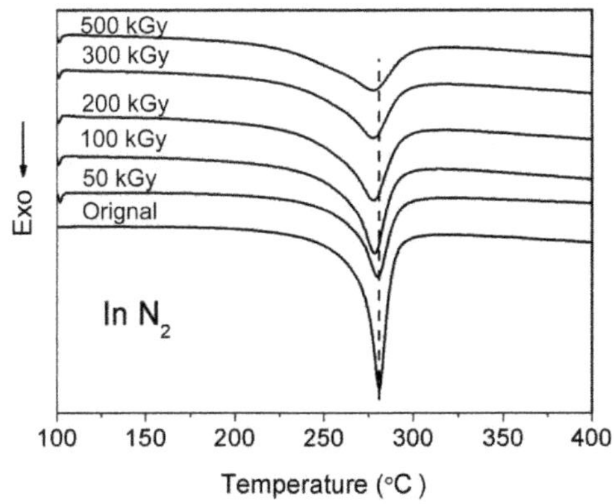

Figure 2.10 DSC curves of PAN fiber samples irradiated with various doses under N_2 atmosphere.[4] *DSC*, Differential scanning calorimetry; *PAN*, polyacrylonitrile.

Table 2.1 Changes in T_i, T_f, and T_P of the exothermic peaks of PAN fibers before and after irradiation

Dose (kGy)	0	50	100	200	300	500
^1T_i (°C)	271	268	261	254	249	237
^2T_f (°C)	296	297	298	298	300	302
^3T_P (°C)	280	280	278	278	278	278
ΔT (°C)	25	29	37	44	51	65
ΔH (J g^{-1})	536	516	497	490	485	480

Notes: 1. Initiation temperature of peak; 2. Termination temperature; 3. Peak temperature.

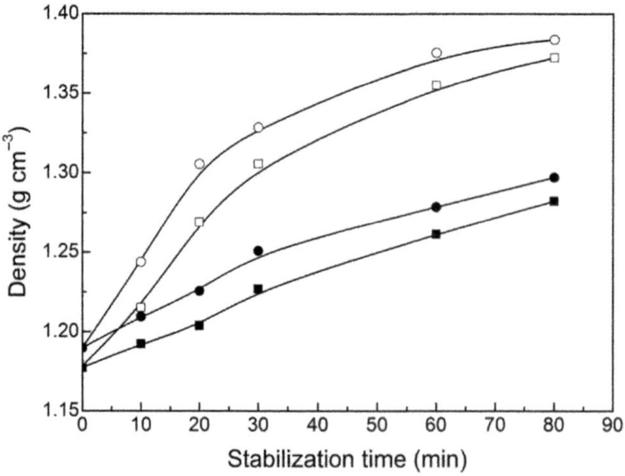

Figure 2.11 Density of preoxidized PAN fibers as a function of preoxidation time (curves 1 and 3: fibers irradiated at 200 kGy; curves 2 and 4: original fiber). *PAN*, Polyacrylonitrile.

produced by irradiation were studied via ESR. It was found that the modification effect was not caused by the trapped free radicals, but rather by the functional groups (allyls) formed during the irradiation process. The allyl groups can initiate the preoxidation reaction during thermal treatment.

The free radicals produced by electron beam (EB) irradiation in PAN fibers can also lead to cyclization of nitrile groups. In addition, the free radicals produced by irradiation are very stable at room temperature and can be retained for a long time without attenuation. This provides the opportunity for the use of free radicals to further initiate reactions. The effect of EB irradiation on the thermal properties of PAN fiber was investigated via DSC, thermogravimetric analysis (TGA), and other means. The results show that the effect of EB irradiation is similar to that of γ-ray irradiation on PAN fiber. The EB irradiation pretreatment reduces the initial temperature of the reaction, mitigating the exothermic reaction which increases the carbon yield significantly. The results of Fourier infrared transform and UV spectroscopy indicate that irradiation induces the formation of a conjugate structure, which improves the thermal properties of PAN fiber. In addition, the activation energy of the cyclization reaction decreases after EB irradiation. For example, the activation energy decreased from 302 to 260 kJ mol^{-1} after irradiation at 200 kGy. Radiation treatment also accelerates the cyclization reaction and improves the degree of cyclization.

Tarakanov studied the effect of LGN-70ICO_2 laser pretreatment on PAN fiber and obtained similar results to γ-ray irradiation.[9] Laser treatment causes the accumulation of energy on the chemical bond of PAN fibers and has a destructive effect on the chemical bond, causing a decrease in fiber strength. The effect of pretreatment time with UV irradiation on the preoxidation of PAN fiber was also investigated. It was found that UV irradiation initiated cyclization reaction at room temperature and promoted the cyclization reaction during the preoxidation process. In addition, UV irradiation treatment can induce the effective cross-linking of melt spinning PAN copolymer. CF was then obtained by thermal oxidation and carbonization, but the mechanical properties of the CFs were poor. While the performance of final PAN-based CF can be promoted via optimization of the PAN copolymer structure (copolymer composition, relative molecular mass, etc.) and melt spinning.

To summarize, irradiation modification of PAN fiber can alleviate the exothermic behavior of the preoxidation process, improve carbon yield, reduce the initiation temperature, and accelerate the preoxidation process. With the development of radiation technology and reduction in radiation processing cost, radiation technology has potential for application in the PAN-based CF production. This has been demonstrated by China in an online CF manufacturing line. It was found that the preoxidation time can be cut at least by 50% when the PAN fiber is irradiated prior to preoxidation.

2.1.1.4 Radiation effect of polyacrylonitrile on the performance of final carbon fiber

The preoxidation process can be improved via irradiation pretreatment of PAN fibers, and preoxidized fibers with excellent structure and performance can be obtained. The structure and properties of preoxidized fibers play an important role in the structure and performance of the resulting CF. Therefore, irradiation can improve the performance of the final CF by improving the preoxidation process. For γ-ray-irradiated PAN fibers, CFs are successfully prepared by continuous preoxidation and carbonization. The tensile strength of CFs is significantly increased by optimizing the preoxidation conditions. Radiation pretreatment not only alleviates the exothermic behavior but also accelerates the rate of preoxidation. Therefore, the strength of the CFs obtained from irradiated PAN fiber is not substantially reduced by shortening the preoxidation time by 40%. The strength of CFs produced from irradiated PAN fibers is obviously higher than that of CFs obtained from original PAN fibers, as shown in Fig. 2.12. Therefore, irradiation modification of PAN fibers can improve the performance of the final CFs, and shorten the preoxidation time.

2.1.1.5 Effect of radiation treatment on the structure and performance of carbon fibers

CFs are used in composites mainly due to their high strength, high modulus, and other excellent performance. Usually, the performance of composite material depends on the mechanical properties of CF. Although CF has excellent mechanical properties, the strength is far lower than its theoretical values. Therefore, CFs are also modified in many ways to improve their properties, in addition to modification of their precursor. During irradiation, the substance absorbs ionizing radiation energy and can be ionized or

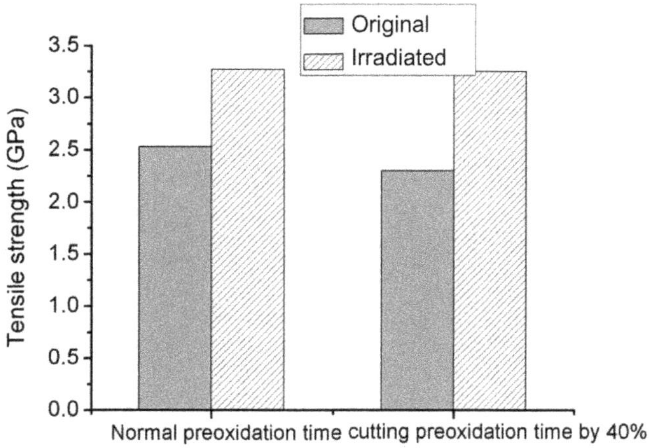

Figure 2.12 Tensile strength of carbon fibers from original and irradiated PAN fibers. *PAN*, Polyacrylonitrile.

excited, and then a series of chemical reactions are induced. Radiation modification does not need any initiator and uniform modification of material can be achieved, due to the high penetration depth by radiation.

The mechanical properties of CFs can be effectively improved by γ-ray irradiation.[10] The strength and Young's modulus of CF significantly increase with an absorbed dose within the low dose range. The strength increases by 17%, from 4.6 to 5.4 GPa while the Young's modulus increases by up to 50% after irradiation with an absorbed dose of 30 kGy (refer to Fig. 2.13). Although the tensile strength of CF decreases with the increase of an absorbed dose, the strength is still relatively high at a dose of 300 kGy. Therefore, the mechanical properties can be obviously promoted at a lower dose, while defects may be introduced into the CFs, leading to a decrease in CF strength at higher absorbed doses.

The increase in CF strength induced by irradiation may result from the reduction of the first-class defects in CFs, while an increase in the degree of graphitization leads to an increase in Young's modulus. The enhanced mechanical properties of CF are also considered to be due to the formation of cross-linking covalent bonds, as shown in Fig. 2.14. It is reasonable to assume that the irradiation of CFs by

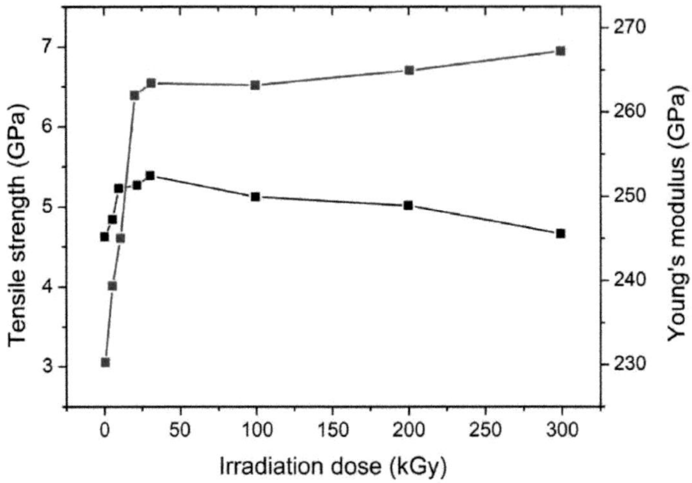

Figure 2.13 Influence of γ-ray irradiation on mechanical properties of carbon fibers[10] (the error bars for values correspond to the standard deviation of measurements from at least 30 filaments).

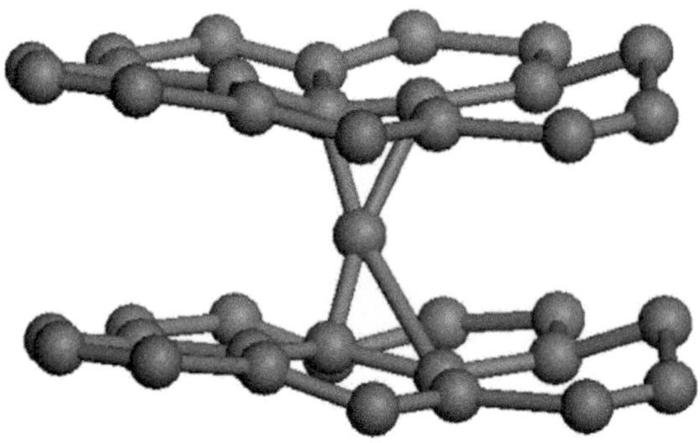

Figure 2.14 View of cross-linking between graphene layers.[10]

high energy γ-ray generates free radicals on the graphene layers and produces covalent cross-linking (bond defects) between the graphene layers of CFs. The density of CF increases with the absorbed dose, as shown in Fig. 2.15. The reduction of interlayer spacing may be responsible for this result during γ-ray irradiation. The formation of covalent bonds between the graphene layers induced by irradiation shortens their distance. Within the absorbed dose range of 0–300 kGy, the interlayer spacing is reduced from 0.3514 nm

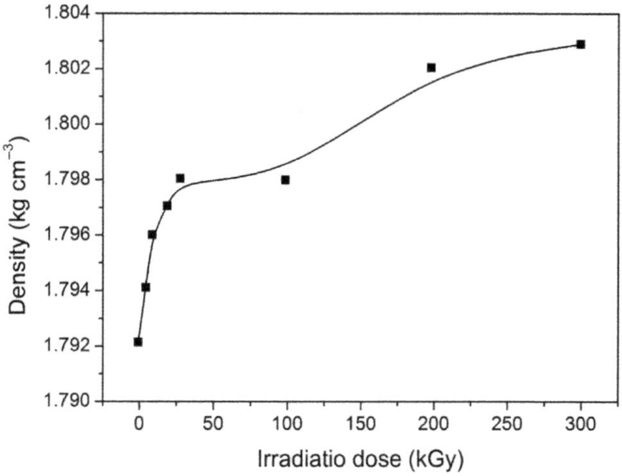

Figure 2.15 Influence of γ-ray irradiation on fiber density (the error bars for values correspond to the standard deviations of measurements from five small bundles).

gradually to 0.3480 nm. The interlayer spacing moves gradually close to the hexagonal graphite layer spacing of 0.3354 nm, indicating a higher degree of graphitization after irradiation.

2.1.2 Silicon carbide fibers

2.1.2.1 Preparation of silicon carbide fibers from polycarbosilane precursors

SiC fiber, a ceramic fiber shown in Fig. 2.16, is an excellent material for high-temperature applications, such as for gas turbines. Limited by its brittleness, its practical application is mainly as an important reinforcing fiber for composite materials. SiC fiber shows good compatibility with metal, ceramics, and resin. Therefore, it can be used to prepare metal, ceramic, and resin-based composite materials, which can be used in the aerospace, defense industry, and potentially nuclear reactors.

The preparation methods of SiC fibers have attracted attention from researchers all over the world. To date, the methods mainly include chemical vapor deposition, CF conversion, auxiliary sintering, and precursor pyrolysis, of which the precursor pyrolysis

Figure 2.16 Photo of SiC fibers.

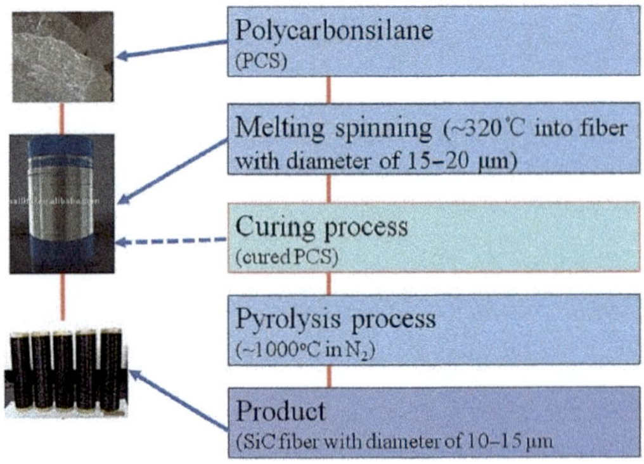

Figure 2.17 Scheme for preparation of SiC fibers.

method is more suitable for industrial production due to its lower cost and higher production efficiency. Therefore, the precursor pyrolysis method has been widely studied and used. The SiC fiber was invented by Yajima in 1975.[11] The processing is by the pyrolysis of a polymer of polycarbosilane (PCS) as the precursor of ceramic fiber. The precursor fiber is obtained using the melt spinning method and thermally treated at high temperature. Thus the organic fibers are transformed into inorganic ceramic materials. The preparation of SiC fibers usually involves the synthesis of organic precursor PCS, spinning, curing, and high-temperature treatment (Fig. 2.17). The mechanical strength of PCS fiber is rather low due to the lower molecular weight of PCS and weak intermolecular force. Therefore, the PCS fibers must be cured prior to pyrolysis by certain methods such that the fiber macromolecules are cross-linked to produce a nonmolten structure. They can then retain the fiber shape under high-temperature treatment. The curing treatment can make PCS fibers transform from thermoplastic into thermosetting polymer, and avoid the fusing of fiber to retain the fiber shape in the process of high-temperature sintering. The curing of PCS can also improve the ceramic yield and reduce defects in the fibers. The curing methods are thermal oxidation, EB curing, thermal cross-linking and chemical cross-linking, of which thermal oxidation and EB curing are the main methods presently.

The thermal oxidation curing of PCS fibers is usually carried out in air within the temperature range of 200–350°C, which is similar to the preoxidation process of CF production. In the thermal oxidation process, oxygen is introduced onto PCS molecular chains and then cross-linking of PCS fibers occurs at high temperature. In the last century, the first generation of SiC fibers (Nicalon: The trademark of the SiC fiber of Nippon Carbon Co. Ltd.) was obtained using this curing method in Japan. A sufficient amount of oxygen is needed during thermal oxidation curing. As the silicon atom has a high affinity to oxygen, the oxygen supplied to the PCS fiber during the curing process remains in the resultant SiC fiber. The oxygen content in the SiC fiber is 10%–12% by weight. However, the heat resistance temperature is limited to around 1200°C for the first generation of SiC fibers. The amorphous phase of the Si–O–C phase in the SiC fibers is decomposed with the emission of SiO and CO gases above 1200°C.

To reduce the oxygen content, radiation curing (cross-linking) of PCS fibers is applied to replace the thermal oxidation process. For EB irradiation, a high energy EB causes the chemical bond breakage of PCS fibers and then the active particles (free radicals) are formed. The free radicals further induce the formation of a cross-linking structure. The cross-linked PCS fibers will not melt and retain the fiber shape during the pyrolysis process. The mechanism of radiation curing is shown in Fig. 2.18. Compared to other polymers, the relative molecular weight of PCS is low. Therefore, a

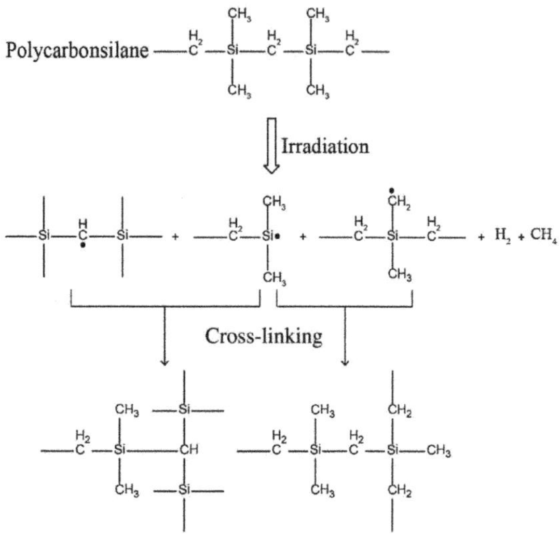

Figure 2.18 Radiation curing mechanism of PCS fibers.[12] *PCS*, Polycarbosilane.

much higher absorbed dose (≥10 MGy) is necessary to achieve cross-linking. Radiation curing is carried out in an inert atmosphere without the introduction of other chemicals or oxygen. The oxygen content of SiC fibers is obviously reduced and then the heat resistance properties are improved. The curing process of PCS fibers plays a key role in the structure and properties of the final SiC fiber. EB radiation curing has been used in the manufacture of SiC fibers with lower oxygen contents (Hi-Nicalon; Hi-Nicalon-S: The trademarks of SiC fibers of Nippon Carbon Co. Ltd.). The high-temperature performance of SiC fibers obtained from thermal oxidation curing and radiation curing respectively can be seen in Fig. 2.19.[12]

2.1.2.2 The properties of silicon carbide fibers obtained from radiation curing

Okamura[13] used EB radiation curing to obtain SiC fibers with oxygen content as low as 0.5 wt%, that is, the second generation of SiC fiber. PCS fibers were irradiated in an inert atmosphere to make the PCS macromolecular chains cross-linked and they were then sintered at high temperature. The high-temperature properties

Figure 2.19 Scanning electron microscope images of SiC fibers after thermal treatment at 1773k in Ar.[12]

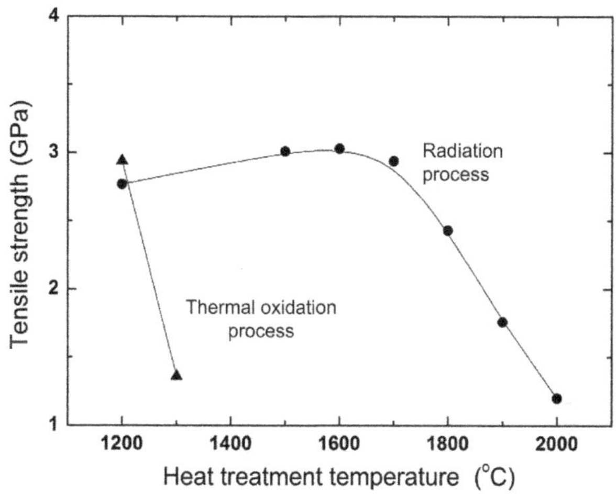

Figure 2.20 High-temperature properties of SiC fibers obtained by thermal oxidation curing and radiation curing.[13]

of SiC fibers are obviously improved after reducing the oxygen content, as can be seen in Fig. 2.20.[13] The residual tensile strength of SiC fibers after thermal treatment in Ar for 10 h at 1500°C is still 2 GPa. When the fibers are treated in Ar for 1 h at 2000°C, they still retain their shape. The second generation of SiC fibers exhibits excellent high-temperature performance.

Although the second generation of SiC fiber has high tensile strength and high-temperature resistance, it contains free carbon. SiC microcrystalline recrystallization and radiation-assisted oxidation of SiC fibers occur when SiC_f/SiC composites are exposed to high reflux of neutron irradiation. The SiC fibers shrink significantly while the SiC matrix swells. Then the SiC fibers are separated from the SiC matrix after neutron irradiation. The mechanical properties are significantly reduced, limiting its application in the field of nuclear energy. Therefore, the third generation of SiC fibers was developed by removing the free carbon in the second generation of SiC fibers. The heat resistance of the fiber is further improved. When the fibers are thermally treated at 1600°C in Ar for 10 h, the tensile strength and modulus are still very high. The Si/C ratio of the third generation of SiC fibers is near the stoichiometric ratio (1:1). The tensile strength of SiC_f/SiC composites containing the third generation SiC fibers remains almost unchanged after neutron irradiation. The use of composites made with the third generation SiC fibers will greatly enhance the security of existing nuclear reactors and drive the development of fourth-generation nuclear reactor technology.

The great improvement in performance of the second and third generation SiC fibers is closely related to the curing method. Both use EB irradiation to achieve the nonmelting of the precursor PCS fibers, which significantly improves the properties of SiC fibers. EB irradiation curing is usually carried out under He atmosphere and a high absorbed dose ($>$10 MGy) is needed for nonmelting of PCS fibers. EB irradiation can cause the temperature to increase quickly. Therefore, the fibers must be cooled by circulating water or gas during the irradiation process to avoid a significant temperature rise. The EB irradiation curing system is shown in Fig. 2.21.[14] The fiber should be annealed within the temperature range of 300–500°C for 30 min to quench the free radicals generated during the curing process to avoid oxidation. Cured PCS fibers are pyrolyzed at high temperatures (1300–1500°C) in an inert atmosphere (Ar or N_2) to prepare SiC fibers.

The extent of cross-linking of PCS fibers is usually evaluated by the gel fraction measurement of PCS fibers in tetrahydrofuran (THF) solution. The gel contents in THF at various absorbed doses are shown in Fig. 2.22. There is no gel formation below 5 MGy

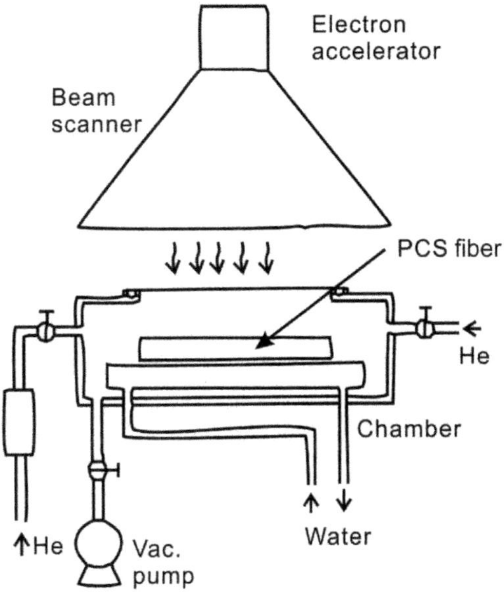

Figure 2.21 EB irradiation curing system.[14]

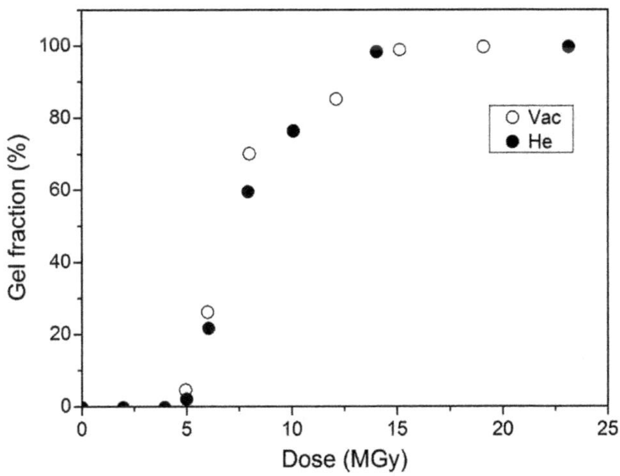

Figure 2.22 Gel fraction of PCS fibers versus absorbed dose by EB irradiation under Vac and in He gas at room temperature.[14] *PCS*, Polycarbosilane, *Vac*, vacuum.

due to the low molecular weight of PCS fiber. With the increase of the absorbed dose, the gel content increases dramatically, eventually being close to 100%. Generally, when the gel content is more than 75%, PCS fibers can retain their shape during the later pyrolysis process into SiC fiber at high temperature. Therefore, the required absorbed dose for curing process is >10 MGy.

2.2 RADIATION-INDUCED SURFACE MODIFICATION IN ADVANCED FIBER COMPOSITES

2.2.1 High-performance fibers and composites

Compared with traditional natural fibers such as cotton, linen, and man-made fibers such as polyester and polypropylene (PP) fibers, high-performance fibers have some excellent properties, such as extremely high strength modulus, high thermal stability, and good acidic and caustic corrosion resistance. Kevlar fiber, ultrahigh molecular weight polyethylene (UHMWPE) fiber, and CF are three representative high-performance fibers. High-performance fibers are not only widely used in national defense, aerospace, and other high-end fields, but also play an important role in civilian applications including transportation, construction, and sports goods.

For fiber reinforced composites, the adhesion strength between reinforcing fibers and the resin matrix determines the properties of composites. A strong interfacial strength can make the load on the composite transfer more effectively between the fibers and the resin matrix. As high-performance fibers are mostly constructed of rigid macromolecular chains and possess high crystallinity, the fiber surface is very smooth and lacks chemically active groups, leading to poor interface adhesion strength between fibers and the resin matrix. Consequently, the load transfer is inefficient, limiting the application of composites. It is reported that the strength conversion ratio of aramid fiber reinforced composites is only 68%–76%, which is even lower than that of glass fiber (>80%).[15] In order to improve the interfacial bonding ability between the fiber and the matrix resin, the surface of high-performance fibers must be modified by physical or chemical methods. Physical modification methods include surface coating, plasma, and EB and γ-ray irradiation. The chemical methods include surface activation and surface grafting. Although the surface of high-performance fibers can be modified using many methods, there are still some limitations. Surface coating has the advantages of simple operation, easy control, and no damage to the fiber, but the effect is insignificant; plasma treatment can introduce polar functional groups on the fiber surface, and improve surface wettability, but it is difficult to achieve in an online continuous operation. Surface grafting by chemical methods not only increases the compatibility between the high-performance fiber and the resin matrix, but also introduces active groups on the surface of the fiber. This method can significantly enhance the adhesion strength of the fiber and the resin matrix. However, chemical surface grafting often takes a long time and the activity of chemical reaction is difficult to control. The application of high energy irradiation (γ-ray and EB) in the modification of polymer materials has proven to be a good choice for improving the surface compatibility of high-performance fibers. Since irradiation generates a large number of free radicals in the fibers, these radicals can trigger the grafting polymerization of acrylate monomers on the fiber surface, leading to the complete surface modification of high-performance fibers (Fig. 2.23).

2.2.2 Radiation modification of aramid fibers

2.2.2.1 Aramid fibers

Aramid fibers are aromatic polyamide fibers, containing more than 85% amide bonds directly connected with the two aryls. This kind of fiber has the following advantages: lightweight, high-strength, high-modulus, high-temperature resistance, and excellent corrosion resistance. The strength and modulus of aramid fibers is five to six times and two to three times, respectively, higher than that of steel wires with the same diameter, while the weight is only approximately one-fifth of the steel wire. The decomposition temperature is 560°C. There are two types of aramid fibers, namely,

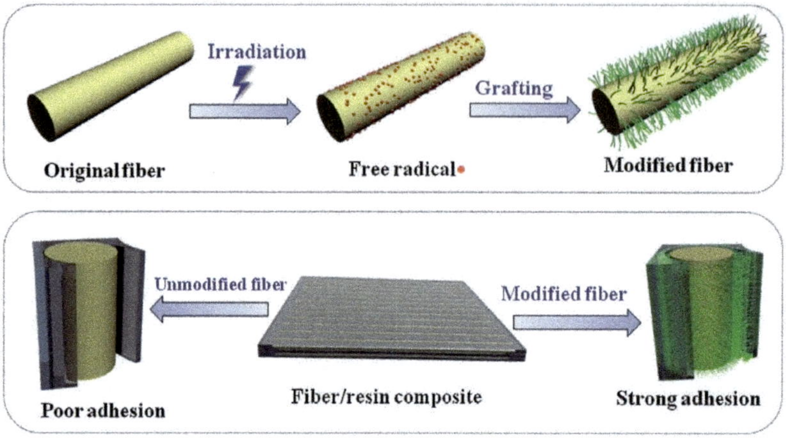

Figure 2.23 Radiation-induced surface modification of high-performance fibers and its application in composites.

meta-aromatic and *para*-aromatic polyamide. For *meta*-aromatic polyamide, the angle between the amide plane and the phenylene plane is 30 degrees, resulting in the most stable structure. The rotation energy barrier of the phenyl–amide bond and the carbon nitrogen (C–N) bond is high, leading to fully extended chains. Furthermore, the existence of strong hydrogen bonding can further enhance the stability of the chemical structure of aramid fibers. The crystalline structure of *para*-aromatic polyamides is pseudo-orthorhombic, and the macromolecular chains are fully extended in the crystalline region. The angle of HN–O is 160 degrees, which is favorable for the formation of intermolecular hydrogen bonding. These unique structures give excellent physical and chemical properties, but also bring some problems: a high degree of crystallinity, a smooth surface, and a lack of chemically active groups, resulting in poor adhesion strength between the aramid fiber and the resin matrix. Therefore, it is necessary to modify the fiber surface for practical applications.

2.2.2.2 Radiation surface modification of aramid fibers

Radiation modification of aramid fiber has been also carried out in China. For example, Huang et al.[16] studied the influences of air and N_2 on the properties of Armos aramid fiber and its composite under γ-ray irradiation. The results showed that γ-ray irradiation could induce cross-linking of aramid fiber macromolecules. When the absorbed dose exceeded 600 kGy, degradation was dominant but there was no significant impact on the fiber performance. The interlaminar shear strength (ILSS) of composite increased with the increase of absorbed dose by γ-ray irradiation, and then decreased with a further increase of the absorbed dose. After fiber surface treatment, ILSS increased by 17.7% in N_2 medium and 15.8% in air medium (Fig. 2.24).[16] Armos fibers were modified by simultaneous radiation grafting in mixtures of phenol-formaldehyde and ethanol at different ratios. It was found that the ILSS of the composite was improved by 25.4% for the fiber irradiated with an absorbed dose of 500 kGy.[17] Furthermore, γ-ray irradiation of Kevlar-49 fiber can increase the oxygen content on the fiber surface, and introduce new functional groups, in order to improve the surface activity of Kevlar fiber. Compared with untreated fibers, composites based on the treated fibers showed an increase of flexural strength, modulus and shear strength, but a decrease in impact strength.[15] Qiu et al.,[18] investigated the effect of γ-ray irradiation (500 kGy) on Apmoc fibers. After treatment, the tensile strength of the multifilament increased by 8.2% while the transverse tensile strength increased by 16.1%. Furthermore, the transverse tensile strength of the composites increased by 16.1%.[18] Although some progress has been made, there are still many challenges to overcome for industrial applications.

2.2.3 Radiation modification of carbon fiber

CF is usually used as a reinforced fiber, and CF reinforced materials are the most widely used. However, due to the lack of active functional groups on the surface of CF, its surface energy is low, and the surface inertia of CF is relatively large. To make full use of the high performance of CF, it is necessary to improve its surface properties and enhance the wettability and cohesiveness between the fiber and the matrix resin. CF surface modification

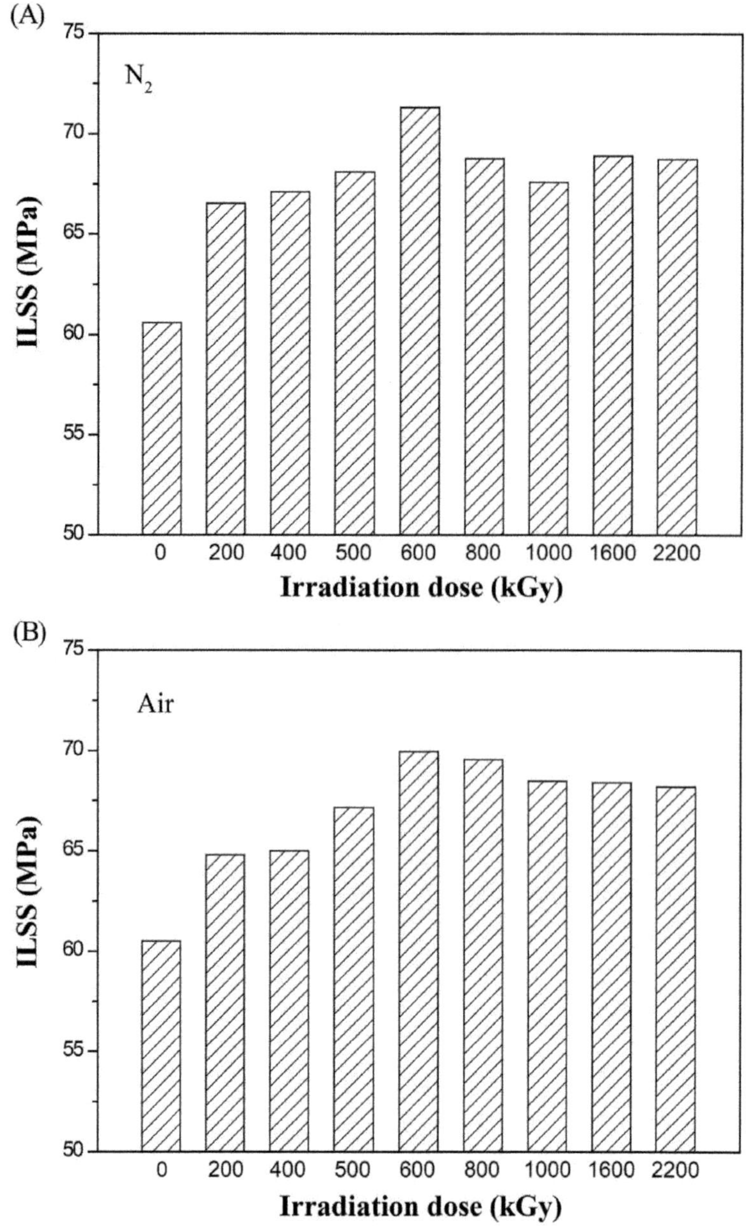

Figure 2.24 ILSS of Armos/epoxy composites untreated and treated with different irradiation doses.[16] *ILSS*, Interlaminar shear strength.

methods can be mainly classified into two categories: (1) oxidation methods, such as electrochemical oxidation and chemical oxidation (2) nonoxidizing methods, such as whisker growth, surface coating, and radiation modification. Radiation modification is an efficient, environmentally friendly and very promising modification method.

High energy irradiation can increase the surface roughness of CF by atomic displacement. Meanwhile, some oxygen-containing functional groups, such as hydroxyl groups, ether bonds and carboxyl groups, can be introduced on the surface, and the mechanical properties of CF are not impaired. The carboxyl groups can effectively improve the interface binding between the CF surface and the matrix resin, thereby enhancing the strength of the composite. The CFs were irradiated in air or epoxy solution and the changes in structure and properties of CFs were further studied using X-ray photoelectron spectroscopy (XPS), torsion brazing analysis, micro-off testing, and other methods.[19] The results showed that the ILSS of CF reinforced composites increased by 37%. The improvement in the ILSS of CF/epoxy composites could be attributed to the enhancement of interfacial adhesion strength of the fiber and matrix after radiation grafting.

In addition to the direct radiation modification of CF, radiation grafting is also an effective modification method that can not only improve the surface roughness of CF, but also increase the surface activity, and improve the surface energy of the CF surface by introducing a large number of reaction functional groups. Radiation grafting also promotes the interface reactions between the CF and matrix resin, and enhances the interfacial strength of the composite. An epoxy resin matrix composite were made by γ-ray irradiation grafting of epoxy resin, epoxy chloropropane, and acrylic acid (AA) onto the surface of CF (Fig. 2.25). Compared with untreated CF epoxy composites, the ILSS of grafted CF epoxy composite was raised by 21.8%, 18.1%, and 17.5%, respectively.[20]

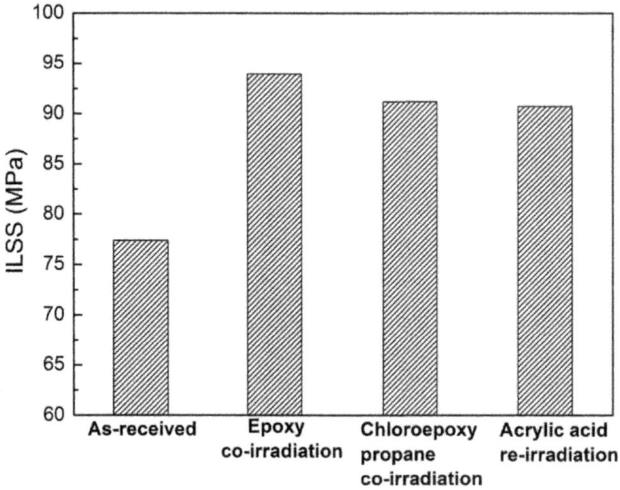

Figure 2.25 Effect of γ-ray irradiation grafting on ILSS of carbon fiber/epoxy.[20] *ILSS*, Interlaminar shear strength.

2.3 RADIATION GRAFTING AND APPLICATION OF ULTRAHIGH MOLECULAR WEIGHT POLYETHYLENE FIBER

2.3.1 Radiation effects on ultrahigh molecular weight polyethylene fiber

UHMWPE fibers have excellent mechanical properties, chemical corrosion resistance, and weathering resistance. They are extensively used in military armor, aeronautical and astronautical devices, fishing nets, marine ropes, and sports equipment. Although UHMWPE consists of the methylene ($-CH_2-$) groups such as conventional PE, some properties are quite different and superior, such as the wear resistance and self-lubricity; in addition, its impact resistance is the highest of all engineering plastics. In contrast, due to the lack of a reactive group of a PE backbone and low surface energy, the surface is hydrophobic and lacks adhesive properties and dyeing affinity, which greatly limit its applications. In order to develop new functional materials based on UHMWPE fiber, surface treatments have been used to modify this polymer.

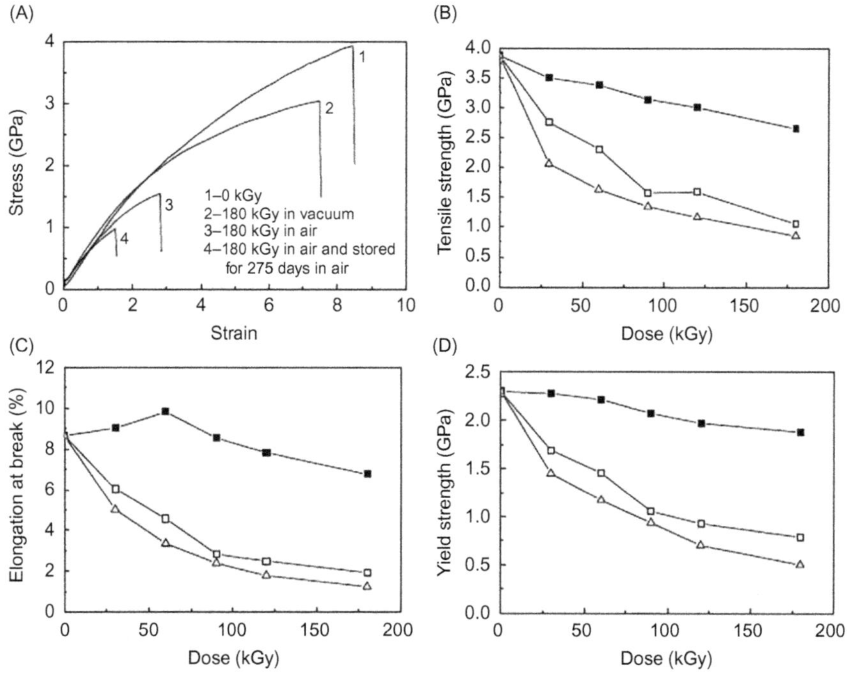

Figure 2.26 Change of tensile properties of UHMWPE fiber by irradiation in vacuum and in air ((A) stress–strain curves of initial fiber and irradiated at 180 kGy; (B) tensile strength; (C) elongation at break; (D) yield strength (■: vacuum; □: air; △: irradiated in air and stored in air for 275 days)).[21]

The investigation of radiation-induced radical species and the subsequent reactions in UHMWPE fiber is important for radiation modification. Radiation effects include the following aspects: the transition mechanism of free radicals, changes of chemical (cross-linking and cracking) and crystal structures, and changes in material performance caused by radiation-induced free radicals. Due to some of the characteristics of UHMWPE fiber such as its high strength, high degree of crystallization and orientation, radiation modification can improve some characteristics of a material while decreasing other properties at the same time (Fig. 2.26).

As shown in Fig. 2.27, the ESR spectra have higher resolution due to the special structure of the UHMWPE fiber.[21] Because the UHMWPE fibers are produced by the gel-spinning processing

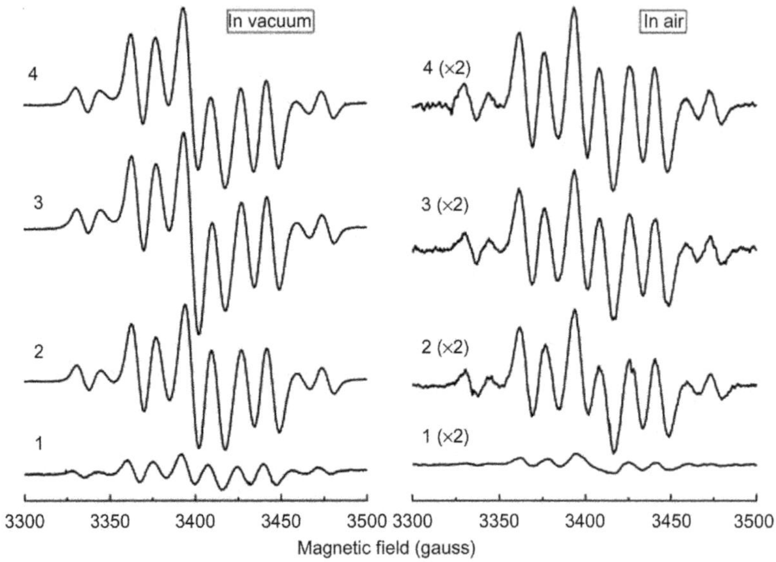

Figure 2.27 ESR spectra of the UHMWPE fiber irradiated (dose rate: 0.73 Gy s^{-1}) in vacuum and in air at room temperature (1: 10 kGy; 2: 60 kGy; 3: 130 kGy; 4: 200 kGy; ×2: the magnification of spectra by a factor of two).[21] *ESR*, Electron spin resonance.

with an ultrahigh draw ratio, the macromolecular chains align with the fiber axis and have a highly regular structure. The degree of crystallinity is higher than 80% and the orientation of crystallites (*c*-axis) is uniform along the fiber direction.

Radiation-induced oxidization of UHMWPE fibers significantly improves the mechanical properties of styrene—butadiene rubber (SBR) composites, and the tensile modulus of the SBR composites with the oxidized fibers at an absorbed dose of 100 kGy can be increased from 1 to 5 MPa,[22] as shown in Fig. 2.28. Xing et al.,[23] studied the radiation-induced grafting reaction of AA, acrylonitrile (AN), and other monomers on UHMWPE fibers. After radiation grafting, the roughness of the surface of fiber is greatly improved. The increase of polarity and roughness is beneficial to the mechanical properties of composites. UHMWPE fiber grafted with a small amount of carboxyl group can improve the mechanical properties of SBR, such as its tensile modulus.[24]

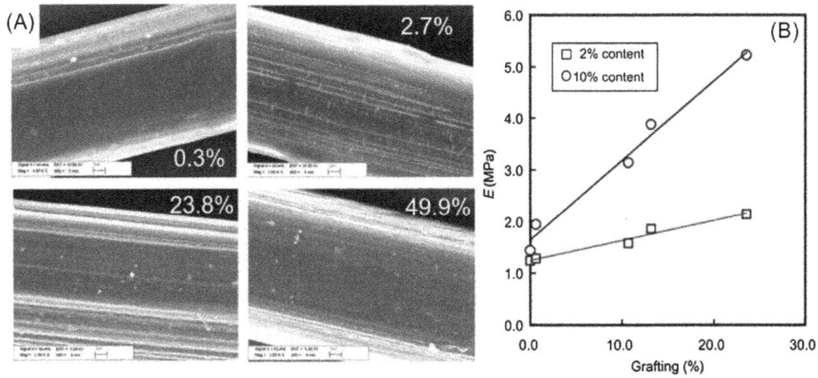

Figure 2.28 Scanning electron microscope micrographs of UHMWPE-*g*-PAA fiber with different DGs (A) and the effect of degree of grafting (DG) of UHMWPE-*g*-PAA fiber on mechanical properties of SBR composites (B).[24] *PAA*, Poly(AA).

2.3.2 Preparation of amidoxime-based UHMWPE fibrous adsorbent for extraction of uranium from seawater

Extensive research on the recovery of uranium from seawater has been conducted to replace uranium locally deposited as terrestrial ore with uranium uniformly dissolved in seawater. At present, there are three major obstacles to using this virtually limitless reservoir as a source of uranium. First, the uranium is in a strongly complexed form ($UO_2(CO_3)_3^{4-}$) at extreme dilution in the presence of relatively high concentrations of other ions. Second, it is very difficult to efficiently contact the extraction agent with very large volumes of seawater, which would be involved during the process of extraction. Most importantly, the extractant must work effectively at the seawater pH and ionic strength, and must be virtually insoluble. Third, deterioration of the extractant due to fouling in the marine environment has to be considered. There are two main sources of fouling: deposition of gelatinous iron oxides onto the adsorbent material surface and marine biofouling (such as shellfish and seaweed).

Although a number of other uranium extraction methods have been investigated, amidoxime (AO)-based adsorbents are recognized by many groups as the most promising material. It has been demonstrated that radiation-induced grafting of AO functional groups onto PE fibrous adsorbents has a rate of adsorption of uranium and vanadium from seawater 100 times higher than that of commercialized resins.[25] Under a memorandum of understanding between the Chinese Academy of Science (CAS) and the US Department of Energy, the research groups at Shanghai Institute of Applied Physics (SINAP), and Oak Ridge National Laboratory (ORNL) parallel-developed and compared several adsorbents based on AO groups in recent years. Nevertheless, real ocean adsorption experiments must still be promoted. In the early days, the Japan Atomic Energy Research Institute carried out two types of marine experiments with AO-based PE fibers prepared by radiation-induced grafting polymerization: (1) a uranium collection system using AO-adsorbent stacks and (2) a uranium collection system using AO braid adsorbent, which can stand on the bottom of the sea and the ocean current can easily force seawater through the braid-type adsorbent. The stack adsorption system involved a floating frame stabilized with ropes and adsorption beds for assembling adsorbent sheets, and uranium collection experiments with adsorbent stacks were performed in the Aomori prefecture of Japan from 1999 to 2001. After a total submersion time of 240 days in the ocean, ~1 kg of uranium was collected by the stack system.[26] Despite some achievements, the researchers found that serious biofouling and the high collecting cost were major problems with the stack adsorption form. Therefore, the braid collection system was developed, and adsorption tests were carried out in Okinawa, Japan. The adsorption ability of the braid system was ~2 times higher than that of the stack form.[26,27] To date, it is the most successful method of uranium collection from marine environments reported in the literature. Recently, several countries have conducted marine tests with natural seawater. The Marine Sciences Laboratory (MSL) of the Pacific Northwest National Laboratory (PNNL) has established a system for estimating the uranium adsorption performance of adsorbents with natural seawater from Sequim Bay, Washington. Kim et al. conducted column experiments at the MSL using an AO-based polymeric adsorbent developed at the ORNL, and the maximum uranium uptake from seawater was 3.3 mg g^{-1} after 8 weeks of contact with filtered seawater.[28] Alexandratos et al., studied the adsorption by different fibers in natural seawater at PNNL for 20.8 days, and the results showed that the adsorption capacities of the fibers were much lower than that in simulated seawater.[29] However, investigations of the adsorption performances of adsorbents in natural seawater are still rare.

We selected UHMWPE fiber as a substrate polymer for radiation grafting because of the following advantages. First, it has ultrahigh tensile strength and corrosion resistance. After immersion in seawater for 13 months, the tensile strength decreases by ~14%, even when soaked in sulfuric acid for 30 min prior to immersion in seawater (Fig. 2.29).[30] In addition, as illustrated in Fig. 2.30, free radicals generated in the fibers after irradiation are stable and have a long half-life (26 days in vacuum, 13 days in air),[21] which is favorable for preirradiation graft polymerization to obtain a high DG. For example, the DG of γ-ray-irradiated UHMWPE fibers with an

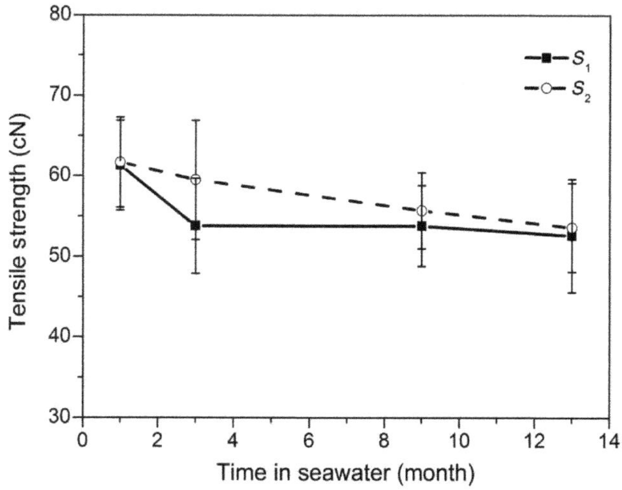

Figure 2.29 Tensile strength of UHMWPE fibers irradiated with an absorbed dose of 50 kGy immersed in simulated seawater for a definite time (S_1 was soaked in sulfuric acid for 30 min prior to immersion in simulated seawater).[30]

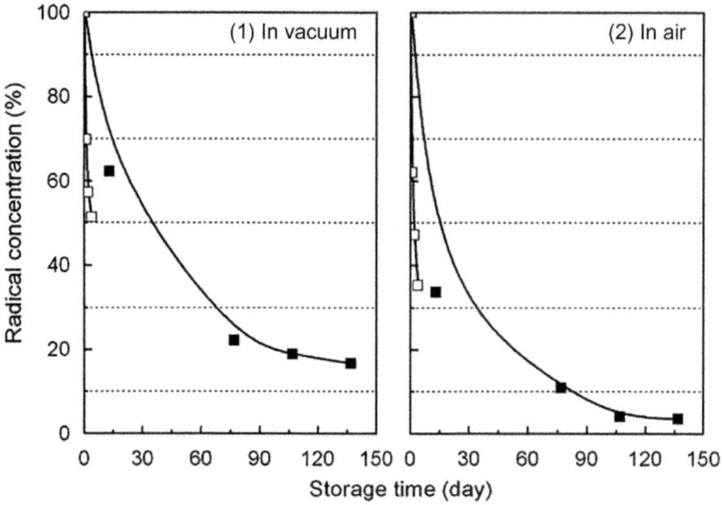

Figure 2.30 Radical concentration versus storage time for UHMWPE fibers with an absorbed dose of 90 kGy (■: original UHMWPE fiber; □: heat-treated UHMWPE fiber at 180°C).[21]

absorbed dose of 50 kGy reaches 360% after 3 h of graft polymerization. Moreover, the UHMWPE fiber has high tensile strength and corrosion resistance even after radiation graft polymerization. In the study by Tamada et al.,[31] PE fiber as a trunk polymer was irradiated with an EB at 200 kGy in N_2 gas, and the DG reached 150% after 4 h of graft polymerization. A higher absorbed dose not only leads to high energy consumption but also impairs the mechanical properties of the fibers, which influences the lifetime and recyclability of the adsorbents. Therefore, UHMWPE fiber is at least equal to or better than PE fiber as a trunk polymer for uranium adsorption.

The process of preparing uranium adsorption material by radiation grafting includes the following steps: (1) irradiation of UHMWPE fiber; (2) postradiation graft polymerization of AN and AA; and (3) amidoximation of the grafted fiber. The reaction process is outlined in Fig. 2.31. Co-grafting of AA with AN onto UHMWPE fibers can enhance the uranium adsorption of the AO-based adsorbents.[28]

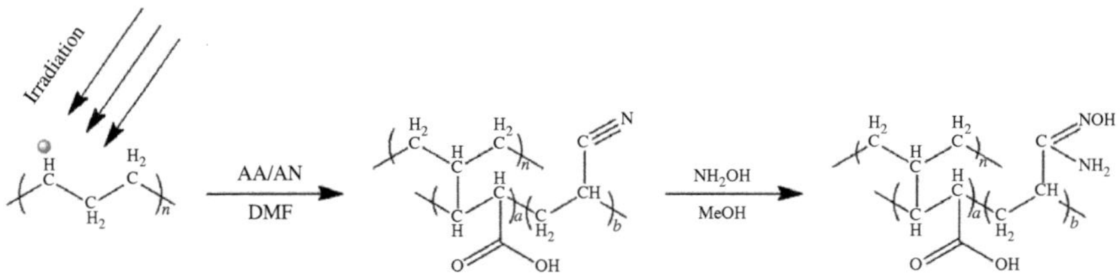

Figure 2.31 Preparation of AO-UHMWPE adsorbents by preirradiation-induced graft copolymerization of AN and AA, and subsequent amidoximation. *AO*, Amidoxime; *AN*, acrylonitrile; *AA*, acrylic acid.

AO-adsorbent fiber prepared using radiation-induced graft polymerization has the following advantages: (1) sufficient loading capacity and selectivity; (2) simple delivery, recovery, and regeneration of adsorbent; (3) suitable for production on a large scale; and (4) less damage to the marine environment.

To assess and characterize the adsorption performance of AO-based UHMWPE adsorbents, seawater adsorption tests were performed in batch adsorption, flow-through adsorption, and marine adsorption modes:

2.3.2.1 Batch adsorption experiment

Batch adsorption experiments were performed using 5 L plastic tanks of simulated seawater. The concentrations of uranium and coexisting ions were quantitatively monitored by adding a standard solution to the sea salt solution to create an initial concentration 100 times higher than that of natural seawater. After adsorption in the simulated seawater for 42 days, the color of the AO-UHMWPE adsorbents changed from light yellow to brown, as shown in Fig. 2.32. This indicates that some metal ions were adsorbed together.

2.3.2.2 Marine adsorption experiment

Three coastal marine areas of China (Xiamen, Daishan, and Raoping), were selected to evaluate the performance of the adsorbents. The Xiamen adsorption platform was located in the Taiwan Strait (118.09°E, 24.43°N), a buoy platform of the State Key Laboratory of Marine Environmental Science (Fig. 2.33A). The platform was ~500 m far from the coast.

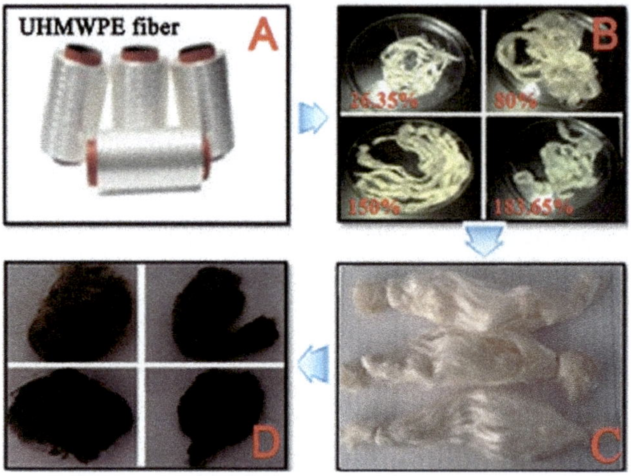

Figure 2.32 (A) Photographs of UHMWPE fiber, (B) grafted fiber, (C) AO-UHMWPE adsorbents before, and (D) after adsorption.

Figure 2.33 Photographs of the three coastal marine adsorption platforms: (A) Xiamen adsorption platform, (B) Daishan adsorption platform, and (C) Raoping adsorption platform.

Adsorbents were strapped and immersed in seawater ~3 m below the water surface. The salinity of the seawater was 31 psu, and the flow velocity was 0.2 − 0.6 m s^{-1}. The pH value of the seawater was 7.5 − 7.9. The adsorption test in this study was carried out at an average seawater temperature of 21.4°C. All metal ion concentrations in the seawater sample are listed in Table 2.2. The Daishan adsorption platform was located in the East China Sea (122.17°E, 30.25°N) at Daishan Lvyuan Desalination Co., Ltd. (Fig. 2.33B). Adsorbents were strapped and immersed in the reservoir ~1 m below the water surface for 15 days. The salinity of the seawater was 27 psu, and the flow velocity was 0.5 m s^{-1}. The adsorption test was carried out at an average seawater temperature of 25.5°C. The concentrations of uranium and vanadium in the seawater sample are listed in Table 2.2. The Raoping adsorption platform was located in the South China Sea (117.08°E, 23.58°N) at Guangdong Zhongkebao Biotechnology Co., Ltd. (Fig. 2.33C). Seawater was pumped to the underground impounding reservoir. The salinity of the seawater was 35 psu. The adsorption test in this study was carried out at an average seawater temperature of 17.1°C. Adsorbents were strapped and immersed in the reservoir. The seawater in the reservoir was almost static with a very low flow velocity less than 0.01 m s^{-1}. The concentrations of elements in the seawater are shown in Table 2.2. AO-UHMWPE fiber adsorbents were freely placed in the seawater (Fig. 2.34). In addition to the impact of competitive ions, biofouling is another important factor influencing the amount of uranyl ions binding to the AO-UHMWPE adsorbent. In Fig. 2.35, it can be seen that sediment and marine life (i.e., seaweed and shellfish) adhere to the surface of the AO-UHMWPE fibers, which significantly affects the diffusion of uranyl ions to the adsorbents.

2.3.3 Preparation of highly durable conductive UHMWPE fibers

Metallization of fibers is widely used in various technological applications because of their desirable properties, including electrostatic discharge, electrical conductivity, radio frequency interference

Table 2.2 Concentrations of various elements in the batch, flow-through, and marine adsorption experiments

Element	Typical seawater conc. (ppb)	Conc. in simulated seawater in batch test system (ppb)	Conc. in simulated seawater in flow-through test system (ppb)	Conc. in seawater at Xiamen (ppb)	Conc. in seawater at Daishan (ppb)	Conc. in seawater at Raoping (ppb)
U	3.3	331	3.6	4.0	2.6	3.1
V	1.5–2.5	150	1.9	0.5	2.9	2.0
Fe	1.0–2.0	141	40.6	14.5	–	48.1
Co	0.05	5.3	0.3	0.02	–	0.06
Ni	1	101	1.1	0.3	–	0.5
Cu	0.6	65.4	5.4	0.1	–	1.2
Zn	4	408	8.2	0.9	–	21.6
Pb	0.03	34.6	31.6	0.07	–	0.7
Mg	1.3×10^6	1.2×10^5	1.2×10^5	0.6×10^5	–	–
Ca	0.4×10^6	0.6×10^5	0.6×10^5	0.2×10^5	–	–

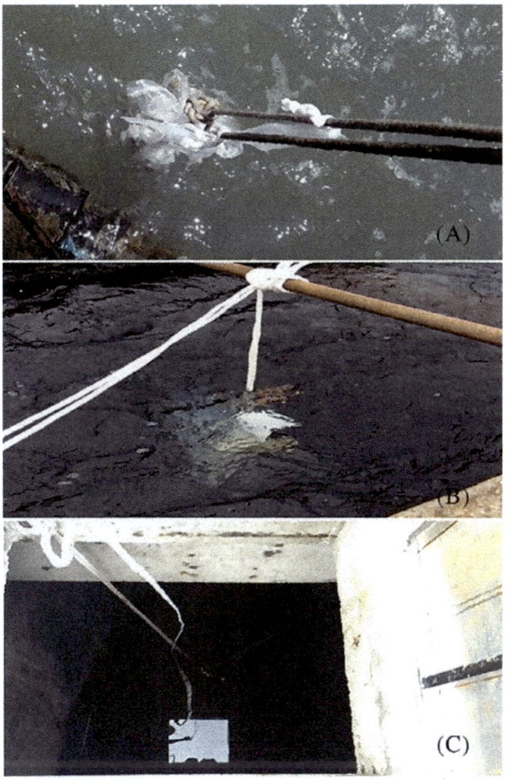

Figure 2.34 Placement of adsorbents in seawater at different marine locations: (A) Xiamen adsorption platform, (B) Daishan adsorption platform, and (C) Raoping adsorption platform.

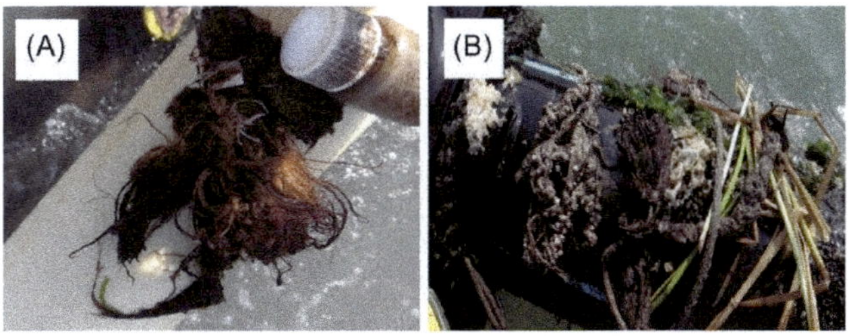

Figure 2.35 AO-UHMWPE adsorbent: (A) after uranyl ion adsorption and (B) contaminated by marine organisms.

protection, and electromagnetic interference protection. Metallization of fibers combines the specific properties of fibers, such as their lightweight, design flexibility, and low cost of manufacturing; and characteristics usually associated with metals, including their reflectivity, electrical conductivity, abrasion resistance, and a variety of decorative effects. The development of conductive fibers is also critical for various applications in wearable electronics and flexible electronics such as wearable displays, solar cells, actuators, data management devices, and biomedical sensors.

The processes that can be used for metalizing polymer can be either in gaseous phase such as physical and chemical vapor deposition, or in solution, as in wet chemical metallization, that is, plating. Plating can be subdivided into electroless plating and electroplating. Electroless plating is an attractive metallization technique because it offers the ability to metalize insulating substrates at low temperature using cost-effective processes and equipment. It is also an important technology for microelectronic metallization, because of its compatibility with the process of Damascene-Cu electrochemical plating and to its ability to form all metal-encapsulated, highly reliable Cu interconnects in an inline and self-aligned manner. Another advantage of electroless plating is the use of localized catalysts for site-selective metallization. The process of electroless plating involves two main steps: (1) surface activation and surface seeding with the catalyst for the next step, (2) an electroless plating bath mainly containing metallic ions (e.g., Ni^{2+} or Cu^{2+}) and a strong reducer. To activate the metal reduction onto the polymer, noble metals Pd, Au, and Ag are usually employed as catalysts to initiate the electroless plating.

At present, there are two main approaches to the surface activation of fiber, which can be classified into surface etching/coarsening and surface addition. Surface etching/coarsening usually adopts chemical or physical means to improve surface roughness or makes active functional groups on the original surface via in situ oxidizing reactions, such as chemical etching, laser ablation, sand blasting, and oxygen plasma. Surface addition refers to add an extra interface layer on the surface of fibers, typically including grafting polymerization, surface silanization, and layer-by-layer deposition of polyelectrolytes. The functions of surface activation are mainly embodied in the following aspects. On one hand, it is able to selectively adsorb and efficiently load the catalyst to decrease the diffusion of catalyst into the plating solution; on the other hand, it can improve the adhesion between the fiber substrate and metal to avoid delamination. The activated surface, especially via surface addition, should be chemically resistant to electroless plating and can further form an interface layer between the original substrate and the deposited metal layer for better adhesion. Surfaces activated by different methods can improve the interface binding force between organic matrix and metal coating, but most are still far from industrial application, either due to complex or environmentally unfriendly processes, or because of the difficulty in large-scale production. For instance, a typical chromium-containing etching agent for surface activation of polymer has been prohibited in Europe because of its harm to the environment, which requires an efficient process to be replaced. A robust interlayer between the substrate surface and the metallic layer might be able to replace the rough interface treated by the acid. Li et al., reported a novel, facile, and versatile approach for preparing highly durable, electrically conductive cotton yarns. 2-(Methacryloyloxy)ethyltrimethylammonium chloride brushes are first grown from cotton surfaces by surface-initiated atom transfer radical polymerization (ATRP).[32] Subsequent electroless deposition of metal particles onto the modified cotton surface to form polymer-bridged metal/cotton hierarchical structures resulted in the formation of conductive yarns. However, in that case, the ATRP process is complex and unsuitable for large-scale preparation. Compared with the ATRP process, radiation graft polymerization is favorable, due to its strong penetration into the polymer trunk, uniform formation of active sites for initiating grafting, and its ability to endow new performance characteristics to a material. Therefore, the radiation graft technique (RIGP) is a suitable methodology for activating the surface of polymer and building a coating which is covalently grafted to the substrate and able to provide a complex catalyst for the growth of the metallic phase which starts inside the host polymer.[33] Moreover, radiation graft polymerization is suitable for large-scale industrial production.

Since UHMWPE fiber exhibits excellent physical and mechanical properties, the metallization of UHMWPE fiber is therefore of great importance with regard to the combined properties of the UHMWPE fiber and copper. The metalized UHMWPE fiber can substitute for copper and steel fibers and act as electromagnetic interference shielding, microwave absorption and reflection, and static charge dissipation materials in coaxial cables and wear protection, which then show wider potential application prospects in the aerospace, telecommunications, and military industries. Despite its many excellent properties, UHMWPE fibers are difficult to plate because of their hydrophobic surface, their chemical inertness, and the lack of functional groups on the surface. To date, how to effectively activate the UHMWPE surface with controllable conductivity and high durability remains a great challenge in the process of electroless plating.

A new kind of UHMWPE fiber-based conductive fiber was synthesized by radiation graft polymerization of AA with subsequent electroless plating of copper onto the AA modified UHMWPE fiber to form polymer-bridged metal/fiber hierarchical structures. The utilization of the RIGP technique and silver activation is very simple and effective to introduce seeds of the catalyst to the fiber matrix and initiate the subsequent electroless plating, which could then be a substitute for the traditional cadmium etching method and prove its applicability in the coating industry. Moreover, thanks to the metallic cation entanglement in grafted poly(AA) chains, mechanical anchoring of the final metallic layer is also obtained by this process.

Because of the high conductivity and robustness, the prepared conductive UHMWPE fibers may be used as conductive wires in electronic circuits. In order to show the conductivity property of the fiber, we constructed a simple circuit by bridging a battery and one electrical contact of a LED with our prepared conductive fiber (Fig. 2.36). When the fiber linked the battery and the LED light,

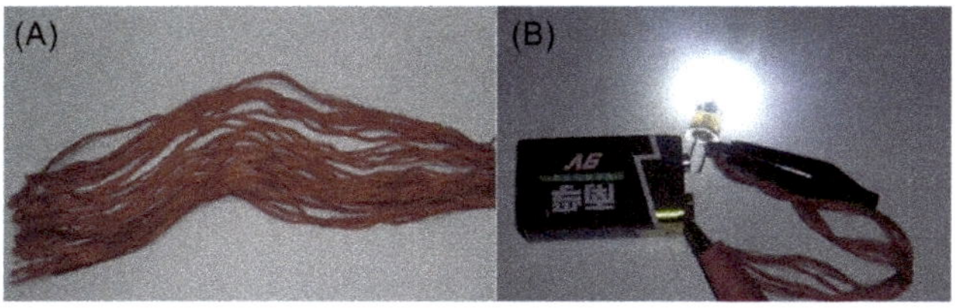

Figure 2.36 Digital images of UHMWPE conductive fiber: (A) prepared by radiation grafting and (B) modified fiber used as electrical wire for powering an LED.

the LED turned on immediately and was illuminated for more than 10 min until it was disconnected. The LED can be repeatedly switched on and off by connecting and disconnecting the contact without loss of fidelity.

2.4 PREPARATION OF FUNCTIONAL TEXTILES BY RADIATION GRAFTING

2.4.1 Types and preparation methods of functional textiles

Textile is closely related to human life, and an essential material for mankind. As early as 2000 years ago, there were many poems about ancients making textiles in *The Book of Songs*, such as the verse, "pool in the east gate is used for retting," which is a description of the process of using hemp fiber for making textiles. For a long time, natural material has been used for textiles: cotton, linen, silk, wool, and so on. In the 20th century, man-made fibers, such as nylon, polyester, and polyacrylic, became an important source of textile materials on the basis of the development of polymer science.

With the improvement of living standards, people are no longer satisfied with meeting their warmth, comfort, decoration, and other needs, and have higher requirements for the functionality of textiles; research and development of functional textiles are becoming increasingly important. At the same time, with the increasingly fierce competition from international textile markets, the demand for functional textiles with high value-added features is also growing. Functional textiles are textiles that have specific physical and chemical properties in addition to the inherent properties of general textiles.

2.4.1.1 Types of functional textiles

2.4.1.1.1 Antimicrobial textiles

Infectious diseases caused by pathogenic microorganisms have always been an important threat to human health. Preventing pathogenic bacteria from harming the human body is an important measure to prevent the occurrence and spread of infectious diseases, and avoiding or reducing the contact between pathogenic bacteria and the human body is an effective means to reduce injury. Ordinary textiles have no inhibiting effect on microorganisms such as bacteria and fungi, and pathogens are prone to grow and reproduce in ordinary textiles, thereby contacting and attacking the human body. Antimicrobial textiles can cut off the pathways of bacteria to prevent the propagation of germs on textiles, so as to effectively avoid the contact between pathogens and the human body, and reduce the risk of human pathogens and cross-infection rates. In addition, antibacterial textiles can also prevent the textiles from being damaged because of microbial erosion, and inhibit the odor caused by bacterial decomposition of the fabric, so as to ensure the health and comfort of the human body.

At present, the production of antimicrobial textiles mainly involves introducing antibacterial agent on textiles. Common antibacterial agents include nanometer metal clusters (such as nano-silver), metal salts (copper and silver salts such as copper sulfate), and some natural polymers with antibacterial functions such as chitosan. For example, Jeong et al.,[34] used the twin-screw extrusion technique to blend Ag nanoparticles into PP fibers, and then produced textiles with good antibacterial properties.[34] However, a shortcoming of conventional production methods is that the combination of antibacterial agents on the textile mainly depends on physical adsorption, because its binding force is weak, and the antibacterial agent may be lost after using and washing, and the textile antibacterial effect will not last. The preparation of long-term antibacterial textiles is expected.

2.4.1.1.2 Antielectromagnetic radiation textiles

Electromagnetic radiation has biothermal and nonthermal effects on biological energy. Electromagnetic radiation in the human living and work environment has always been endlessly debated in relation to the health hazards for human beings. Taking safety into consideration, antielectromagnetic radiation has aroused great interest, and also has a great market. There are three methods for the preparation of antielectromagnetic radiation textile: First, use electromagnetic shielding fiber for spinning or blend and process it with other fiber; second, carry out the antiradiation function process for the general textile; and third, add a wire mesh sandwich in ordinary textiles. At present, the most mature and the most widely used is the blending of metal fibers and ordinary fibers, and the shielding rate of textile is between 20 and 35 dB, which basically can shield the vast majority of electromagnetic radiation exposed in daily life. In addition, the fabric feels good, wears comfortably, and can be washed in water, although long-term use will cause the metal wire breakage to impact the shielding effect. Coating method was also used to introduce conductive macromolecules and magnetic nanoparticles coated on ordinary textiles, in order to prepare textiles with good antielectromagnetic radiation properties. However, the effectiveness of the bond between the coating and the textile, and the durability of the modified fabric are still problems to be solved.

2.4.1.1.3 Antiultraviolet textiles

Excessive exposure to UV rays causes harm to human skin, and is one of the important factors causing skin cancer. Anti-UV textiles can effectively block UV rays and protect skin. The preparation of anti-UV fabric mainly depends on the after-finishing method, and there are two main types of finishing agents: One is coating inorganic particles with stronger reflection to UV ray, such as zinc oxide and titanium dioxide; the other is finishing with an absorbent that can transform the high energy UV into heat or harmless low-energy UV radiation. Using the wet chemical method, ZnO nanoparticles can be immobilized onto the surface of fabric to endue its good anti-UV effect.

2.4.1.1.4 Waterproof and moisture-permeable textiles

Waterproof and moisture-permeable textiles can ensure that water does not penetrate into the fabric under a certain pressure, while also allowing water vapor to pass through the fabric quickly and achieving the effect of being waterproof and moisture-permeable, which is known as "breathable fabric." The common processing methods are laminating and coating. At present, the most common is laminated fabric, which is mainly composed of multiporous Teflon film, PU film, and fabric composite lamination. Gore-Tex fabric is a more mature product, and has a very high market share of fabrics for outdoor sportswear. Directly grafting hydrophobic groups onto cotton fabrics can introduce the cotton waterproof function while maintaining the original comfortable and breathable characteristics of cotton, and has good application prospects. A research group at the SINAP, CAS, has carried out extensive investigations in this area.

2.4.1.1.5 Flame-retardant textiles

Flame-retardant fabric with fire-retardant properties is widely used in industrial products. The flame-retardant property of textiles can be obtained by two methods: First, for the finishing of flame-retardant textiles, the defect is that the flame retardant is weakly bonded to the textile, and the flame-retardant property will gradually decrease with the increase of the service life and washing times; second, the flame-retardant fiber is synthesized first, and then woven into the flame-retardant textile. Then, its flame-retardant property can be retained for a long time. For example, a layer-by-layer self-assembly method was used to process the surface of cotton in order to obtain branched polyethylene imides and sodium montmorillonite clay coating on the surface of cotton sheet.[35,36] A residual weight of 15% can be obtained after heating to 500°C in the TGA. Based on the analysis of fiber morphology of the coating samples after vertical combustion, it was found that the fiber shape and weaving structure were maintained, indicating that modified cotton has better flame-retardant properties.[35,36]

2.4.1.1.6 Antistatic textiles

The static electricity accumulated on textiles is not only harmful to human health, but also has potential dangers. The surface resistance of antistatic textiles is much smaller than that of ordinary textiles, which can effectively accelerate the dissipation of charge and leakage of static, and reduce the damage caused by the accumulation of static on the surface of textiles. The method of obtaining antistatic fabric mainly involves embedded conductive fiber and the fabric surface finishing: the former is knitted with ordinary fiber and wire to obtain a fabric with lasting antistatic effects; the latter is to cover a layer of antistatic resin on the fabric surface using antistatic resin, in order to increase the conductivity of the fiber and have the antistatic effect. However, the persistence of this method is expected to be improved. Zhou et al.,[35] modified the polyvinyl chloride substrate with ZnO whiskers to prepare the composite interface. It was found that the composite coating could effectively disperse the charge on the surface of the fabric and had good antistatic function.[37]

2.4.1.2 The preparation methods of functional textile

The preparation methods of functional textiles are divided into two categories. One is to prepare the functional fiber first and then weave it into textile. Through the method of blending and composite spinning, the functional additives of uniform dispersion will be added to the fiber to obtain functional fiber; the surface of conventional fiber is processed, for example, to introduce a resin with a specific function on the surface of the fiber, in order to achieve fiber functionalization. The functional fibers are woven to obtain functional textiles. The other is to modify ordinary textiles with functional group. The physical finishing process includes three methods: impregnation, padding, and coating, and its main drawback is that the bonding force is weak between the finishing agent and textiles. However, its strength is more durable, and functionality can be maintained for a long time. The chemical finishing method involves grafting a functional monomer onto a polymer substrate, to obtain a new functional textile. The advantage is that its functionality can be maintained for a long time. Biological ecological finishing is a finishing method which has emerged in recent years, and it adopts biological enzymes with biological activity in the finishing of textiles.

2.4.2 Advantages of preparing functional textiles by radiation grafting

At present, the main drawback of conventional functional textile preparation methods is that the combination of functional materials (or groups) on textiles is insecure, resulting in poor functional durability. The energy of ionizing radiation (γ-ray or EB) is several orders of magnitude higher than covalent bond energy (3–4 eV). Therefore, it is possible to generate free radicals conveniently on textiles through the irradiation method. Low absorbed dose might be able to initiate the graft polymerization of vinyl monomers, forming a covalent bond between the textile and the graft segments, to achieve permanent functionalization.

2.4.3 Current status of preparation of functional textiles by radiation grafting

Due to the advantages of radiation grafting in the preparation of functional textiles, it has also attracted wide attention from many researchers. Gupta et al.,[38] carried out radiation grafting of *N*-isopropylacrylamide and AA on the surface of polyester fabric to obtain thermally sensitive fabrics. Li's group at the SINAP (CAS) have used the radiation grafting method to prepare various kinds of functional textiles, such as super-hydrophobic textiles,[39–41] antistatic cotton fabrics,[42,43] self-cleaning textiles,[44] and antibacterial textiles.[45,46] The preparation of functional textiles using radiation grafting can be divided into two categories: (1) radiation grafting of organic monomers and (2) radiation grafting of nanomaterials and organic monomers.

2.4.3.1 Radiation grafting of organic monomers

2.4.3.1.1 Direct grafting of organic monomers onto textiles

Vinyl monomers, represented by AA or methacrylic acid and its derivatives, can be easily triggered to polymerization by free radicals. Their self-polymerization reactions are mild and efficient. Therefore, they are the most commonly used functional organic monomers in radiation grafting. In particular, their acid ester derivatives can introduce a variety of specific functional groups without significant impacts on the polymerization performance, and are powerful tools for the preparation of functional textiles.

For example, through the radiation grafting method, superhydrophobic cotton can be prepared by grafting (methyl)acrylic acid ester derivative with the hydrophobic group onto cotton. Superhydrophobic material refers to material contact angle of water of which is greater than 150 degrees, which is a kind of bionic material with a surface water-repellent effect; its surface is a rough structure with micro—nanoscale and low surface energy. Li et al.,[39,40] prepared super-hydrophobic cotton fabric by radiation grafting of poly(perfluorobutyl)ethyl aster on cotton. The results showed that the hydrophobicity of the grafted cotton textiles became stronger with the increase of the DG, and the initial contact angle became larger. When the DG exceeds 10%, stable super-hydrophobic cotton textiles were obtained, and the contact angle was between 150 and 160 degrees. The surface chemical structure of grafted cotton was analyzed by XPS and the F/C element ratio was calculated. When the DG was less than 10%, the F/C ratio linearly increased with the degree of grafting. When the grafting rate was more than 10%, the F/C element ratio was stable at around 1.08 instead of changing with the DG, which confirmed that the super-hydrophobic characteristic was caused by the reduction of the surface energy of the fluorinated polyacrylate graft chain.

Because the graft chain was connected to the cotton fiber macromolecules by covalent bonds, theoretically the combination of low surface energy groups on cotton was strong. The covalent binding force could be investigated through the laundry-proof test. According to the testing standards of the American Association of Textile Chemists and Colorists (AATCC) 61 (2006), 2 A (water temperature of 49, 0.15% of nonfluorescent brightener detergent powder, and 50 steel balls), after 50 times of accelerated laundering (equivalent to 250 times domestic laundering), the super-hydrophobic cotton textiles obtained by radiation grafting still retained good super-hydrophobic characteristics. If washed once weekly, super-hydrophobic cotton textiles can be used for ~5 years. The results showed that the strength of the covalent bonds between the fluoropolymer graft chain and the cotton fiber macromolecules in the super-hydrophobic cotton prepared by the radiation grafting method was higher, which could not only resist the chemical erosion from detergent, but also the mechanical wear of steel balls. Therefore, it is an effective way to improve functional textile laundering durability through covalent bond connecting.

Taking only hydrophobicity into consideration, the surface energy of the long-chain alkanes was low enough. At the same time, considering the cost of fluorine-containing monomer, the researchers used the same method to graft alkyl methacrylate onto cotton, in order to obtain super-hydrophobic cotton.[41] The minimum grafting rate was closely related to the monomer structure, and the longer the alkyl chain, the lower the grafting rate needed to achieve super-hydrophobicity. For example, cotton grafted with methyl methacrylate had no super-hydrophobicity, no matter how high the DG was; for grafted *n*-hexyl methacrylate, super-hydrophobic cotton could be obtained when the grafting rate was over 20%; for grafted lauryl methacrylic (LMA), which contained dodecyl chain, the minimum grafting rate of super-hydrophobicity was only 10%.[41]

The intensity of the covalent bond and the wear resistance of the super-hydrophobic cotton were also investigated. According to the GB/T 3920 (2008) specifications, under the pressure of 44.8 kPa, after the cotton-*g*-PHMA and the cotton-*g*-PLMA were rubbed 8000 and 9000 times, respectively, by a copper pillar, the super-hydrophobicity was damaged (contact angle is less than 150 degrees); after ironing using steam (~200°C), the super-hydrophobicity was restored. The mechanism of ironing repair is due to the penetrating action of irradiation, and the hydrophobic graft chains are evenly distributed on the surface and interior of the fiber, while when the grafted chains of the surface are worn, the internal grafted chains are transferred to the surface by ironing, thereby repairing their super-hydrophobicity.

More importantly, if one carries out the periodic ironing of super-hydrophobic cotton at the beginning, such as ironing once every 2000 times of friction, which can greatly extend its life. For example, cotton-*g*-PHMA was destroyed (Fig. 2.37A) after being rubbed 16,000 times, which is twice as much as the unironed type, while the abrasion resistance of LMA grafted cotton (Fig. 2.37B) is more than 24,000 times.[41] The above studies confirmed that the strength of covalent bonds formed by radiation grafting between the textile and functional grafting chains was sufficient to provide long-term protection.

2.4.3.1.2 Grafting of reactive group-containing monomers and subsequent functionalization

In addition to simple grafting, the use of some vinyl monomers containing reactive functional groups can obtain technical textiles through the polymer chemical reaction by radiation grafting,

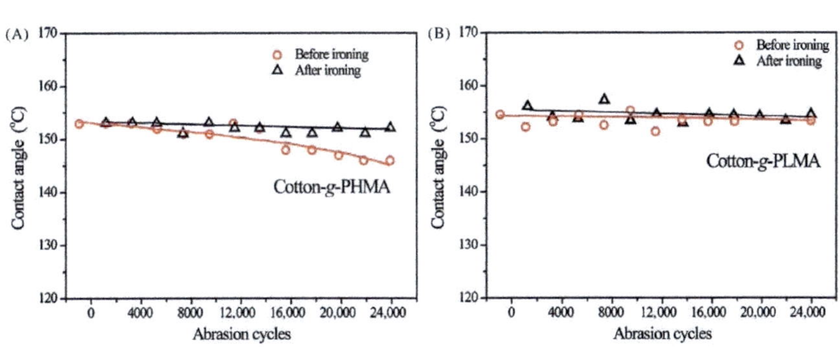

Figure 2.37 Abrasion-resistant and ironing repair performance of ultra-hydrophobic cotton: (A) cotton fabric grafted with HMA and (B) cotton fabric grafted with LMA.[41] *HMA*, *n*-Hexyl methacrylate; *LMA*, lauryl methacrylic.

thereby achieving specific functions. For example, AN is grafted to PE nonwoven fabric or UHMWPE before reacting with hydroxylamine nitrile, which can convert the nitrile group into the AO group. Since the AO group is a good coordinating group, the obtained nonwoven fabric or fiber with polyamidoxime grafted segments can complex with a number of transition metals.

Similarly, grafting of glycidyl methacrylate (GMA) and other monomers with epoxy groups by irradiation, and then reacting with various organic amines to open the rings of epoxy groups, makes it possible to prepare a large amount of amine-grafted textiles which can complex with metal ions. In addition, all levels of amine have not only coordination ability, but also different reducibility, and the metal ions adsorbed on the textiles can be reduced by amine groups. Ma et al.,[47] prepared a tertiary amine group-containing textile by the grafting of GMA on the surface of PE/PP nonwoven fabric and the subsequent ring-open reaction of diethylamine and GMA. This special textile could be coordinated with chloroauric acid and restore gold nanoparticles. Morphology-controlled gold nanotubes with a minimum wall thickness of 280 nm and an external diameter of 2.3 μm could be prepared by calcination. In this method, the tertiary amine group-containing textile not only acted as ligands and reducing agents, but also played the role of template during the calcination process.[47]

Conductive textiles can be prepared by grafting GMA on textiles as well as the subsequent reactions. As stated above, conventional conductive textile is generally blended with ordinary fibers through metal wire, and its durability is insufficient. Polyaniline (PANI) is a typical conductive polymer, and if PANI is grafted onto the textile, conductive textiles can be prepared. However, conventional irradiation induces radical polymerization and fails to achieve the grafting of PANI. Therefore, a three-step method is required to introduce PANI chains on the surface of cotton: (1) grafting GMA on cotton and introduction of epoxy groups; (2) using the reaction of epoxy groups to obtain aminophenylamine, and introduction of the aniline group into the cotton; and (3) immersing the cotton fabric into the aniline solution to initiate the

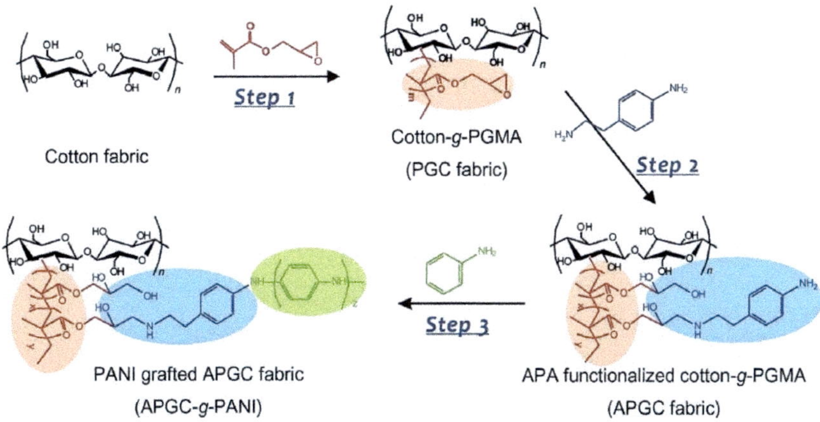

Figure 2.38 Three-step method used for achieving grafting polymerization of aniline on cotton and preparation of antistatic cotton.[43]

redox polymerization of aniline, then the PANI chains can be grafted onto the cotton, as shown in Fig. 2.38.[42,43] The surface resistance of cotton modified in this way could be reduced from $10^{11}\,\Omega\,\square^{-1}$ to $10^{8}\,\Omega\,\square^{-1}$, and the modified cotton presented good antistatic effects. After 40 accelerated laundering cycles according to the testing method of the AATCC 61 (2006) standard (equivalent to 200 conventional domestic launderings), the surface resistance of the modified cotton was almost constant. This proved that the combination of PANI graft chain and cotton was very strong and the resulting conductive cotton had good durability.

2.4.3.2 Radiation grafting of nanomaterials and organic monomers

Inorganic nanomaterials have abundant sources and varied functions, and have been hotspots of materials science research in recent decades. Functional textiles whose surface is coated with inorganic nanomaterials can bring more and richer functions and applications for textiles. How to make a connection between the inorganic nanoparticles and textiles is still a problem. Since the chemical structure of carbon nanomaterials such as graphene and its derivatives is close to that of organic macromolecular material, and graphene oxide (GO) has been reported to be antibacterial, GO and functional monomers were grafted onto cotton, and chemical initiators were used to initiate graft polymerization in the same system, in order to prepare durable antibacterial cotton. The GO-grafted cotton had good antibacterial properties, and its long-term properties were also tested according to AATCC standards. The results showed that the antibacterial rate of GO-grafted cotton was still above 80% after washing 100 times. The antibacterial rate of the cotton prepared by chemical grafting method significantly declined after washing 50 times. After washing 100 times, the antibacterial rate was reduced to below 30%.[45]

However, other metal oxide inorganic nanomaterials are difficult to form covalent bonds directly with cotton. The C—C double bond structure can be first modified and introduced on the surface of the metal oxide inorganic nanoparticle, and then co-grafted with acrylate monomer. Taking titanium dioxide nanoparticles as an example (Fig. 2.39), the C—C double bonds were introduced by using the esterification reaction of the hydroxyl group on the surface of titanium dioxide with maleic anhydride, and then the modified titanium dioxide nanoparticles were coated on the cotton after blending with hydroxyethyl acrylate. The titanium dioxide nanoparticles could be co-grafted on the surface of cotton under irradiation.[44]

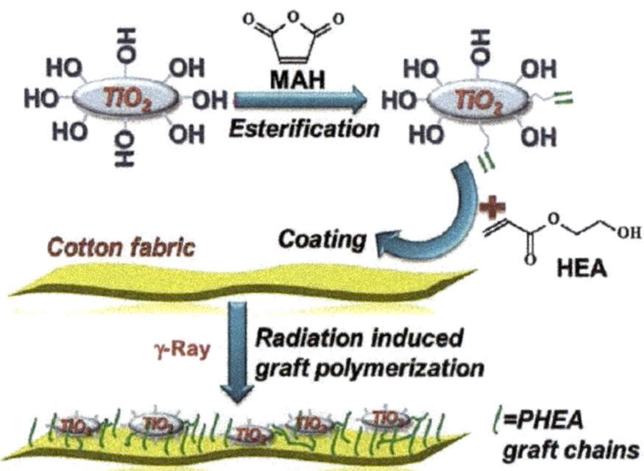

Figure 2.39 Radiation grafting of carbon—carbon double bond modified titanium dioxide nanoparticles.[44]

With the introduction of super-hydrophilic titanium dioxide nanoparticles and hydrophilic graft segments on the surface, cotton hydrophilicity was enhanced after grafting. At the same time, due to the photocatalytic function of titanium dioxide, the oil red dye on the surface of the modified cotton could be degraded under UV light, indicating that the titanium dioxide nanoparticles were linked to the cotton by the poly(2-hydroxyethyl acrylate) grafted chains through the covalent bonds, while their surface active sites were still exposed. After 30 accelerated laundering cycles (equivalent to 150 times domestic washing), the grafted cotton still had photocatalytic effects, and scanning electron microscope observation showed that a large number of titanium dioxide nanoparticles were still firmly bonded to the surface of cotton fiber. This indicated that although only 2% of the area on the surface of the titanium dioxide was covered by maleate, the introduction of double bonds was sufficient to allow it to participate in the graft polymerization and have some intensity. Although the laundering durability was slightly worse compared with the simple monomer graft modified cotton (as described above for super-hydrophobic cotton), 150 times ordinary washing could meet general use requirements.

Metal nanoparticles can be fixed on textile by using simultaneous radiation grafting of organic polymer and radiation reduction of metal ions. This enables the achievement of macromolecular graft chains to encapsulate the metal nanoparticles and their connection to the surface of the textile by covalent bonds. For example, when the cotton was placed in a mixed solution of silver nitrate and ethyl methacrylate, metal organic composite nanoparticles with pomegranate-like structures could be obtained on the surface of cotton by irradiation (Fig. 2.40). The as-obtained functional cotton had strong bacteriostatic properties, and could remain antibacterial property after many launderings. At the same time, due to the encapsulation and protection of polymer chains, the loss of silver nanoparticles was less during the washing process, and the concentration of silver ions in the washing solution was also lower. Therefore, it had little impact on the environment.[46]

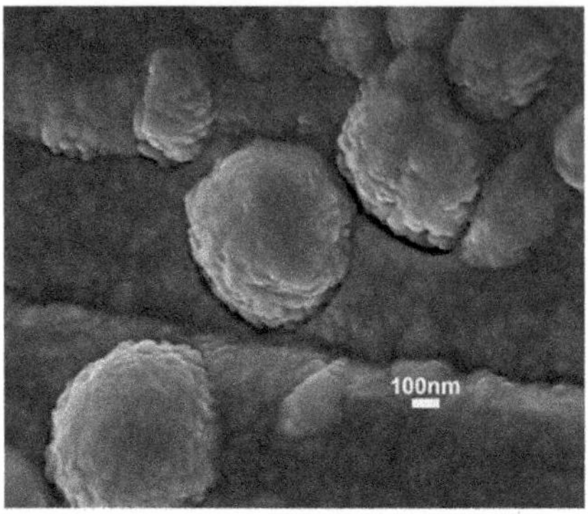

Figure 2.40 Radiation reduction to form the pomegranate-like structure nanoparticles (composited silver and polymer chains) on the surface of the cotton.[46]

REFERENCES

1. Henrici-Olivé G, Olivé S. Molecular interactions and macroscopic properties of polyacrylonitrile and model substances. *Adv Polym Sci* 1979;**32**:123–52.
2. Warner SB, Uhlmann DR. Oxidative stabilization of acrylic fibers. Part 3: morphology of polyacrylonitrile. *Mater Sci* 1979;**14**(8):1893–900.
3. Hill DJT, Land AP, Odonnel JH, Pomery PJ. The effects of gamma radiation on polyacrylonitrile. *Polym Degrad Stabil* 1992;**38**(3):193–203.
4. Liu WH, Wang MH, Xing Z, Qi Y, Wu G. Radiation induced crosslinking of polyacrylionitrile fibers and the subsequent regulative effect on the preoxidation process. *Radiat Phys Chem* 2012;**81**(6):622–7.
5. Liu WH, Wang MH, Xing Z, Wu G. Radiation oxidation and subsequent thermal curing of polyacrylonitrile fiber. *Radiat Phys Chem* 2014;**94**:9–13.
6. Fitzer E, Froh W, Heine M. Optimization of stabilization and carbonization treatment of PAN fibres and structural characterization of the resulting carbon fibres. *Carbon* 1986;**24**(4):387–95.
7. Dossantos LGC, Kawano Y. Degradation of polyacrylonitrile by X-ray-radiation. *Polym Degrad Stab* 1994;**44**:27–32.
8. Tarakanov BM. Effect of γ-radiation on the structure and thermal properties of polyacrylonitrile fibers. *Fibre Chem* 1995;**27**:150–4.
9. Tarakanov BM. The effect of laser radiation on the structure, thermal, and optical properties of poly(acrylonitrile). *Vysokomolekulyarnye Soedineniya Seriyaa & Seriya B* 1997;**39**(6):972–6.
10. Xiao H, Lu YG, Wang MH, Qin X, Zhao W, Luan J. Effect of gamma irradiation on the mechanical properties of polyacrylonitrile based carbon fiber. *Carbon* 2013;**52**:427–39.
11. Yajima S, Hayashi J, Omori M. Continuous silicon carbide fiber of high tensile strength. *Chem Lett* 1975;**9**:931–4.

12. Okamur K, Shimoo T, Suzuya K, Suzuki K. SiC based ceramic fibers prepared via organic to inorganic conversion process a review. *Cera Soc Jpn* 2006;**114**(6):445–54.
13. Seguchi T. New trends of radiation application to polymer modification irradiated in oxygen free atmosphere and at elevated temperature. *Radiat Phys Chem* 2000;**57**:367–71.
14. Ichikawa H. Development of high performance SiC fibers derived from polycarbosilane using electron beam irradiation curing—a review. *Ceram Soc Jpn* 2006;**114**(1330):455–60.
15. Peng T, Ye G. Development of *p*-aromatic polyamide fiber and its reinforced composite material. *China Synth Fiber Ind* 2004;**27**(06):46–9.
16. Zhang Y, Huang Y, Liu L, Wu L. Surface modification of aramid fibers with γ-ray radiation for improving interfacial bonding strength with epoxy resin. *J Appl Polym Sci* 2007;**106**(4):2251–62.
17. Zhang YH, Huang YD, Liu L, Cai KL. Effects of γ-ray radiation grafting on aramid fibers and its composites. *Appl Surf Sci* 2008;**254**(3):153–3161.
18. Qiu J, Liu L, Zhang Z, Huang Y. Effect of radiating APMOC fibers on tensile strength of its reinforced composite. *Aerosp Mater Technol* 2000;**30**(03):34–7.
19. Xu ZW, Huang YD, Zhang CH, Liu L, Zhang Y, Wang L. Effect of γ-ray irradiation grafting on the carbon fibers and interfacial adhesion of epoxy composites. *Compos Sci Technol* 2007;**67**(15):3261–70.
20. Sharma M, Gao SL, Edith M, Sharma H, Wei LY, Bijwe J. Carbon fiber surfaces and composite interphases. *Compos Sci Technol* 2014;**102**(4):35–50.
21. Zhao Y, Wang M, Tang Z, Wu G. ESR study of free radicals in UHMW-PE fiber irradiated by gamma rays. *Radiat Phys Chem* 2010;**79**(4):429–33.
22. Kawashita M, Itoh S, Araki R, Miyamoto K, Takaoka GH. Surface structure and apatite-forming ability of polyethylene substrates irradiated by oxygen cluster ion beams. *Biomed Mater Res A* 2007;**82**(4):995–1003.
23. Xing Z. *Study on modification of ultra-high molecular weight polyethylene fiber by radiation-induced graft polymerization [D]*. Shanghai: Shanghai Institute of Applied Physics, Chinese Academy of Sciences; 2013.
24. Kondo Y, Miyazaki K, Yamaguchi Y, Sasaki T, Irie S, Sakurai K. Mechanical properties of fiber reinforced styrene–butadiene rubbers using surface-modified UHMWPE fibers under EB irradiation. *Eur Polym J* 2006;**42**(5):1008–14.
25. Seko N, Tamada M, Yoshii F. Current status of adsorbent for metal ions with radiation grafting and crosslinking technique. *Nucl Instrum Methods Phys Res, Sect B* 2005;**236**:21–9.
26. Saito T, Brown S, Chatterjee S, Kim J, Tsouris C, Mayes RT, et al. Uranium recovery from seawater: development of fiber adsorbents prepared via atom-transfer radical polymerization. *J Mater Chem A* 2014;**2**(35):14674–81.
27. Seko N, Katakai A, Hasegawa S, Tamada M, Kasai N, Takeda H, et al. Aquaculture of uranium in seawater by a fabric-adsorbent submerged system. *Nucl Technol* 2003;**144**(2):274–8.

28. Kim J, Tsouris C, Oyola Y, Janke CJ, Mayes RT, Dai S, et al. Uptake of uranium from seawater by amidoxime-based polymeric adsorbent: field experiments, modeling, and updated economic assessment. *Ind Eng Chem Res* 2014;**53**(14):6076–83.
29. Seko N, Tamada M, Yoshii F. Current status of adsorbent for metal ions with radiation grafting and crosslinking techniques. *Nucl Instrum Methods Phys Res, Sect B* 2005;**236**(1):21–9.
30. Hu J, Ma H, Xing Z, Liu X, Xu L, Li R, et al. Preparation of amidoximated ultrahigh molecular weight polyethylene fiber by radiation grafting and uranium adsorption test. *Ind Eng Chem Res* 2016;**55**:4118–24.
31. Tamada M. *Current status of technology for collection of uranium from seawater.* Erice, Italy: Japan Atomic Energy Agency; 2009.
32. Liu X, Chang H, Li Y, Huck WTS, Zheng Z. Polyelectrolyte-bridged metal/cotton hierarchical structures for highly durable conductive yarns. *ACS Appl Mat Interfaces* 2010;**2**(2):529–35.
33. Gao Q, Hu J, Li R, Pang L, Xing Z, Xu L, et al. Preparation and characterization of super-hydrophobic organic-inorganic hybrid cotton fabrics via γ-radiation-induced graft polymerization. *Carbohydr Polym* 2016;**149**:308–16.
34. Jeong SH, Yeo SY, Yi SC. The effect of filler particle size on the antibacterial properties of compounded polymer/silver fibers. *Mater Sci* 2005;**40**(20):5407–11.
35. Horrocks AR, Kandola BK, Davies PJ, Zhang S, Padbury SA. Developments in flame retardant textiles—a review. *Polym Degrad Stabil* 2005;**88**(1):3–12.
36. Wu H, Zhang M, Jiang S. The development and application of flame resistant fabrics. *Adv Text Technol* 2002;**10**(2):37–9 [in Chinese].
37. Zhou Z, Chu L, Tang W, Gu L. Studies on the antistatic mechanism of tetrapod-shaped zinc oxide whisker. *J Electrostat* 2003;**57**(3–4):347–54.
38. Gupta B, Mishra S, Saxena S. Preparation of thermosensitive membranes by radiation grafting of acrylic acid/*N*-isopropyl acrylamide binary mixture on PET fabric. *Radiat Phys Chem* 2008;**77**(5):553–60.
39. Deng B, Cai R, Yu Y, Jiang H, Wang C, Li J, et al. Laundering durability of superhydrophobic cotton fabric. *Adv Mater* 2010;**22**(48):5473–7.
40. Cai R, Deng B, Jiang H, Yu Y, Yu M, Li L, et al. Radiation induced graft polymerization of a fluorinated acrylate onto fabric. *Radiat Phys Chem* 2012;**81**(9):1354–6.
41. Wu JX, Li JY, Deng B, Jiang H, Wang Z, Yu M, et al. Self-healing of the superhydrophobicity by ironing for the abrasion durable superhydrophobic cotton fabrics. *Sci Rep* 2013;**3**:2951.
42. Wu BT, Zhang BW, Wu JX, Wang Z, Ma H, Yu M, et al. Electrical switchability and laundering durability of conductive textile. *Sci Rep* 2015;**5**:11255.
43. Wu BT, Zhang BW, Wu JX, Ma H, Yu M, Li L, et al. Preparation of antistatic textile with covalently grafted polyaniline by a three-step process. *J Radiat Res Radiat Process* 2015;**33**:040302 . [in Chinese].
44. Yu M, Wang ZQ, Liu HZ, Xie S, Wu J, Jiang H, et al. Laundering durability of photocatalyzed self-cleaning cotton fabric with TiO_2 nanoparticles covalently immobilized. *ACS Appl Mater Inter* 2013;**5**(9):3697–703.

45. Zhao JM, Deng B, Lv M, Li J, Zhang Y, Jiang H, et al. Graphene oxide-based antibacterial cotton fabrics. *Adv Healthc Mater* 2013;**2**(9):1259–66.
46. Liu HZ, Lv M, Deng B, Li J, Yu M, Huang Q, et al. Laundering durable antibacterial cotton fabrics grafted with pomegranate-shaped polymer wrapped in silver nanoparticle aggregations. *Sci Rep* 2014;**4**:5290.
47. Ma HJ, Yao SD, Li JY, Cao C, Wang M. A mild method of amine-type adsorbents syntheses with emulsion graft polymerization of glycidyl methacrylate on polyethylene non-woven fabric by pre-irradiation. *Radiat Phys Chem* 2012;**81**(9):1393–7.

CHAPTER THREE

Radiation Cross-Linking and Its Application

Cong Zhang[1], Maojiang Zhang[1], Guozhong Wu[1], Yuhai Wang[2,3] and Lin Zhang[2,3]
[1]Shanghai Institute of Applied Physics, Chinese Academy of Sciences, Shanghai, P.R. China
[2]EVE Rubber Institute Co., Ltd., Qingdao, P.R. China
[3]Qingdao Key Laboratory for Advanced Tire Materials, Qingdao, P.R. China

3.1 RADIATION CROSS-LINKED WIRE/CABLE AND HEAT-SHRINKABLE MATERIALS

3.1.1 Radiation cross-linked wire/cable

Since cross-linking polyethylene (PE) induced by irradiation is found, extensive efforts have been made on the research and commercialization of PE cross-linking by radiation. The rapid development of high energy radiation devices such as electron beam (EB) accelerators and ^{60}Co isotope sources has provided important support to the industrialization of radiation cross-linking technology. Radiation cross-linking has become the most industrialized and sector with the highest production value in radiation processing industry.

In the cable industry, there are three kinds of cross-linking, namely, chemical, silicon, and radiation. In recent years in China, with the rapid development of EB accelerator manufacturing (1.5—5 MeV), medium and large-sized cables such as nuclear cables and marine cables can be produced by radiation cross-linking. According to our estimation, more than 400 installations of EB accelerators are in operation for industrial applications, mainly for cross-linking of cables and wires.

Radiation Technology for Advanced Materials.
DOI: https://doi.org/10.1016/B978-0-12-814017-8.00003-2

Radiation cross-linked wires and cables have a number of advantages. Since radiation cross-linked PE cables and wires have larger current-carrying capacity and stronger resistance to overload, the insulation thickness can be reduced and therefore the material costs and cable/wire weight can be reduced. In addition, such cross-linked products have good resistance to low temperature and weather, acid and alkali resistance, abrasion resistance and cracking resistance, and improved mechanical strength and flexibility, good flame retardancy, and longer service life.

Radiation cross-linked cables and wires are the two most popular commercialized products of radiation processing. Radiation cross-linked cables include 1 and 10 kV PE-insulated overhead cables, 1 kV PE-insulated power cables (replacements for polyvinyl chloride (PVC)-insulated power cables), PE-insulated control cables, PE airport lighting cables, solar photovoltaic (PV) cables, and K2 and K3 cables for nuclear power stations and marine environments. Radiation cross-linked wires include 125°C polyolefin (POF)-insulated locomotive oil-resistant wires, 105°C PVC-insulated wires (UL class), 105, 125, and 150°C POF-insulated wires (UL class), automotive low-voltage wires, and some aviation low-voltage wires.

Radiation cross-linked cables for nuclear power plants in China are mainly K1-type (within the island) cables intended for third-generation nuclear power. In 2015, China restarted its nuclear power program and was expected to construct more nuclear power units every year during the 13th 5-year plan period. These approved projects are all third-generation nuclear power generation units with better safety performance. Although nuclear power cables have promising prospects, the number of manufacturers is quite small, and most are only capable of producing K3-type wires and cables intended for second-generation nuclear island periphery. To date, Jiangsu Shangshang Group is the only enterprise that can produce K1-type cables for nuclear islands. Shell cables, produced by a local Chinese company and intended for third-generation nuclear power plants AP1000, have passed the International Electrotechnical Commission (IEC) test, the thermal aging test, the normal radiation test, accidental radiation test, the loss of coolant accident (LOCA) test, and the long-term immersion test.

The differences of cables between second-generation and third-generation nuclear power stations include the following: service life extension from 40 to 60 years; higher insulation performance (insulation resistance constants under 20 and 90°C are 3760 and 3.76 MΩ km, respectively); halogen-free flame retardant property. Moreover, greatly improved radiation resistance of nuclear power cables is required, apart from the ^{60}Co γ-ray radiation resistance test, for example, AP1000 and CAP1400 nuclear power cables are also required to pass a low-dose rate β irradiation resistance test.

Tests required on the first area cable (inside the containment) of AP1000 nuclear power plants include electrical performance, mechanical performance, halogen-free flame retardancy, and cable 60-year thermal life assessment. After 60-year life evaluation by ^{60}Co γ-ray irradiation testing, a β-ray radiation test at a dose rate and total absorbed dose (Figs. 3.1 and 3.2) follows. The LOCA test is normally carried out at a temperature of 251.7°C and a pressure of 427.5 kPa. After the LOCA test, the cables are immersed in 2428 ppm aqueous boric acid solution at 85.2°C for 10 months to conduct the insulation resistance test (Fig. 3.3).

Figure 3.1 ^{60}Co source irradiation resistance test diagram of nuclear power cable.

Figure 3.2 β-Ray irradiation resistance test diagram of nuclear power cable.

Figure 3.3 Sheath immersion test chart of Jiangsu Shangshang Company's cables after LOCA test. *LOCA*, Loss of coolant accident.

The Commercial Aircraft Corporation of China Ltd. has successfully developed the ARJ21 series and model C919 large commercial airplane. The development of domestic large airplanes has provided a market for the development of aviation cables in China. Currently, aviation wires are mainly made from radiation cross-linked poly(ethylene-tetrafluoroethylene) (XL-ETFE), commonly known as F40. Aviation Wire 55 made by the Tyco-Raychem

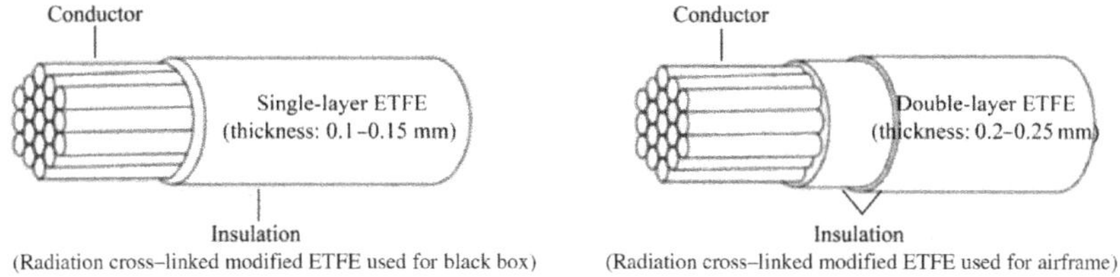

Figure 3.4 ETFE aviation wire made by US Tyco-Raychem Company.

Company of the United States is dominant internationally. At present, only companies like DuPont of the United States and Daikin of Japan can produce ETFE, and this kind of material is quite expensive. To make XL-ETFE wire fit for aviation uses, the protection of inert gas and a relatively high temperature during the cross-linking radiation process should be ensured. In addition, to remove the remaining hydrogen fluoride as well as the accumulated radicals during radiation cross-linking, the XL-ETFE wire should be post-treated under an inert gas atmosphere and at a relatively high temperature (Fig. 3.4).

3.1.2 Radiation cross-linked heat-shrinkable materials

Radiation cross-linked heat-shrinkable material, also called polymer shape-memory material, is a kind of intelligent polymer material cross-linked by irradiation. The development of heat-shrinkable materials started in the 1950s. In 1959, the Raychem Company of the United States applied for the first patent for PE heat-shrinkable tube. By the 1990s, the Raychem Company had developed into a large multinational business with 8000 employees, more than 3000 heat-shrinkable product specifications, and an annual output value of up to $3 billion. The Changchun Applied Chemistry Institute, Chinese Academy of Sciences, began research on radiation cross-linked heat-shrinkable materials in the 1960s, and successfully applied it in China's space satellite. Since the 1980s, many institutions and universities in China have been involved in the development of radiation cross-linked heat-shrinkable cables and films. By 2013, it was estimated that there were more than 180 enterprises in China that produced radiation cross-linked heat-shrinkable materials and products at scale, with over 36,000 employees and over 7 billion Yuan sales income.

Currently, heat-shrinkable materials developed in China are mainly applied in electronics, electric power, telecommunications, oil pipelines, automobiles, high-speed rail, and the military industry. China has become a major producer of heat-shrinkable materials. However, the technical level of our products is still of medium and lower grade, while high-grade products are monopolized by foreign companies such as the American Tyco Company. Domestic enterprises have also developed a variety of high-performance heat-shrinkable materials, including halogen and phosphorus-free heat-shrinkable materials used for electronics, and heat-shrinkable materials for nuclear power plants and aviation, high-performance heat-shrinkable materials for automobiles and high-speed railways, and high-performance cable accessories.

3.1.2.1 Halogen/red phosphorus-free heat-shrinkable materials for electronics

Heat-shrinkable materials are most widely used for electronics. Chinese companies can produce heat-shrinkable materials free of halogen flame retardant, but red phosphorus flame retardants are normally used in heat-shrinkable materials to pass the VW-1 vertical burning test under the UL224 standard. Therefore, only black, iron red, and dark-color heat-shrinkable tubes are available on the market. Red phosphorus decomposition may also occur in its formation and burning processes. Flame retardants containing red phosphorus are harmful to both the environment and human health. Many large international electronics companies already explicitly refuse to use red phosphorous heat-shrinkable tubes. At present, there are two types of heat-shrinkable materials free of red phosphorus flame retardant. One is the only use of inorganic flame retardants that do not contain any phosphorus element. But this kind of material is difficult to pass the American UL224 standard VW-1 vertical burning test. The other is the use of phosphorus/nitrogen-containing flame retardants and inorganic flame retardants or other halogen-free flame retardants, which are relatively easy to pass the standard VW-1 vertical burning test.

3.1.2.2 Heat-shrinkable materials used for nuclear power stations

The Tyco-Raychem Company in 2001 successfully developed the WCSF series heat-shrinkable tubes (Fig. 3.5) and accessories with a service life of 40 years, which are used for nuclear power plants and passed the LOCA/main steam line break (MSLB) test. However, this series of materials used halogen-containing flame retardants. The test steps for this material included: equivalent simulation of 40 years of thermal aging life at 90°C (thermal aging at 150°C for 1500 h, γ-ray irradiation of 2.0–2.9 MGy (samples at a voltage of 1000 V and rated current to simulate LOCA/MSLB accident); 30 days of chemical spray (6200 ppm boric acid, 50 ppm hydrazine, pH 10.5 sodium phosphate buffer)).

China has now built many third-generation nuclear power stations. Its nuclear power plant test life has increased from 40 years for the second generation to 60 years for the third generation; low-smoke halogen-free flame retardant materials for both cables and accessories are adopted. Several of China's enterprises have

Figure 3.5 Heat-shrinkable tube for nuclear power stations made by Tyco-Raychem Company.

Figure 3.6 Low-dose β-ray resistance test on heat-shrinkable tubes for nuclear power stations by SINAP, CAS.

attempted to develop third-generation K1, K2, and K3 heat-shrinkable materials in recent years, with the following technical features: (1) the assessment of thermal life of 60 years, (2) the γ-ray radiation resistance test, and (3) the LOCA/design basis event (DBE) test. At present, there are only a few research reports on the test evaluation of the low-dose β-ray resistance of heat-shrinkable materials for third-generation nuclear power plants. The Shanghai Institute of Applied Physics (SINAP) of Changchun Institute of Applied Chemistry (CAS) has established a low-dose β-ray test facility for the evaluation of heat-shrinkable materials in nuclear power plants (Fig. 3.6). The dose rate of the EB accelerator is very low; hence it is easy to evaluate the radiation oxidation of material.

3.1.2.3 High-performance heat-shrinkable tubes for automobiles and high-speed trains

Heat-shrinkable tubes dedicated to high-speed railways (Fig. 3.7) are mainly used in electrified railway rebars for insulation. Its characteristics include high breakdown strength and tensile strength, strong weather resistance, and good aging resistance. It is also an environmentally friendly product. Heat-shrinkable tubes dedicated to high-speed railway Borg boards, designed and manufactured according to DERAY-MC 225 technical requirements, can be applied in the world's cutting-edge high-speed railways which reach

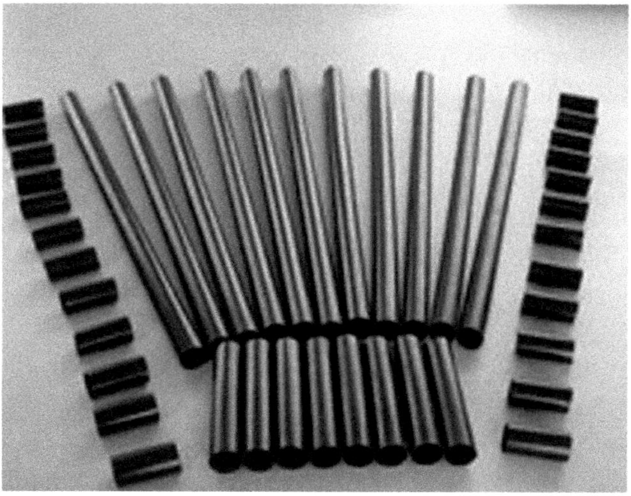

Figure 3.7 Heat-shrinkable tubes intended for high-speed railways.

400 km h^{-1}. The mechanical, thermal, chemical, electrical, and other performance indicators are made to meet American Society for Testing and Materials (ASTM), IEC, and Verband Deutscher Elektrotechniker standards with the working temperature of −55 to 105°C.

3.1.2.4 Heat-shrinkable material for aviation

In comparison with common heat-shrinkable materials, the heat-shrinkable materials used for aviation have several advantages: thin-walled, light weight, and high performance to meet the flight requirements. The current production includes the following categories:

1. Thin-walled high transparent heat-shrinkable tube. Material: transparent XL-PVDF, working temperature: −55 to 150 or 175°C, with high strength, strong flame retardancy, and high chemical stability.
2. High-temperature and oil-resistant heat-shrinkable tube. Material: radiation cross-linked fluorine rubber, working temperature: −40 to 200°C, excellent resistance to oil, hydraulic oil and lubricants and other corrosive oil liquids at relative high temperatures.

3. High-performance cable accessories: at present, the middle and low end of the heat-shrinkable cable accessories are 100% made domestically, such as low-end first-generation semiconductive/insulated composite tube of pure heat-shrinkable composite tube (Fig. 3.8), with the following technical characteristics: semiconductive heat-shrinkable polymer used in the outer layer, and insulating heat-shrinkable polymers used in the inner layer. Middle-end second-generation semiconductive/insulated composite tube of heat-shrinkable/chilling-shrinkable composite tube (Fig. 3.9), with a semiconductive heat-shrinkable polymer used in the outer layer and insulated elastomeric polymer in the interior. The high-end third-generation semiconductive/insulated composite tube of heat/chilling-shrinkable composite tube, with semiconductive heat-shrinkable polymer in the outer layer, insulated heat-shrinkable polymer in the first inner layer and insulated elastomeric polymer in the second inner layer.

For the heat-shrinkable materials, the development work needs to be strengthened in two directions: reducing costs, and improving the performance to meet the new needs of the aviation field.

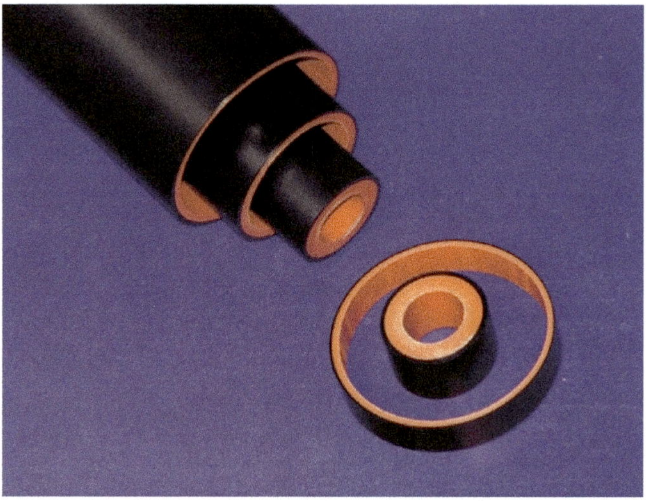

Figure 3.8 Double-walled heat-shrinkable tube for high-performance cable accessories.

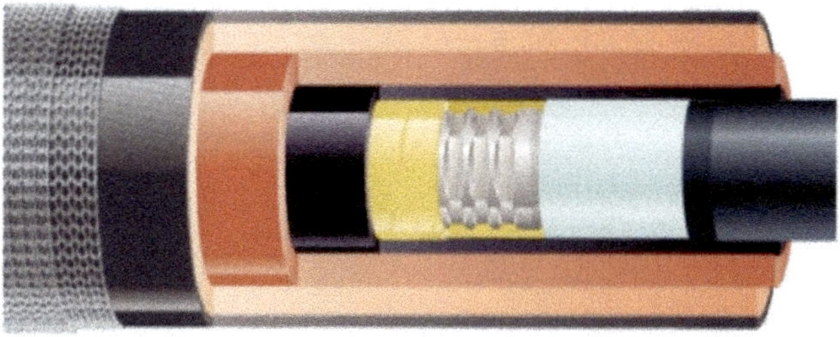

Figure 3.9 Heat/chilling-shrinkable composite tube for high-performance cable accessories.

3.2 RADIATION CROSS-LINKED POLYOLEFIN SHRINK FILM

Shrink film is a material made of polymer plastic, which shrinks tightly over whatever it is covering when heat is applied. Shrink film is commonly used in the outer packing of various items, and widely used in packaging small pieces of goods, automatic packaging, assembly packaging, and other products. A variety of commercial shrink films, such as PVC shrink film, PE shrink film, polypropylene (PP) shrink film, polyvinyl dichloride shrink film, and polyethylene terephthalate (PET) shrink film are produced by multilayer co-extrusion processing.

POF shrink film is one of the fastest growing environmentally friendly packaging shrink films. Standard POF shrink film is generally composed of three to seven polymer layers in which the core layer comprises linear low-density PE (LLDPE) and each skin layer comprises co-PP or cross-linked PE. POF shrink film (Fig. 3.10) offers many advantages such as nontoxicity, high shrinkage rate, good toughness, and economic cost. It is a biaxially oriented multilayer co-extruded POF film. It is recognized that the performance of POF shrink film is better than single PE, PVC, or PP shrink film.

Despite its many excellent properties, standard POF shrink film also has some drawbacks, such as mechanical strength, sealing strength, compatibility, and transparency. Many techniques have

Figure 3.10 Photograph of standard POF shrink film. *POF*, Polyolefin.

been adopted to address each of these shortcomings. For example, some strategies, such as chemical additives, polymer blending, corona discharging, and ultrafast cooling, have been employed to improve the performance of standard POF shrink film. Differing from the above methods, radiation cross-linking is effective in enhancing the properties of POF shrink film. POF macromolecular chains can be cross-linked by irradiation to form a more durable structure.

The central effect of irradiation on the chemical structure of POF shrink film is cross-linking, due to the reactions of free radicals with vinyl bonds produced during irradiation. Cross-linking induces the formation of a three-dimensional network of C—C bonds (Fig. 3.11), making the POF thermoset. Cross-linking can be enhanced with the addition of a cross-linking agent, which reduces the absorbed dose for a certain degree of radiation cross-linking. Relatively low-energy EB irradiation provides significantly enhanced physical properties of POF shrink film to meet the desirable performance of shrink film for packaging.

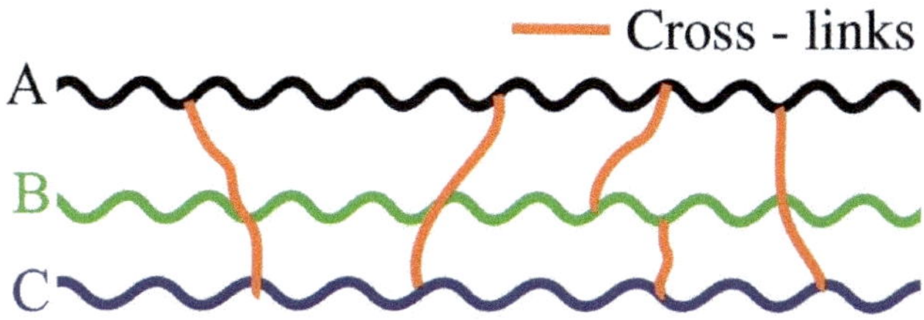

Figure 3.11 Schematic diagram of free radical reaction involving macromolecular chains of POF shrink film via covalent bonds during irradiation.

Figure 3.12 Photograph of multilayer co-extrusion (POF) dual bubble-blown film machine (designed by Jinming Machinery Co., Ltd). *POF*, Polyolefin.

The manufacture of POF shrink film requires a dual bubble-blown film machine, including extrusion, die head, heater, cooling device, winder, and additional EB accelerator unit when radiation cross-linking is desired, as shown in Fig. 3.12. The process of blowing film is mainly divided into two parts: a primary tube (first bubble) is extruded by either blowing or casting the tube onto an external or internal mandrel, respectively, using water for cooling at this point, then the primary tube is reheated and inflated into a second bubble using infrared heating technology. The orientation temperature range of POF is usually below the POF melting temperature. After being stretched, the film is promptly cooled to quench it, thus freezing the macromolecules of film along the stretching direction. When heat is applied, shrink film will shrink tightly over whatever it is covering.

In the past few decades, only a few companies have used radiation cross-linking technology to produce high-performance POF shrink film. Both medium-energy and low-energy EB accelerators can be used for radiation cross-linking, depending on the manufacturing process (Fig. 3.13). POF shrink film can be

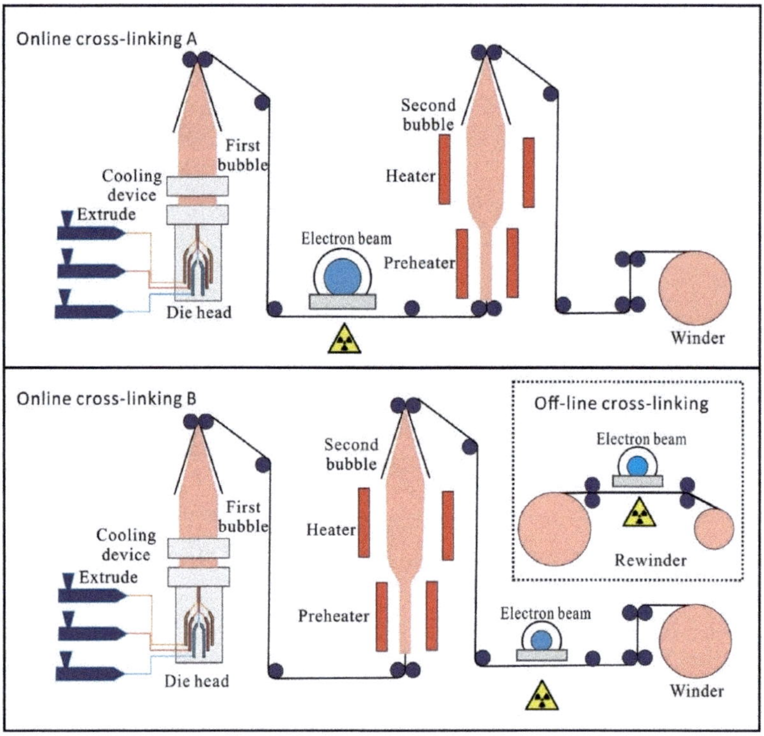

Figure 3.13 Schematic of POF shrink film irradiated after blowing the first bubble (online cross-linking A) and the second bubble (online cross-linking B or off-line cross-linking) by EB accelerator.

produced by irradiation of the primary POF tube using a 0.5–2 MeV EB accelerator, and the following biaxial stretching and orientation, known as "online cross-linking A" (Online-A). With the development of very low-energy EB accelerators with high beam current, POF shrink film can be produced by radiation cross-linking after the second bubbles. This is known as "online cross-linking B" (Online-B). In addition, radiation cross-linking can also be achieved by the irradiation of POF shrink film (e.g., PP/PE/PP multilayer POF shrink film) with a low-energy EB accelerator in a separate rewinding process, known as "off-line cross-linking" (Off-line).

The advantages of radiation cross-linked POF shrink film are as follows:

1. *Stronger seal strength*: the seal seams are the weakest point in a shrink-wrapped package. Cross-linked POF shrink film has higher seal strength than standard POF shrink film, reducing the probability of damaged packages.
2. *Less residue*: cross-linked POF shrink film also leaves less solid residue on sealers to promote longer sealing blade and wire life, which is suitable for packaging with high-speed machines and high output operations.
3. *Increased clarity*: the ability to use a thinner cross-linked POF shrink film with comparable strength increases the exhibition clarity of the film.
4. *Excellent shrinkage*: the shrinkage rate of cross-linked POF film is higher, and the range of shrinkage temperatures is wider. The reduction of shrinkage is lower than that of standard POF shrink film, and cross-linked POF shrink film can be stored for longer.
5. *Good compatibility*: irradiation is applied to POF shrink film to improve compatibility, and it is easy to solve the problem of delamination.

When irregular items are packaged using standard POF shrink film (Fig. 3.14A) or radiation cross-linked POF shrink film (Fig. 3.14B) under the same heating conditions, the original sample shows obvious breakage with a poor packaging effect, but the cross-linked one shrinks tightly over the item it is covering.

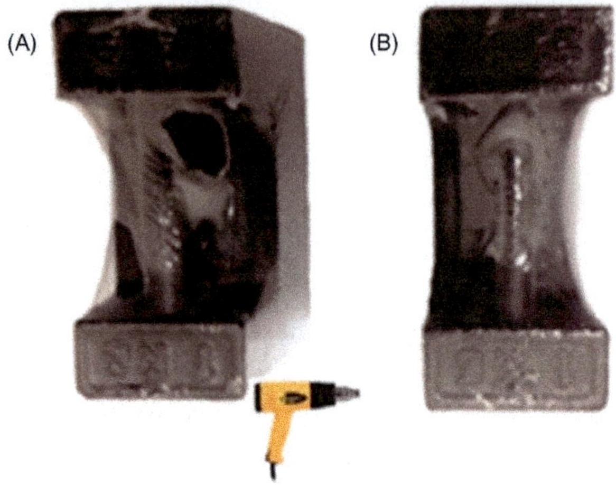

Figure 3.14 Photographs of (A) standard POF shrink film and (B) radiation cross-linked POF shrink film used to pack items under the same conditions.

Table 3.1 Comparison of online cross-linking A, online cross-linking B, and off-line cross-linking

Item	Online-A	Online-B	Off-line
Thickness	>300 μm	<40 μm	
EB	0.5–2 MeV	<300 keV	
	High voltage	Low voltage	
Stability of second bubble	Good	Normal	
Degree of cross-linking	Restriction	Flexibility	
Security	High requirement	Low requirement	
Continuous operation	Continuous production	Continuous production	Intermittent production

However, cross-linked POF shrink film, produced by EB accelerator irradiation after blowing the first bubble (Online-A) or blowing the second bubble (Online-B or Off-line), has some differences in properties, as shown in Table 3.1. For example, LLDPE and PP have different radiation effects, which lead to difficulty in that the second bubble is biaxially stretched and oriented. Therefore, the layers of Online-A POF shrink film all consist of LLDPE, while those of Online-B or Off-line POF shrink film can consist of LLDPE or both LLDPE and co-PP. In Online-A, the composite primary tube has a three-dimensional network structure, which reduces the probability of breaking the film. The thickness of the second bubble film is more uniform and has higher transparency when it is biaxially stretched and oriented. However, the absorbed dose for Online-A POF shrink film needs to be limited compared to that of Online-B or Off-line POF shrink film, otherwise the cross-linked primary tube is difficult to make a second bubble. Furthermore, radicals may be quenched during annealing in the second bubble, improving the stability of POF shrink film. For Online-B or Off-line, the absorbed doses can be adjusted without any strict restraint, which is beneficial for the production of products with different degrees of cross-linking. However, product stability is relatively less, because the cross-linking does not have a heating process with annealing. Moreover, since Online-A using medium-energy accelerators requires complex equipment with high voltage, security is a major problem. Online-B or Off-line using low-energy accelerators is more secure, but it is relatively expensive. For Online-B or Off-line, the film-blowing machine is slightly modified compared with that of Online-A, and this technology can reduce costs. Online-A, Online-B, and Off-line each have many advantages, utilizing radiation technology to develop high-performance shrink film. All are necessary and complement each other with their different advantages.

As a popular environmentally friendly heat-shrinkable packaging material, the market for radiation cross-linked POF shrink film is still in the fast growing stage. With the development and reduction of cost of low-energy EB accelerators, more companies will adopt online or off-line cross-linking techniques to produce POF shrink film with a variety of properties.

3.3 APPLICATION OF RADIATION VULCANIZATION IN RUBBER AND TIRES

The application of radiation for the vulcanization and modification of rubber is a prospective technology in polymer processing. This research began in the middle of the 20th century, and many world-famous tire manufacturers had applied radiation vulcanization technology in their plants by the end of the 20th century. At present, more than 93% of radial tires made in Japan are treated by low-energy accelerators for prevulcanization. In recent years, EB radiation vulcanization has become gradually recognized and accepted in the rubber and tire industry in China. Some tire manufacturers have already established production lines with the adoption of radiation vulcanization.

The following is the introduction of radiation vulcanization, the effect of radiation on rubber, and the advantages of radiation vulcanization compared with chemical vulcanization. In addition, there is a summary of the application and development of EB irradiation in the rubber and tire industry at home and abroad. Modified off-line irradiation equipment for the prevulcanization process and application examples of this technique is also described.

EB radiation prevulcanization has many advantages, including reducing the cost and improving the performance of tires. This technology is certainly important for the manufacture of high-performance radial tires. It has an enormous potential market in the future.

3.3.1 Radiation vulcanization of rubber

3.3.1.1 Radiation vulcanization and prevulcanization of rubber

Free radicals are formed when rubber macromolecules are subjected to high energy rays such as EB and X-ray at normal temperatures and pressures, causing cross-linking of main chains to form three-dimensional networks, and resulting in the same level of vulcanization compared with chemical vulcanization. The process is defined as radiation vulcanization. Radiation prevulcanization means a certain degree of weak cross-linking between main macromolecular chains by adjusting the irradiation parameters.

3.3.1.2 Radiation effects on rubber

The formation of cross-linked bonds and the degradation of polymer main chains happen simultaneously during the course of polymer irradiation. The radiation cross-linking *G* value of common rubber is called the cross-linking yield. The *G* value is an important index of rubber radiation cross-linking.[1] Table 3.2 displays the

Table 3.2 Radiation cross-linking *G* values of common rubber

Type	Natural rubber	Butadiene rubber	Nitrile-butadiene rubber	Butadiene styrene rubber	Chloroprene rubber	Ethylene-propylene rubber	Silicone rubber
G value	1.3–1.8	2–5.8	1.4	3.8	9.6	1.8	2.2

G values of common rubber. The vast majority of alkene rubbers are cross-linkable. However, many factors, including the chemical structure of polymer main chains, polymer morphology, irradiation conditions and characteristics and the amounts of accessory ingredients and sensitizers in rubber can significantly affect the formation of rubber cross-linking bonds.[2]

3.3.1.3 Advantages of radiation vulcanization

After radiation vulcanization, the tearing, aging and ozone resistance properties of rubber are improved. By radiation vulcanization, full vulcanization can be achieved rapidly at normal temperature and pressure with many advantages, including speed, flexible production, energy saving, and low pollution. Once the setup parameters of radiation vulcanization are established, the EB can be evenly distributed throughout the sample, leading to a uniform cross-linking density and a single network structure. Radiation vulcanization can overcome the drawbacks of chemical vulcanization, including pollution, energy consumption, time consumption, and nonuniform vulcanization.

3.3.1.4 Radiation vulcanization of common rubber in tire industry

A variety of rubber are used in the tire industry include natural rubber (NR), isoprene rubber (IR), butadiene styrene rubber (SBR), butadiene rubber (BR), ethylene-propylene diene terpolymer (EPDM), butyl rubber (IIR), chlorinated IIR (CIIR), and brominated IIR (BIIR). Radiation vulcanization of these types of rubbers is introduced briefly in the following sections.

3.3.1.4.1 Natural rubber and isoprene rubber

NR is widely used in all components of tires, with the largest consumption share in the tire industry, and it can be prevulcanized by

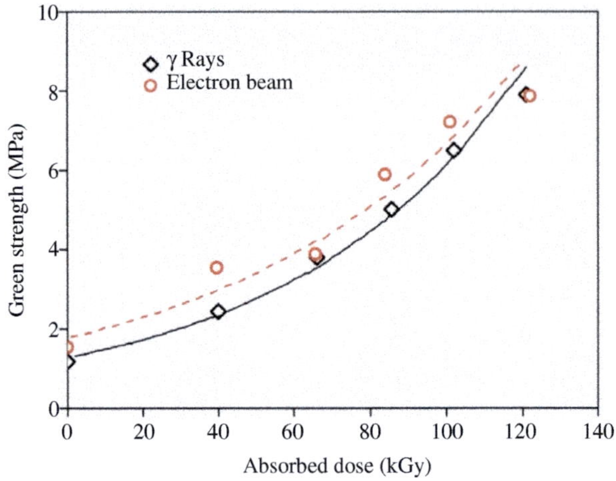

Figure 3.15 Effect of absorbed dose on green strength of NR.[3] *NR*, Natural rubber.

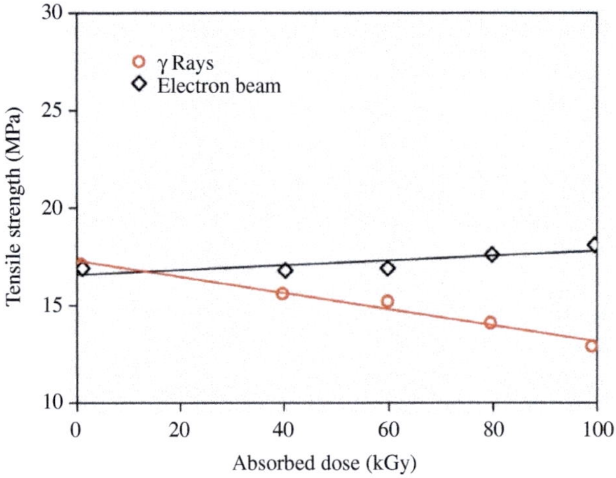

Figure 3.16 Effect of absorbed dose on tensile strength of NR.[3] *NR*, Natural rubber.

EB and γ-ray. Fig. 3.15 shows that the green strength of radiation-prevulcanized NR increases with absorbed dose. Figs. 3.16 and 3.17 indicate that the properties of NR do not change significantly before and after radiation prevulcanizing.[3] Table 3.3 shows the properties of radiation vulcanization and sulfur vulcanization for IR compound made according to the method of ASTM D3184

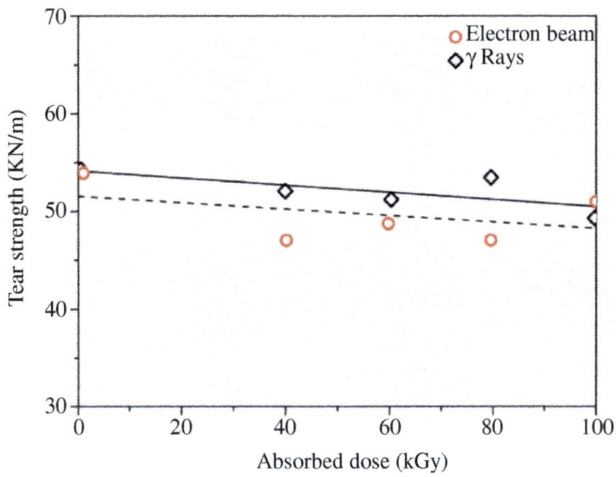

Figure 3.17 Effect of absorbed dose on tear strength of NR.[3] *NR*, Natural rubber.

Table 3.3 Mechanical properties of irradiation vulcanizate and sulfur vulcanizate for IR[4]

Item	Irradiation	Sulfur
Shore A hardness	55	50
Tensile strength (MPa)	29.6	28.0
Elongation (%)	690	620
Tensile strength at 100% (MPa)	1.8	1.6

(1995). The Young's modulus and tan δ of both vulcanizates are similar, as shown in Figs. 3.18 and 3.19, while the tensile strength, elongation at break, and tensile strength at 100% of radiation vulcanizate are higher than the performance of sulfur vulcanizate. This indicates that radiation vulcanization can obtain vulcanizate with good properties.[4]

3.3.1.4.2 Styrene-butadiene rubber and butadiene rubber

SBR is a copolymer with styrene and *cis*-1,4 butadiene, and is widely used in the tread of passenger car tires and carcass compounds. It has the largest consumption share of all synthetic rubbers. *Cis*-1,4 polybutadiene is the most important blended rubber in the tread formula, polymerized by *cis*-1,4 butadiene. The wear

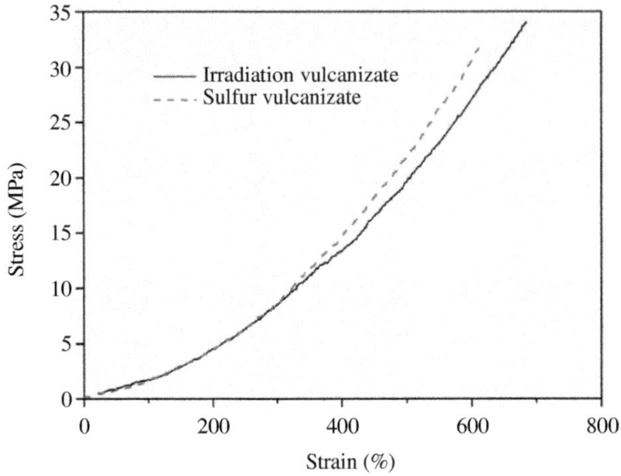

Figure 3.18 Curves of stress–strain of irradiation vulcanizate and sulfur vulanizate.[4]

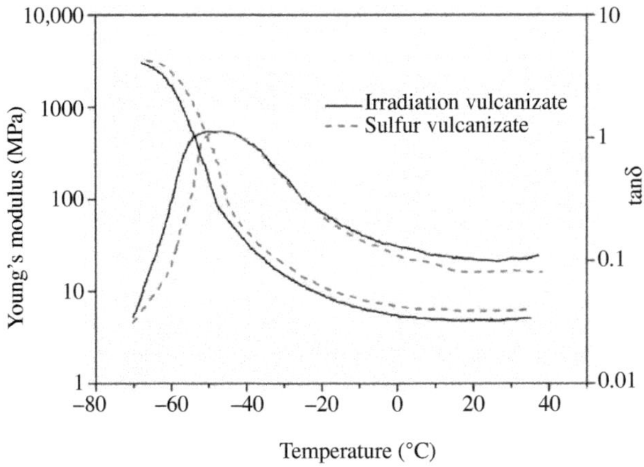

Figure 3.19 Spectrogram of dynamic mechanical thermal analysis (DMTA) of radiation vulcanizate and sulfur vulcanizate.[4]

resistance of BR is also the best of the common rubbers. SBR and BR can be cross-linked by irradiation, and the degradation rate for both is close to zero when they are irradiated in vacuum. The mechanical properties of SBR compound measured according to ASTM D3185 (2006) for rubbers vulcanized by irradiation and sulfur are shown in Table 3.4. The tensile strength, elongation at break, tensile strength at 100%, and hardness of radiation

Table 3.4 Comparison of mechanical properties of radiation vulcanizate and sulfur vulcanizate for SBR

Item	Radiation vulcanizate	Sulfur vulcanizate
Shore A hardness	75	67
Tensile strength (MPa)	21.8	19.7
Elongation at break (%)	398	360
Tensile strength at 50% (MPa)	2.5	1.6
Tensile strength at 100% (MPa)	4.6	2.5
Tensile strength at 200% (MPa)	9.0	7.2
Tensile strength at 300% (MPa)	16.3	14.7
Tear strength ($N\ m^{-1}$)	4.0	4.7

vulcanizate are superior to those of sulfur vulcanizate, and the dynamic properties of both are similar.

3.3.1.4.3 Ethylene-propylene diene terpolymer

EPDM with unsaturated bonds is produced by the terpolymerization of ethylene, propylene, and a small amount of alkadiene. This is applied in tire sidewalls and can be cross-linked by irradiation. As shown in Fig. 3.20, the tensile strength of EPDM increases with the absorbed dose until 120 kGy. When the absorbed dose increases above that level, the tensile strength decreases to a larger extent. The elongation at break decreases with the increase of absorbed dose. Radiation cross-linking leads to an increase in the rigidity, and hence an increase in the tensile strength at 100% and Shore A hardness, as shown in Fig. 3.21.[5]

3.3.1.4.4 Butyl rubber, brominated butyl rubber, and chlorinated butyl rubber

IIR, BIIR, and CIIR are widely used in the inner liners of tires due to their excellent pressure tightness. The main chain of IIR is isobutene. IIR is a radiation-degradable rubber, and cannot be vulcanized by irradiation. The properties of BIIR and CIIR are similar to those of IIR, BIIR, and CIIR rubbers and can be vulcanized by irradiation at a quite low absorbed dose, and they are degraded by irradiation with a higher absorbed dose. The mechanical properties

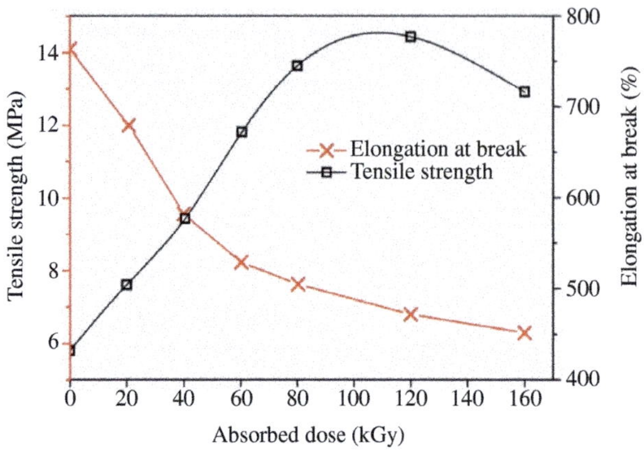

Figure 3.20 Effect of absorbed dose on tensile strength and elongation at break for EPDM.[5] *EPDM*, Ethylene-propylene diene terpolymer.

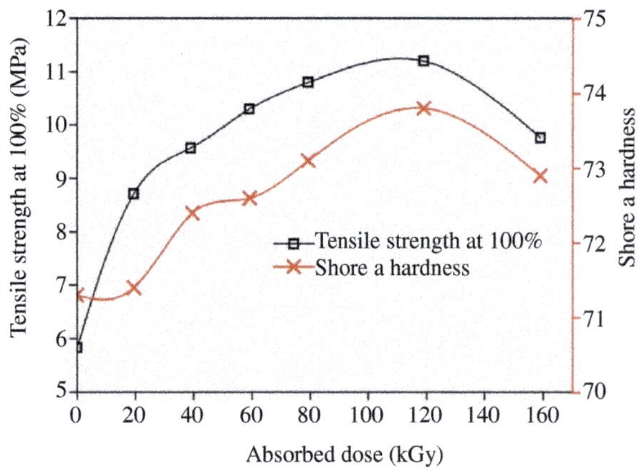

Figure 3.21 Effect of absorbed dose on tensile strength at 100% and shore A hardness for EPDM.[5] *EPDM*, Ethylene-propylene diene terpolymer.

of CIIR can be improved markedly when it is vulcanized by irradiation at 30 kGy with the addition of polyfunctional monomers.

3.3.2 Application of radiation technology in tires

3.3.2.1 Applications of radiation technology in tire industry

The application of EB irradiation in tires started in the late 1950s. In 1957, a whole tire was successfully cured by Goodrich by ^{60}Co irradiation. However, to date, there have been few open publications on the radiation curing of whole tires. In the tire industry, the main purpose of irradiation is to prevulcanize parts of tires, in order to reduce costs and improve product quality. In the 1980s, the American Firestone Company built the world's first radiation prevulcanization production line for tires, and then America's largest tire company Goodyear also adopted the same technology. This technology was promoted and developed successively in tire companies.

Table 3.5 shows the product technical indices of the compounds of tire parts after finishing the heat vulcanization with and without radiation prevulcanization. It can be seen that the green strength and the percent recovery of the rubber increase greatly after radiation prevulcanization.[6]

The application of radiation prevulcanization in tire components has several advantages. First, it can improve the green strength (GS) of rubber, to ensure tire components in the process of forming dimension stability. Second, it can reduce the thickness of various components, to increase the dosage of synthetic rubber and reduce the cost. Third, it can speed up the curing speed, save energy consumption, and improve the degree of uneven curing during the vulcanization process, which is caused by the temperature difference between the inside and outside of the tire. EB radiation prevulcanization can increase the rubber adhesive strength, reduce the rubber hysteresis loss and heat build-up, hereby improving tire performance.

3.3.2.2 Current situation

The first set of radiation prevulcanization equipment for tires was installed in Japan in 1977. Up to 2005, more than 60 low-energy accelerators were installed in the tire industry of Japan. More than 93% of radial tires are prevulcanized by irradiation in Japan. European and American tire companies have also applied EB radiation prevulcanization in the tire industry, but the number of production lines is smaller.

Research on radiation vulcanization of rubber started in the 1960s in China, but no EB accelerator equipment was available for rubber at that time, which restricted the industrial application of

Table 3.5 Properties of each part of the tire with and without radiation prevulcanization

Component	Rubber	%	0 (kGy)		50 (kGy)		100 (kGy)		150 (kGy)		200 (kGy)	
			G.S.	P.R.	G.S.	P.R.	G.S.	P.R.	G.S.	P.R.	G.S.	P.R.
Inner liner	IIR	20	3	8	15.3	31.5	20	42.5	22.9	80	33.8	—
	SBR	35										
	NR	45										
Flipper	NR	80	22.8	9	35.1	39	42.1	48	52	51	58.2	—
	SBR	20										
Sidewall	NR	45	8.2	10	17.6	35.5	20	113.5	21.2	110	—	—
	SBR	30										
	EPR	25										
Tread	EPR	20	13	10.5	17.5	35.5	21.7	71	26	118.5	—	—
	NR	45										
	SBR	35										

IIR, Butyl rubber; *SBR*, butadiene styrene rubber; *NR*, natural rubber; *EPR*, ethylene propylene rubber.

radiation prevulcanization in tires. In 2004, the National Development and Reform Commission initiated a special support program for "nonpowered nuclear technology industrialization of high technology," and radiation vulcanization of rubber was selected as one of the radiation technologies supported by the program. With government support, this technology was successfully applied in one company, and it was confirmed that radiation vulcanization could improve the stability of tires and reduce the thickness of tire carcasses. In 2009, Henan Isotope, Aeolus Tire Co., and CAS jointly developed "radiation prevulcanization application in production of truck bus radial tyre," and finished trial production of an all-steel heavy truck radial tire (12.00R2018 154/149 K).[7,8] In addition, some local makers of EB accelerators began to provide equipment for radiation prevulcanization in tire production lines.[9,10] At present, the practical application of radiation prevulcanization is still very limited, and only very few leading enterprises have adopted this technology.

3.3.2.3 Equipment for radiation prevulcanization of tire parts

3.3.2.3.1 Basic structure of tires and the manufacturing process

The basic construction of a radial tire is shown in Fig. 3.22. The tire manufacturing process is described below. First, in the tire building machine the semifinished products are assembled into green tires according to the tire building process, and then the green tire is made into a tire by vulcanizing with the vulcanizing

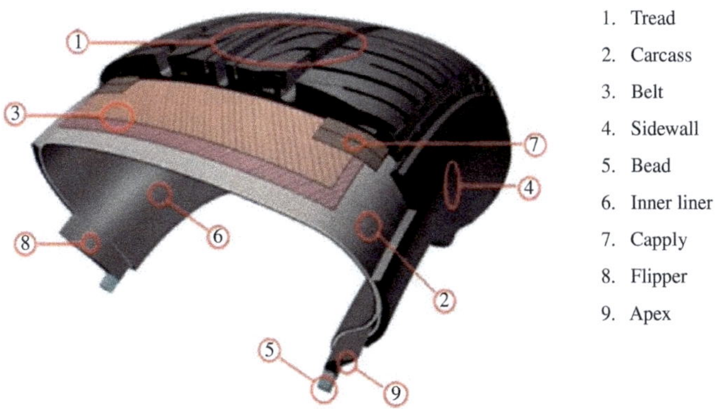

Figure 3.22 Basic structure chart of tire.

machine at a certain temperature, pressure, and time. In the process of building, the tire carcass and inner liner parts are deformed, resulting in defects such as uneven distribution of polyester cord and cord migration. These problems can be overcome by increasing the thickness of the carcass and inner liner, but the material cost is higher.

According to the tire parts recipe composition, structure and shape, radiation prevulcanization can be implemented to tire parts: carcass, belt, sidewall, tread, and flipper.

3.3.2.3.2 Radiation prevulcanization of tire carcass

EB radiation of a tire carcass for prevulcanization is shown in Fig. 3.23. First, the tire carcass is opened for radiation prevulcanization on the conveyor by EB accelerator, and the carcass is taken up after the irradiation, and then the prevulcanizing process is completed.

3.3.2.4 Application of online radiation prevulcanization

3.3.2.4.1 The retrofit scheme of tire rolling production line

In order to realize online EB irradiation, an EB irradiation system was added to the original fiber rolling production line, which does not affect the normal run of the original production line. Considering the demands of the original production line layout and space, the added EB accelerator was installed in the coiling location. In order to guarantee continuity of production, the forward cord conveyor frame, the double location coiling, and some drive motors were installed in the corresponding positions, and the control system and communication system were upgraded to ensure the stability of the production process, and to achieve uniform carcass radiation. Fig. 3.24 shows a diagram of the new online radiation prevulcanization process with two EB accelerators. This technology can make full use of the EB energy, improve production efficiency, and ensure the uniformity of absorbed dose on carcasses.

3.3.2.4.2 Difference in dose distribution between off-line and new online radiation prevulcanization

EB irradiation is a basic energy transfer process, and its penetration is dependent on the wavelength of the irradiation and accelerating

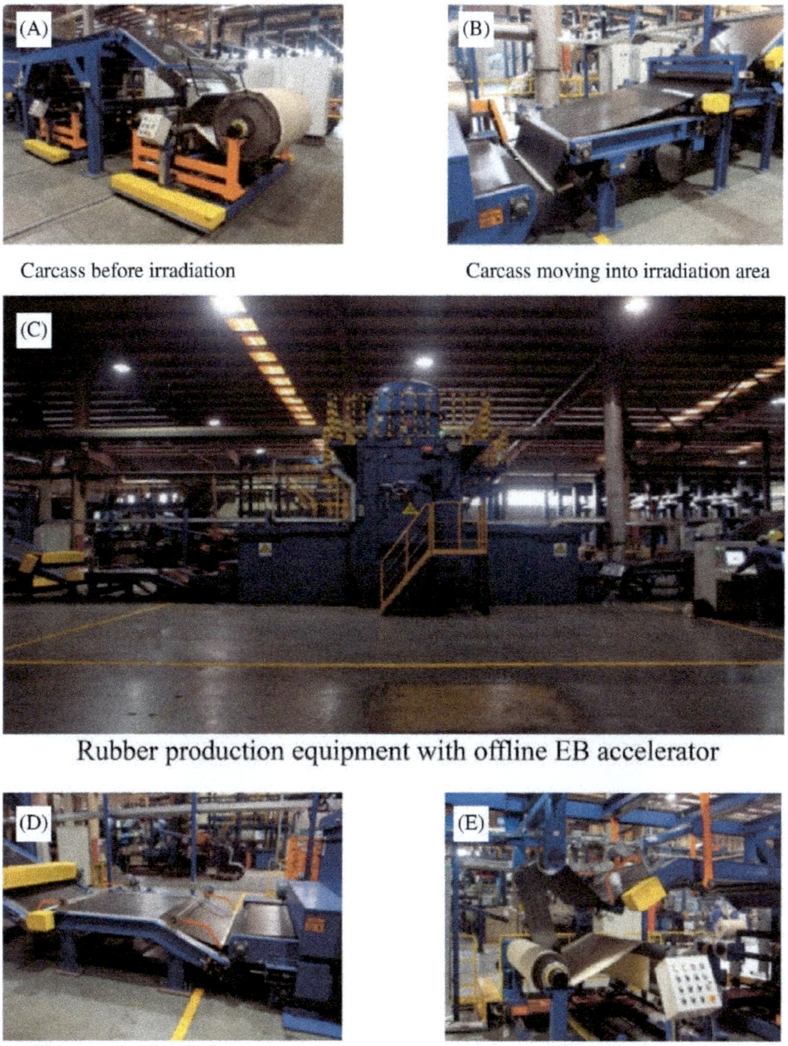

Figure 3.23 Off-line EB accelerator irradiation production line (from A to E).

voltage. The penetration depth of EB generated by a low-energy electron accelerator in material is very shallow. In order to increase the penetration depth, the method of irradiation on both sides is usually used, and the penetration depth is 2.5 times thicker than that of irradiation on one side. Compared with off-line radiation prevulcanization, the distribution of the absorbed dose is more

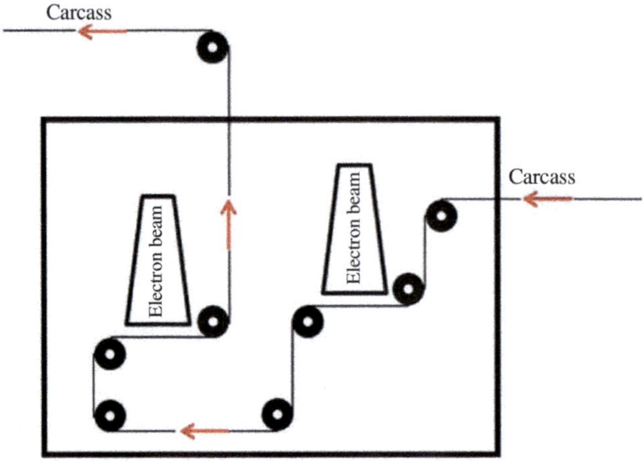

Figure. 3.24 Diagram of new online radiation prevulcanization process.

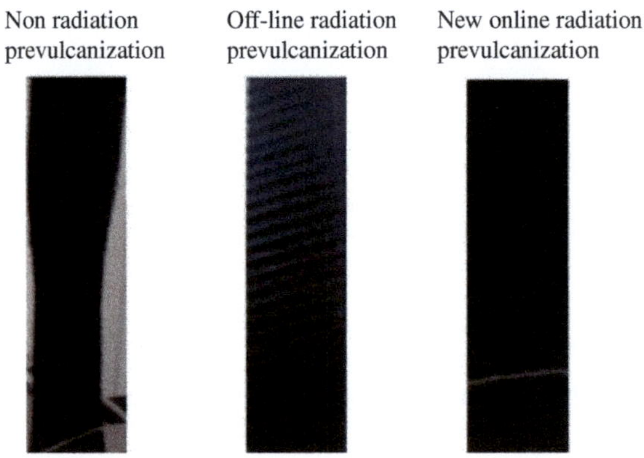

Figure 3.25 Comparison of tensile properties of tire carcasses produced by different radiation prevulcanization processes.

even for a carcass with the same thickness by online radiation prevulcanization, and production efficiency is higher.[11,12]

3.3.2.4.3 Tensile properties of tire carcass by radiation prevulcanization

Tension testing with the same tensile deformation for carcasses produced by different technology (as shown in Fig. 3.25), shows that the tensile deformation of carcasses without irradiation is large but there is almost no deformation for the carcasses produced by off-line radiation prevulcanization. However, the cords are not perpendicular with the rolling direction for off-line radiation prevulcanization. For the carcasses produced by online radiation prevulcanization, there is also no deformation and the cords are still perpendicular to the rolling direction, which can ensure the shape and dimensional stability of carcasses during the expansion process of tire building.

3.3.2.4.4 Carcass appearance

As shown in Figs. 3.26 and 3.27, there are many pleats on the carcass produced by off-line radiation prevulcanization, caused by

Figure 3.26 Off-line radiation prevulcanization process.

Figure 3.27 New online radiation prevulcanization process.

uneven stress in the guide uncoiling process. This problem was solved by the new online radiation prevulcanization process, in which the stress on the carcass is even.

3.3.2.4.5 Reduction of the amount of waste and number of bad tires by the application of new online radiation prevulcanization

Fig. 3.28 shows the amount of waste and number of bad tires produced by the new online irradiation prevulcanization process. Since many different types of tires are produced in the factory, the carcasses produced by the new online radiation prevulcanization are used step-by-step in the implementation of the online radiation prevulcanization process, lasting ~15 days. The amount of waste and number of bad tires in this process decreased markedly, and remains at a low level.

After irradiation, the defective rate of tire appearance caused by cord exposure decreased from 0.04% without irradiation to 0.02%, and the qualified rate of product increased from 98.25% to 99.58%. After the application of the online radiation prevulcanization technology, the production efficiency and quality of semifinished and finished products increased markedly, and the defective rate of semifinished and finished products and labor costs decreased significantly, compared with the off-line radiation prevulcanization system.

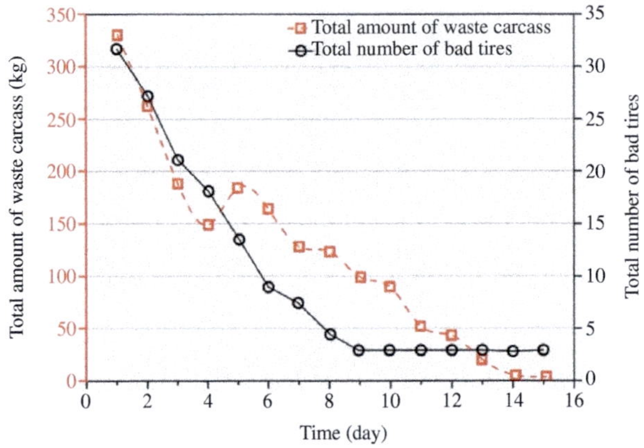

Figure 3.28 Changes in amount of waste and number of bad tires for new radiation prevulcanization process.

The implementation of the new online radiation prevulcanization system further improved the dimensional stability of rubber sheet, and decreased the effect of tension on rubber sheet, and the carcasses became thinner, which reduced the material and production costs. This technology improves the quality of tire production and will be widely applied in China in the future.

3.4 RADIATION VULCANIZATION OF NATURAL RUBBER LATEX AND SYNTHETIC RUBBER LATEX

NR is a milky fluid (Fig. 3.29) flowing out of rubber trees, that is, NR latex, and it falls into the category of thermoplastic resin, with an average size of latex particles of 0.4 μm. Natural latex features high elasticity, good film-forming performance in the adherence and flexibility of its film, which therefore has outstanding flexing resistance, seismic resistance, and creep resistance, and is applicable for part adhesion and adhesion between materials of different thermoelastic coefficients under dynamic conditions. Composite latex is obtained from various kinds of monomers through polymerization. Its compact microstructure contributes to features such as elasticity, insulation, and water- and air-proofedness after processing. NR has excellent elasticity within a broad temperature range, that is, −50 to 150°C, as well as good fatigue resistance, electrical insulation, chemical resistance, and wearability. Therefore it is widely used for the production of tires, rubber hoses, adhesive tapes, insulation materials, and rubber shoes.

Figure 3.29 Natural rubber latex collected from rubber trees.

3.4.1 Radiation vulcanization of natural rubber latex

3.4.1.1 Natural rubber latex

NR latex, a natural macromolecular compound, is mainly comprised of polyisoprene, that is, rubber hydrocarbon, which accounts for 91%–94%, plus other nonrubber materials such protein, fatty acid, ash, sugar, and so on. Rubber latex without impurities is transparent and slightly yellow. NR has excellent mechanical strength, good viscosity, and dielectric properties, as well as good electrical insulation performance. NR is the most widely used general rubber and falls into the category of recyclable natural resources. NR latex, with its comparatively strong chemical reaction ability, is liable to aging under the influence of light, heat, ozone, radiation, ions, and metals like bronze and manganese. An antiaging agent is normally added to improve the thermal stability of NR latex. Rubber latex is mainly produced in a few Asian countries, including Thailand, Indonesia, Malaysia, China, India, and Sri Lanka, and also in some African countries, including Nigeria. After the United States, China is the second largest rubber consumption country.

3.4.1.2 Radiation vulcanization of natural rubber latex

Latex particles are the general term for rubber particles and nonrubber particles in latex. Among them, rubber particles, accounting for 20%–50% of the volume of latex, are aggregations of rubber molecules, and the rubber within the particle is water insoluble. The external surface of rubber particles is comprised of protein and fatty acid soap. The particles bear negative charges and present Brownian movement in latex.

Chemical vulcanization, with the use of nitrosamine in the vulcanization process, may pose a serious threat to the health of human beings. Radiation cross-linking not only greatly improves the mechanical properties of products, but also has features such as high purity, no waste, and high production efficiency. Radiation cross-linking of natural latex can be realized at room temperature by γ-ray and EB irradiation,[13] and the scheme is shown in Fig. 3.30.

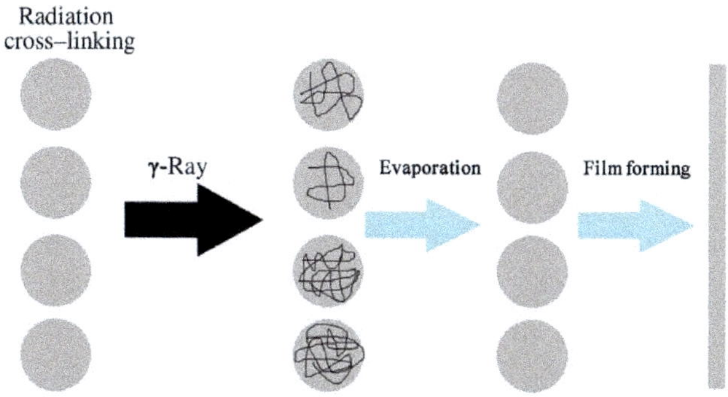

Figure 3.30 Process of radiation cross-linking with natural latex.

As an important industrial raw material, rubber latex is widely used in multiple industries. The stability of latex particles can be increased through radiation cross-linking and grafting settlement. At present, cross-linking is mainly applied to improve the properties of rubber latex, that is, cross-linking of remaining double bonds within rubber particles through the induction of free radicals, and cross-linking reactions within latex particles not only reduce the adhesion effect between latex particles, but also increases their stability. Grafting can change the features of the surface of latex particles and cover the surface with another polymer layer.

^{60}Co γ-ray can be utilized for radiation vulcanization of NR latex, due to its high penetration depth. This technology is adopted in Indonesia, India, Malaysia, and Thailand, and a pilot plant has been established in Malaysia. Taking the radiation establishment of the Malaysian Nuclear Agency as an example, the whole device consists of loading and unloading zones, dosing zone, and latex distribution and irradiation rooms. Latex is transported into snake-shaped stainless-steel pipes placed on the two sides of the plate through a metering pump, and the absorbed dose is controlled by pump-in flow, which is between 12 and 15 kGy normally. If necessary, the absorbed dose can be reduced to $\sim$5 kGy. Industrialization of natural latex with ^{60}Co radiation has been realized in Malaysia. Such radiation-processed rubber latex has three distinguishing features: no protein, no ammonium nitrate-type chemical materials, and longer shelf life (9 months).

There are also reports on the impact of radiation conditions on cross-linking of natural latex. For example, Bohm et al.[14] studied the impact of temperature on the degree of cross-linking and degradation of NR latex, and found that the degree of cross-linking of latex at room temperature was 0.9 and the ratio between degradation degree and cross-linking degree was 0.05; when the temperature was reduced to 77 K, the cross-linking degree changed only slightly, while the ratio between degradation degree and cross-linking degree rapidly increased to 0.16. In addition, the O_2 atmosphere can increase the degradation rate of rubber latex and decrease the cross-linking rate.

Radiation also leads to the decomposition of protein in NR latex. Natural latex contains a considerable amount of protein, which easily leads to the degeneration of hygroscopicity, conductivity, and heat build-up of natural latex products, and may cause contact hypersensitivity. Hence, it is necessary to reduce the protein content as much as possible. Research results showed that, when the mass fraction of residual soluble proteins in natural latex products was less than 110×10^{-6}, they would not cause allergic reactions.[15] However, the mass fraction of residual soluble protein in natural latex currently produced by centrifugal concentration is far larger than 110×10^{-6}. Therefore, an allergy caused by natural latex gloves is common. The main method for the reduction of the protein content in natural latex is decomposition with the use of enzymes. In comparison, the utilization of radiation technology to remove the protein from natural latex is much safer.

3.4.2 Radiation vulcanization of synthetic rubber latex

Synthetic elastomeric polymer is also called synthetic elastomer. Its production volume is only third behind plastic and synthetic fiber. Since natural latex is usually expensive, synthetic latex can be produced on a large scale and its cost is lower, depending on the monomers used. Of course, synthetic latex also has disadvantages. For example, it has a relatively poor stretching effect, lower tearing strength, and mechanical strength. Nevertheless, compared with

Table 3.6 Narpove grade and application

Brand no.	Name	Particle diameter	Purpose
VP-121	Styrene-Butadiene UFPR	~100 nm	Simple and effective preparation of fully vulcanized thermoplastic elastomer
VP-301	Acrylate UFPR	100–150 nm	Toughened polyester, POM, nylon, AS, PV-9C, PMMA, and other resins, preparation of rubber blending material
VP-401F	Nitrile-butadiene UFPR	~100 nm	Preparation of thermosetting plastics like high-performance friction material and modified phenolic resin
VP-501	Carboxylic nitrile-butadiene UFPR	100–150 nm	Preparation of toughened PVC and other plastics, fully vulcanized thermoplastic elastomer, and rubber blending material
VP-101B	Composite nucleating agent		Used in PP, being α crystal nucleating agent, and capable of increasing PP crystallinity, and improving manufacturing performance and mechanical performance of product
VP-101T	Composite nucleating agent		Used in PP, being β crystal nucleating agent, and capable of increasing PP crystallinity, and improving manufacturing performance and mechanical performance of product
VP-801E	Composite nucleating agent		Used in PP, and capable of increasing the transparency of PP
VP-101KJ	Composite antibacterial agent		Used in plastics like PP and PE, assigning antibacterial and antifungal properties to them

Note: Narpove is a kind of product developed and registered by the Beijing Chemical Industry Research Institute of SINOPEC; POM, polyoxymethylene; AS, acrylonitrile-styrene copolymer; PMMA, polymethyl methacrylate.

natural latex, materials made from synthetic latex are cheaper and more widely used.

In a similar manner, synthetic latex can also be cross-linked by irradiation. If a higher degree of cross-linking is achieved, powder rubber can be obtained by drying the cross-linked synthetic latex. Ultrafine fully vulcanized powder rubber (UFPR), independently developed by the Beijing Chemical Industry Research Institute of SINOPEC, has been widely used in toughening modification of plastics and the preparation of thermoplastic dynamic vulcanized alloys, etc. (Table 3.6).[16] Furthermore, this institute has applied for over 70 patents both at home and abroad for the production of a series of UFPRs by radiation cross-linking.

In addition, UFPR can effectively improve the overall performance of thermosetting resins such as epoxy, phenolic, and unsaturated resins.[16] Epoxy resin adhesive and phenolic resin plastics modified with UFPR have been widely used in hydropower projects, leak-proofing and stoppage of large-scale water dams, and manufacturing industries like the military, electric appliances, electronics, mechanics, and so on.

REFERENCES

1. Gao Z, Yang W. Radiation vulcanization of rubber. *Nucl Tech* 1987;**10**(11):38–42 [in Chinese].
2. Xu Y. Radiation vulcanization—a new way to cure synthetic rubbers. *China Synth Rubber Ind* 1997;**20**(5):318–20 [in Chinese].
3. Bao M, Kang X, Jiao Y, Zhang Y, Liu J, Guo Z. Comparison study on irradiation precuing NR of γ-ray and electron beam. *Spec Purp Rubber Prod* 2008;**2**(28):17–19.
4. Xu Y. Radiation vulcanization of *cis*-1,4-polyisoprene containing fillers. *China Synth Rubber Ind* 1995;**18**(3):139–41 [in Chinese].
5. Hu G, Liu B, Du J, Gao T, Zhai ML. Effect of radiation vulcanization on the mechanical properties of ethylene propylene dine terpolymer rubber. *J Radiat Res Radiat Process* 2012;**30**(2):81–6 [in Chinese].
6. Hunt JD, Alliger G. Rubber—application of radiation to tire manufacture. *Radiat Phys Chem* 1979;**14**(79):39–53.
7. Li LX. MESNAC electronic radiation pre-vulcanization line of tire won the second prize of the science and technology progress of the petrochemical federation. *Rubber Ind* 2011;**12**(58):746 [in Chinese].
8. Wang YH, Zhou TM. Application of electronic radiation pre-vulcanization in passenger car radial tire production. *Tire Ind* 2012;**12**(32):750–4.

9. Li CQ. Preliminary study on application of radiation pre-vulcanization in radial tires. *China Rubber* 2013;**8**:43–5 [in Chinese].
10. Song TS. Radiation vulcanization technology of rubber with electron beam. *Rubber Plast Res Util* 2005;**3**:10–17 [in Chinese].
11. Mullins L, Tumer DT. Estimation of chain fracture during the cross-linking of rubber hydrocarbons by high-energy radiation. *Nature* 1959;**183** (4674):1547–8.
12. Mullins L, Tumer DT. Radiation crosslinking of rubber III, chain fracture. *J Pol Sci* 1960;**43**(141):35–47.
13. Zhou HL, Ao LJ, Chen M, Qian HL, Wang C. Effect of sandy-clay dispersion on the property of clay-rubber master batch. *J Chin Electr Microsc Soc* 2001;**20**(4):379–80 [in Chinese].
14. Böhm GGA, Tveekrem JO. The radiation chemistry of elastomers and its industrial applications. *Rubber Chem Technol* 1982;**55**(3):575–646.
15. Azizan MR, Shahnz M, Hasma H. Latex protein allergy: a prevalence study of factory workers. *J Natur Rubber Res* 1996;**11**:240–6.
16. Guo ZT, Zhang GM, Shang HZ, Zeng XM, Guo JM. Research and development of radiation vulcanization of rubber material, Science and Technology Information. *Raw Mater* 2006;**22**(7):23–4 [in Chinese].

CHAPTER FOUR

Radiation Cross-Linking for Conventional and Supercritical CO_2 Foaming of Polymer

Chenguang Yang and Guozhong Wu
Shanghai Institute of Applied Physics, Chinese Academy of Sciences, Shanghai, P.R. China

In the early 1980s, Martini of the Massachusetts Institute of Technology first proposed the concept of micro-foam plastic and developed the relevant molding technology.[1] In polymeric micro-foam material, there are a large number of smaller cells than the original defects in the polymer gaps, leading to a reduction in the density of the product and an improvement in its rigidity. The mechanical strength and many other properties of polymer are almost unchanged by the introduction of small cells.

Microporous polymer is a new type of polymer material with a cell diameter less than 10 μm and cell density exceeding 10^8 cells cm^{-3}. Compared with dense polymer, microporous polymer foam has unique properties, including light weight, good mechanical properties, long fatigue life, high impact strength, good toughness, and low dielectric constant and thermal conductivity. Because the traditional hydrochlorofluorocarbon (HCFC) type of foaming agent destroys the ozone layer, the development and application of environmentally-friendly foaming agents have attracted much attention. Supercritical carbon dioxide ($scCO_2$), a nontoxic, non-polluting green foaming agent for polymer microporous foaming, has obvious advantages, including its cost-effectiveness, its ease to obtain, and its high foaming efficiency, and is one of the most promising alternatives to chlorofluorocarbon blowing agent.

Radiation Technology for Advanced Materials.
DOI: https://doi.org/10.1016/B978-0-12-814017-8.00004-4

4.1 INTRODUCTION OF MICROPOROUS FOAMING MATERIAL

Polymer foam technology has a long history. The bakelite bubble was first prepared in the early 1920s.[2] Polyurethane (PU) foam and polystyrene (PS) foam appeared in the 1930s.[3] Polyethylene (PE),[4] polyvinyl chloride (PVC),[5] epoxy resin, and phenolic foams[6] were prepared in the 1940s. In the 1950s, PS and flexible PU foams were available.[7] The pore diameter of conventional macromolecular polymer foams is usually between 0.1 and 1 mm and the cell density is less than 10^6 cm^{-3}. Larger pore size often leads to initial cracking and the reduction in the mechanical properties of material.[2]

In order to achieve both low cost and better performance of plastic products, Martini and his colleagues used inert gases, such as CO_2, N_2, as the blowing agents to prepare micron-grade cellular foam in 1981.[1] This microcellular foam has a smaller cell diameter (0.1–10 μm) and a higher cell density (1×10^9–1×10^{15} cm^{-3}) compared with conventional foams. Due to the smaller cell size of the microcellular foam, the original crack tip of the material is passivated, which helps to prevent the expansion of the crack under stress. Therefore, the mechanical properties are obviously better than those of general foam. Compared with nonfoamed polymer material, the density of microcellular foam can be reduced by 5%–95%. Their fracture toughness, specific strength, rigidity and fatigue life are increased by 3–5 times. Microcellular foam has low thermal conductivity, high thermal stability, and strong absorption impact ability.[1,8,9] It is widely used in many applications, such as packaging, sound insulation, insulation, moving equipment, aviation and automotive parts, molecular filters, biomedical materials, adsorbents and catalyst carriers.[10–12]

Polymer micro-foam molding technology has been studied in many reports, mainly on the micro-foam molding of amorphous and semicrystalline polymers, such as poly(vinyl chloride), polystyrene (PS), polycarbonate, poly(methyl methacrylate), poly (ethylene terephthalate) and polysulfone resin. At present, the methods of preparation of microcellular polymer foams include monomer polymerization, thermally induced phase separation, supercritical fluid (SCF) antisolvent precipitation, and super-saturated gas. N_2 or CO_2 and other chemically-inert small molecules as foaming agent have attracted much attention, since they have no toxicity to human beings and livestock, and cause no pollution to the environment. Especially in recent years, owing to restriction on the use of the foaming agent Freon, the advantages of the production of microcellular foam by super-saturated gas have become more prominent.

The micro-foam formation process consists of three necessary components. The first is the dissolution of SCF (mainly CO_2 and N_2) into the melted polymer and the formation of single-phase polymer/gas solution; then, the thermodynamic instability of the system was caused by controlling the condition of temperature and pressure, so that the gas solubility in solution decreases. Due to the reduction of gas in the equilibrium concentration, a large number of bubble nucleations appear in the polymer matrix, and then gradually develop to generate tiny holes, as shown in Fig. 4.1. Since the uniform and compact microstructure of foam material determines its high mechanical strength, it has broad application prospects.

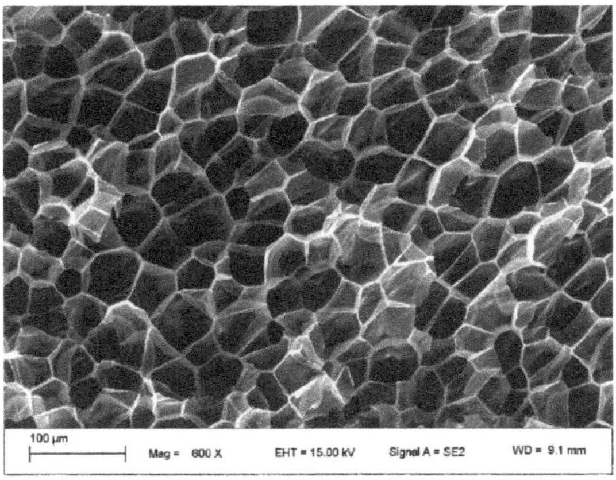

Figure 4.1 Scanning electron microscope (SEM) micrograph of polypropylene (PP) foam material.

4.2 PREPARATION OF MICROPOROUS POLYMERS USING SUPERCRITICAL CO_2

SCF is a noncondensable high-density fluid and the temperature and pressure are higher than or close to the critical point. It is a fluid between gas and liquid, with approximate liquid density, solvent strength and heat transfer coefficient, low viscosity and high diffusivity. From the perspective of interaction between molecules, the fluid state is determined by the intermolecular cohesion and molecular thermal motion balance. On one hand, the intermolecular condensation energy has a limited maximum value; on the other hand, the thermal motion of molecules increases with increasing temperature. When the temperature is above the critical point, no matter if the distance changes between molecules, the thermal motion of molecules plays a dominant role, and this state is called the supercritical state.

The supercritical phenomenon was first discovered by Andrews of Queen's College in 1850.[13] In 1879, Hannay and Hogarth found that SCFs had significant dissolution capacity.[1] Ethylene polymerization was found in the supercritical state in 1933.[2] In the 1970s, a supercritical extraction process was successfully applied to the decaffeination of coffee beans in industrial production.[9] Since the 1980s, many efforts have been made in SCF technology research, covering the food, spices, medical and chemical fields. The characteristics of SCF are as follows:

1. The density is similar to liquid, and is more than two orders of magnitude higher than the general gas. A small change of the temperature or pressure near the critical point causes a significant change in fluid density.
2. The dielectric constant increases with the increasing pressure of the fluid, which is beneficial to dissolve low volatiles. The solubility of the corresponding solutes can be increased by 5–10 orders of magnitude.
3. Viscosity is one order of magnitude smaller than that of liquid, similar to ordinary gas. The diffusion coefficient is two orders of magnitude higher than liquid, and it therefore has good mobility, permeability, and transmission performance.[1]

Table 4.1 Properties of liquid, SCF, and gas[1]

Properties	Liquid	SCF	Gas
Density ($g\ cm^{-3}$)	1	0.1–0.5	10^{-3}
Viscosity (Pa s)	10^{-3}	10^{-4}–10^{-5}	10^{-5}
Diffusion coefficient ($cm^2\ s^{-1}$)	10^{-5}	10^{-3}	10^{-1}

SCF, Supercritical fluid.

Table 4.1 lists a comparison of certain properties of SCFs with gases and liquids. Because SCFs have similar density, dissolution and heat transfer coefficients and near-gas low viscosity and high diffusion coefficients to liquids, SCFs can be used in fields including extraction, synthesis, environmental protection, materials processing, painting and dyeing, and medicine.

The most commonly-used SCFs include CO_2, water, ammonia, ethylene, ethane, propylene, and propane. The most widely-used is CO_2. Fig. 4.2 shows the phase diagram of CO_2, which shows the critical point of CO_2 and the range of each phase. scCO_2 has many unique advantages compared with other SCFs: (1) CO_2 is noncombustible, nontoxic, noncorrosive, safe in operation, and has good environmental compatibility. (2) The lower critical temperature (critical point of 31.1°C) is close to room temperature, the supercritical operation of natural products causes no damage to active substances and heat-sensitive materials. The critical pressure is not high (the critical point is 7.37 MPa) and the processing equipment is not difficult, which is suitable for industrial applications.[14] (3) The mass transfer coefficient is high and the equilibrium concentration can be reached in a short time. (4) scCO_2 has good solubility for many substances. (5) scCO_2 has a higher equilibrium concentration in the polymer, and can swell almost the entire polymer and reduce the glass transition temperature, melting point, and viscosity.[15] (6) CO_2 is cheap and easy to obtain, and is easy to separate and recycle.[12] Based on the above advantages, the application of scCO_2 has been widely investigated.[16–19]

scCO_2 is a clean and versatile solvent that is expected to replace toxic organic solvents and chlorofluorocarbons as a polymerization medium. It has unique advantages, being chemically inert, with no

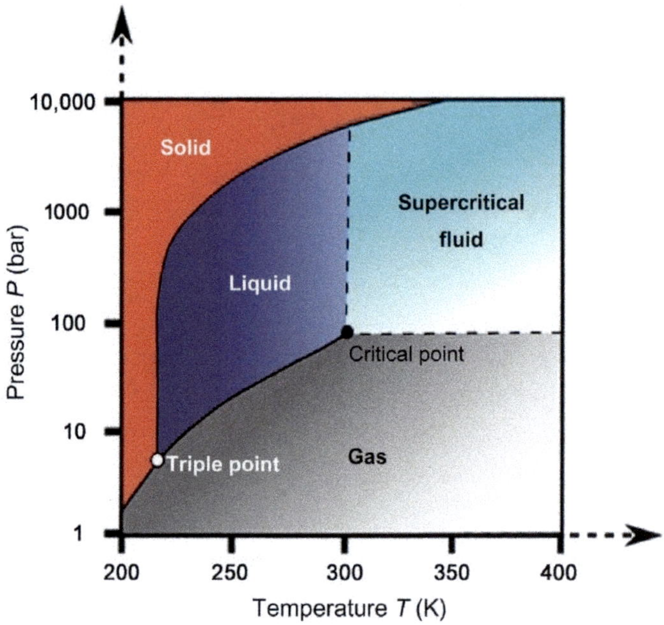

Figure 4.2 Supercritical fluid *P–T* phase diagram.[14]

side effects, and no chain transfer reaction. The dissolving capability can be adjusted by pressure or a small amount of cosolvent and it has a strong swelling and diffusion ability in polymer. In addition, $scCO_2$ can make other small molecules more easily diffuse into the polymer and purify the product easily. Parks et al.[16,20] prepared microporous PU foam by polymerization of diisocyanate with polyhydroxy compounds of various propylene oxide and ethylene oxide in $scCO_2$. Loy et al.[18] made porous silicon materials directly by using $scCO_2$ as the sol-gel polymerization reaction solvent for alkoxysilane. Wood et al.[21] prepared porous polymer beads with $scCO_2$ by suspension polymerization. In the preparation process, they only used water and CO_2 and they also found that changing the pressure of $scCO_2$ could regulate the surface area of the beads, and thus $scCO_2$ acted as a solvent. There are a few steps to prepare the microporous material in this way, and almost no solvent is used. The basic principle of $scCO_2$ as a foaming agent for preparing foam microporous polymer is that $scCO_2$ has many outstanding characteristics, such as a high diffusion rate, and high swelling and plasticizing capacity. A polymer/gas homogeneous saturated system can be easily established under high pressure. Then the phase of the system can be separated by increasing the polymerization temperature or reducing pressure on the over-saturated thermodynamic state, which causes bubble formation and growth, and finally forms microcellular foams. The foaming mechanism is as follows:

1. *Homogeneous nucleation by supercritical CO_2*

   The melted polymer is dissolved by scCO_2, and the scCO_2 can plastify the homogeneous phase system of the mixture. By controlling the pressure and temperature of the mixed system (e.g., by rapid depressurization or rapid heating), the CO_2 is supersaturated, resulting in the formation of a large amount of gas nuclei in the polymer matrix.
2. *Heterogeneous nucleation of micropores*

   Nonhomogeneous nucleation appears near the additives (such as talcum powder, nano SiO_2, clay, metal particles, heterogeneous polymer or polymer aggregation structure) in the melt polymer. At the interface, nucleation sites are formed in the foaming process caused by heterogeneous nucleation, thereby increasing the pore density. The nucleation efficiency of nano filler is theoretically superior to that of ordinary packing.
3. *The external force field applied to strengthen the nucleation of the bubble*

   The foaming process under an external force field can enhance the formation of the cores, reduce the viscosity of the foaming system, and improve the performance of products.[22] For example, higher shear stress can lead to an increase in the nucleation density. The vibration force field in microcellular foam plastics can induce the nucleation of foam, and at the same time, it can make the bubble rupture and form a homogeneous system.

4.3 SUPERCRITICAL CO_2 FOAMING OF RUBBER AND ESTER POLYMER

It has been found that most thermoplastic polymers can be used as raw materials for the preparation of microporous foamed plastics by $scCO_2$.[23,24] Common thermoplastic microporous foaming systems are PET and blends, PS and blends, PP/PE systems, polymethyl methacrylate (PMMA), thermoplastic PU, and other thermoplastic materials and composites. In most $scCO_2$ foaming processes, thermoplastics are used as the matrix. Poly(ethylene-co-octene) (PEOc) foam material was also prepared by $scCO_2$ foaming.[25] The PEOc foam prepared by this process is different from general microcellular thermoplastic, because it has resilience. The technology of $scCO_2$ foaming has been greatly developed since 2000. By combining the $scCO_2$ foaming process with other plastic modification and processing methods (such as graft copolymer foaming, biomaterial foaming, etc.), the application of $scCO_2$ foaming process has continuously attracted attention.

Jacobs et al.[26] studied the foaming process of poly(ethylene-co-vinyl acetate) (EVA) using $scCO_2$. Vinyl acetate rubber is a kind of soft rubber and the diffusion coefficient of CO_2 in EVA can be obtained at normal pressure. In addition, the adsorption pressure and the frequency of temperature determine the cell size and bulk density of polymer foam pores. Control of expansion temperatures below the melting point of EVA can produce perforated microporous foams. Increasing the adsorption pressure after rapid expansion can produce dense foams with microporous dimensions and the additional adsorption temperature has a very important effect on the morphology of the foam. As the temperature increases, microporous foam with a smaller cell size can be obtained. If the temperature expands above the melting point, the foam becomes unstable and collapses. In general, EVA rubber foam prepared using $scCO_2$ requires strict control of the foaming parameters, which are widely used as bio-stents and drug delivery systems. Song et al.[27] also studied the preparation of silicone rubber foams using $scCO_2$. First, methyl vinyl silicone rubber, silica, hydroxyl silicone oil, and dicumyl peroxide were mixed in a mixer (Haake Rheomix 600p) to obtain a rubber mixture. The rubber mixture was then prevulcanized at 125°C for 2–10 min in a press vulcanizer. Rubber mixture sheets were foamed using $scCO_2$ in a special device. The final foamed sample was heated in an oven at a temperature of 150°C for 30 min and then heat-treated at 210°C for 3 h. By controlling the internal cross-linked network structure, cross-linked highly elastic foam plastic was obtained using $scCO_2$ foam technology, and it was found that preplasticizing time was the key factor affecting the foaming process. The microstructure of the lowest density and the best pore size was obtained, as shown in Figs. 4.3 and 4.4.

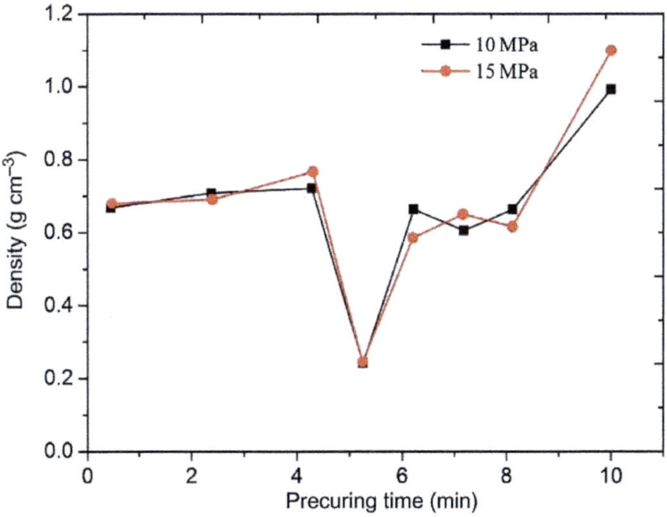

Figure 4.3 Preparation of silicone rubber foams under different conditions. Temperature: 60°C, soak time: 3 h, decompression time: less than 2 s.[27]

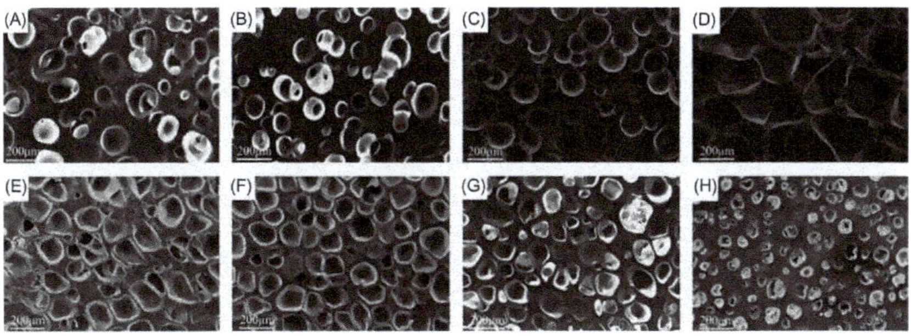

Figure 4.4 SEM micrographs of silicone rubber foam materials at different preplasticizing times (10 MPa): (A) 0 min, (B) 2 min, (C) 4 min, (D) 5 min, (F) 7 min, (G) 8 min, and (H) 10 min.[27]

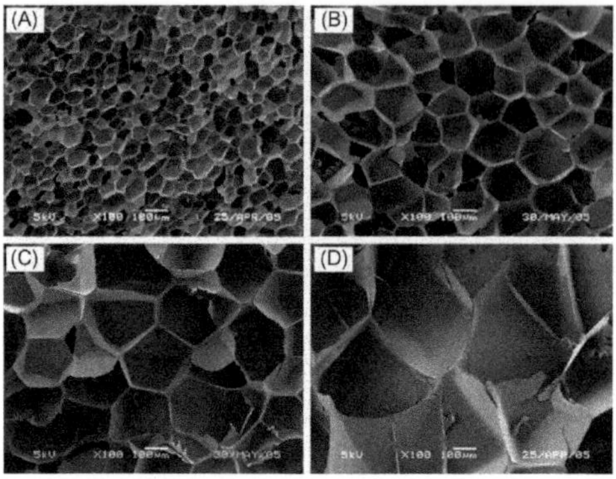

Figure 4.5 SEM micrographs of SMMA foam prepared at different decompression ratios at 118°C [(A) <1 s; (Bb) 35 s; (C) 100 s; (D) 230 s].[28] *SMMA*, Poly(styrene-*co*-methyl methacrylate).

Jacobs et al.,[28] studied the foam morphology parameters of poly (styrene-*co*-methyl methacrylate) (SMMA) produced using scCO_2 and found that the desired foam morphology can be controlled by changing the appropriate foaming temperature and pressure rate. The rapid depressurization rate can make the average diameter value of the prepared foam cells become larger, the cell density become smaller, and the mechanical properties are better (Fig. 4.5). At the same time, a higher temperature resulted in a larger cell size distribution ratio and reduced uniformity. The main reason was that the size of cell reached a limitation, and cells competed to grow, leading to decreased uniformity.

4.4 RADIATION CROSS-LINKING FOAMING TECHNOLOGY

In order to improve the thermal and mechanical properties of polymer foams, a variety of modifications have been carried out, of which the radiation cross-linking is the most commonly used method. The cross-linking of polymer makes it change from a two-dimensional structure to a three-dimensional (3-D) network structure, which significantly improves the polymer mechanical properties, resistance to environmental stress cracking, chemical corrosion and creep resistance, and heat resistance. Radiation cross-linked polymer foam is a semirigid foam between hard and soft, and is light, sound-proof, and earthquake-proof. This material can be used in construction, automobiles, ship-building, and packaging. High energy radiation (γ-ray or Electron beam) is applied to generate macromolecular free radicals, and the subsequent reaction of free radicals leads to cross-linking of polymer.

4.4.1 Radiation cross-linking for the polymer of foaming

Crystalline polymer structure is described by a three-phase model, that is, a crystalline phase, an amorphous phase, and mesosphere (crystalline and amorphous interface). The fact is that radiation cross-linking of the crystalline polymer mainly occurs in the amorphous region.

Due to variation in structural, physical, and other factors, a variety of cross-linking bond structures exist. The cross-linking of T-type or Y-type is dominant in the early stage of irradiation, followed by H-cross-linking, indicating that end radicals have higher activity and higher cross-linking probability than the main chain free radicals. Vilaplana studied the effect of radiation on the chain structure of PE at air atmosphere, and believed that the process of radiation cross-linking and oxidative degradation happened simultaneously.[29] Radiation oxidative degradation is dominant at low dose, and oxidative degradation and cross-linking tend to balance at high dose. Oxidation degradation leads to the formation of sol and cross-linking results in the formation of gel. Lu[30] studied the effects of absorbed dose on the structure and properties of the PE chain under the condition of limited O_2. The results showed that limited O_2 has little effect on the performance of PE at the absorbed doses of 0–300 kGy. The gel content and notched impact strength of PE increased as the dose was increased, but the crystallinity, melting temperature, and the elongation at break decreased with the increase of dose.

Xing et al.[31] prepared radiation cross-linked low-density polyethylene (LDPE) foam by $scCO_2$ foaming. The molecular chain of PE is interrupted by γ-ray irradiation to produce active free radicals[32] and can initiate a series of complex chemical reactions, such as cross-linking, degradation, oxidation, and branching.[29,33] The final results are determined by cross-linking and degradation reactions. The gel content increases rapidly as the absorbed dose increases. This indicates that in the dose range of 0–100 kGy, the irradiation mainly causes the cross-linking of LDPE. The main reaction scheme is as follows:

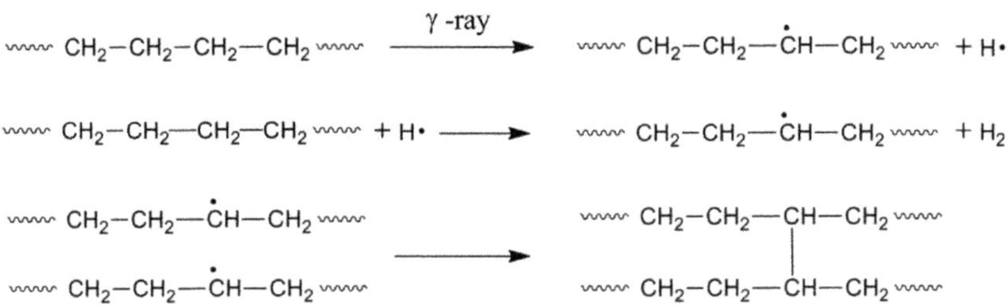

4.4.2 Radiation cross-linking foaming

The production of radiation cross-linked polymer foam is a mature radiation processing technology. Sekisui and Toray began industrial production of radiation cross-linked polymer foams in 1965 in Japan, and at least 15,000 t of polymer foams are now produced annually. The use of radiation cross-linking can improve or adjust the viscoelasticity and viscosity of the plastic in the molten state, so that the bubbles produced during the foaming process are not easily broken. In the radiation cross-linking foaming process of polymer, no gel is formed or the gel content is very low, and the foaming agent is decomposed to produce polymer foam by heating to a higher temperature.

Since radiation cross-linking can improve the melt viscosity of plastics, the foaming temperature window is wider. For PE, a lower absorbed dose is enough to increase the viscosity of the melt and make the foaming process easier. It is believed that PE can be foamed with a gel content of 0.2%–0.4%. For biodegradable

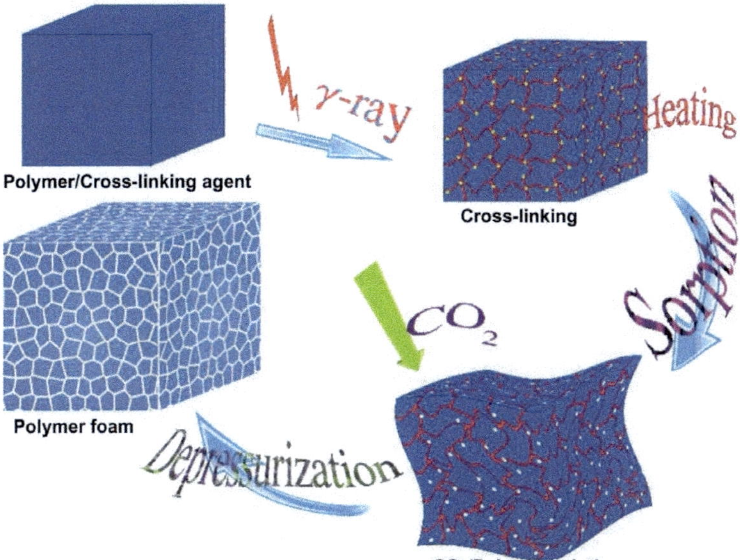

Figure 4.6 Schematic diagram of radiation cross-linking polymer foaming.[34]

succinic acid butanediol, a gel content of 2%–5% is required for foaming. Because PP contains branched chains and is easy to crack by irradiation, it is necessary to add a small amount of radiation cross-linking agent to improve cross-linking and foaming. PP foam material has higher temperature resistance than PE foam, and can be used for heat insulation materials in the roofs of cars. A schematic diagram of the preparation of radiation cross-linked polymer foam by $scCO_2$ is shown in Fig. 4.6[34].

In our study, we investigated the foaming behavior of γ-irradiated LDPE using $scCO_2$ as the blowing agent.[35] It was found that a percentage of gel formed in the LDPE matrix at various absorbed doses, and the degree of cross-linking increased with increasing dose over the dose range investigated in Fig. 4.7 (A). The volume expansion ratio (R_v) of LDPE foam increased initially, and then decreased as the absorbed dose increased in Fig. 4.7 (B). The maximum R_v was observed at 50 kGy, implying an optimum absorbed dose of 50 kGy for the foaming of LDPE. Fig. 4.8 shows the cellular structure of nonirradiated and irradiated LDPE foams.

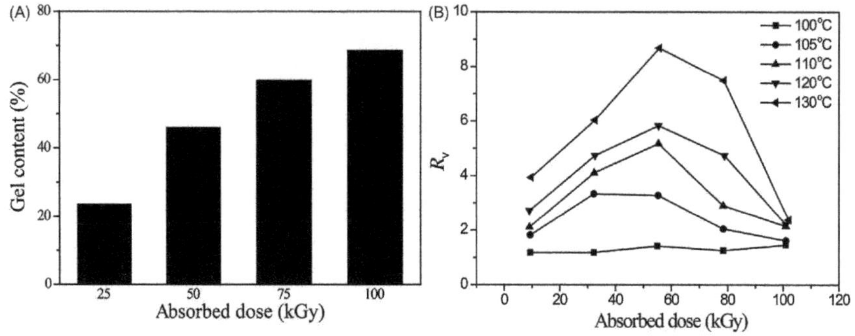

Figure 4.7 (A) Gel content of irradiated LDPE samples and (B) volume expansion ratio (R_v) of LDPE foams prepared at different temperatures.[35]

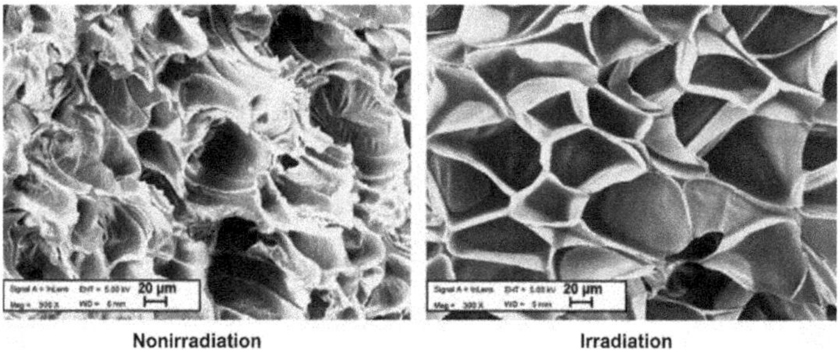

Figure 4.8 SEM micrographs of nonirradiated and irradiated specimens foamed at 130°C.[35]

It can be clearly seen that the irradiated sample possesses an impinging polygonal closed-cell structure.

The mechanical strength and heat resistance of radiation cross-linked polymer are higher than those of conventional polymers. Radiation cross-linked polymer foam has better electrical insulating performance. Radiation cross-linked polymer foams contain no chemicals harmful to human body, and may be used for medicine and food packaging. In addition, this technology has many advantages:

1. Reduced environmental pollution. The radiation cross-linking foaming process is nontoxic.
2. Easier control of the process. Radiation cross-linking foaming is easier to control than chemical cross-linking foaming, and the choice of raw materials, such as base resin or foaming agent, is easier.

3. Higher output of production. In chemical cross-linking, the cross-linking agent is decomposed by heating in a wide temperature range. It is difficult to guarantee the uniformity of the product and the extent of cross-linking cannot be easily controlled. Radiation cross-linking is carried out at room temperature. Therefore, the texture is uniform and the extent of cross-linking is controllable.
4. Quick foaming speed. Since the radiation process can achieve uniform cross-linking of the product at any temperature, the foaming rate is faster than the chemical cross-linking process.
5. Production of various kinds of products. By controlling the absorbed dose and dose rate, it is easy to control the cross-linking degree and produce polymer foams with different pore sizes.

4.4.3 Radiation cross-linking foaming of polypropylene

The main varieties of polymer foams are made from PU, PS, and PE. PU foam has harmful isocyanate residues after the foaming process, and the foaming material cannot be recycled. PS foam is difficult to degrade and is normally called "white pollution". The United Nations Environment Organization has decided to stop using PS foam products. The rigidity of cross-linked PE foam is relatively low, and its maximum service temperature is 80°C. Compared with other plastic foams, PP foam has many advantages:

1. The rigidity of PP is higher than PE; flexural modulus of PP is ~1.52 GPa, much higher than that of PE (207 MPa).
2. PP has better impact resistance than PS, and PP foam is an environmentally-friendly material.
3. PP foam has a higher thermal deformation temperature and can be applied in some high temperature fields. PP foam can withstand thermal treatment at 130°C, which is much higher than the maximum use temperature of 80°C for PE foam. The closed cell structure of PP foam makes its thermal conductivity not affected by humidity, so it can be used as thermal insulation material.

4. Good low temperature properties: products exhibit good performance even at −30°C.
5. Energy absorption: because PP foam products have good energy absorption characteristics, they have excellent compression resistance and energy absorption performance. PP is widely used in automobile bumper energy handling systems and other anticollision energy absorbing parts.[36]
6. Shape restoration stability: PP foam products can be quickly restored to their original shape without repeated permanent deformation after repeated impact and deflection.
7. The density is low and can be used repeatedly: the density of PP foam products can reach a very low level. PP foam has good flexibility, it is not easily broken, but easy for recycling and decomposition.
8. Good surface protection and sound insulation performance: PP foam possesses good hardness, moderate softness, and no scratches, as well as providing good surface protection.

Although PP foam products have good performance and prospects, the production of PP foam at an industrial scale is not easy. At present, many efforts are being made to develop PP foam and replace expanded PS as a green packaging material. Extrusion foaming of PP sheet by the Amut Company in Italy is the latest case of PP foaming on an industrial scale. PP with high melt strength produced by the Montel Company of Belgium (High Melt Strength Polypropylene, HMSPP), combined with PP homopolymer and chemical foaming agent with low-value ozone depletion potential, produces a small pore and uniform distribution PP foam sheet (Expanded polypropylene, EPP) and the density is 0.6 g cm^{-3}. The mass fraction of HMSPP is 12.5%, and it has high economic and environmental significance. Since HMSPP has high tension and strain hardening in the melting process, the cell does not get destroyed, which has a closed cell structure of foam products with high insulation and rigidity.

In Japan, the industrialization of polyolefin foam materials has been realized for many years. At present, the production of radiation cross-linked PE foam is common in China and the mass production technology is mature. However, the technology for the mass production of radiation cross-linked PP foam has not been established to date in China. Radiation grafting or cross-linking has been applied to improve the performance of PP foam, but has not been successfully applied for industrial production.

Wang et al.[37] studied the $scCO_2$ foaming of radiation cross-linked PP by γ-ray. It was found that the melting point of PP decreased slightly after γ-ray irradiation, and the viscosity decreased markedly with the increase of absorbed dose. The foaming behavior of radiation cross-linked PP improved. It is easy to obtain the foam with higher foaming rate by radiation modification of PP. Under the same $scCO_2$ foaming pressure, radiation cross-linked PP needs lower foaming temperature, as can be seen in Tables 4.2 and 4.3 (foaming temperature of 150°C).

The volume expansion rates of PP foams are shown in Tables 4.2 and 4.3. The value of the radiation-modified PP foam is significantly higher than that of the un-irradiated PP foam. The expansion rate of PP foam with a dose of 40 kGy in Table 4.3 is 8,

Table 4.2 Effects of foaming temperature on expansion ratio and average pore diameter for radiation cross-linked PP[37]

Foaming temperature (°C)	Foam density (g cm^{-3})	Foam volume expansion ratio	Average cell diameter (μm)
147	0.59	1.52	8
150	0.45	2.00	11
152	0.36	2.49	49
154	0.20	4.60	82

Table 4.3 Effects of absorbed dose on cellular structure of PP foam (foaming temperature: 150°C)[37]

Absorbed dose (kGy)	Foam density (g cm^{-3})	Foam volume expansion ratio	Average cell diameter (μm)
0	0.40	2.25	10
10	0.30	3.02	18
20	0.20	4.56	25
30	0.20	4.49	40
40	0.11	8.09	96

which is higher than that of the PP foam in Table 4.2. This is due to the cross-linking structure formed in the irradiated PP, and the higher melt strength enhances the growth of the foaming bubble, making the cell wall not easy to break. A blend of isotactic PP with high-density PE was irradiated by γ-ray to induce cross-linking and then foamed using $scCO_2$ as the blowing agent.[38] The results showed that irradiation can enhance foaming ability, improve the cellular structure, and increase the mechanical properties.

Liu et al.[39] prepared PP foam sheets by electron-beam irradiation. It was found that the irradiation atmosphere has a great influence on the gel content. Radiation cross-linking occurred mainly in the amorphous regions of PP, while the dissolved O_2 in the amorphous regions consumed the radiation-induced free radicals, increased the degradation tendency of the system, and reduced the efficiency of the cross-linking reaction, causing the N_2 protection to greatly improve the gel content. At lower doses, the effect of N_2 protection is very obvious. The gel produced by irradiation under N_2 protection is much higher than that produced by irradiation in air at the dose of 1–5 kGy. The main reason is that since free radicals are less consumed under N_2 protection than in the air, the gel content increases. Ugimoto produced high-molecular-weight PP by radiation graft chain modification.[40] Lugão et al.[41] improved significantly the melt strength of PP by radiation cross-linking. Hong et al.[42] studied the effect of gel content on the foaming ability of PP.

Few reports are available on the preparation of radiation cross-linked PP using $scCO_2$ as the foaming agent. Our team compared the micropore structure of un-irradiated PP foam and cross-linked polypropylene/triallylisocyanurate (PP/TAIC) foam at the dose of 30 kGy.[34] The un-irradiated PP foam samples showed serious collapse and fusion, and the cellular structure of cross-linked PP foam was well observed, as shown in Fig. 4.9. Cross-linking between macromolecules happened in irradiated PP, forming a 3D-cross-linked network. When the temperature was raised above the melting point of crystallization, the melt PP could form a highly elastic state, and higher melt strength and viscosity can

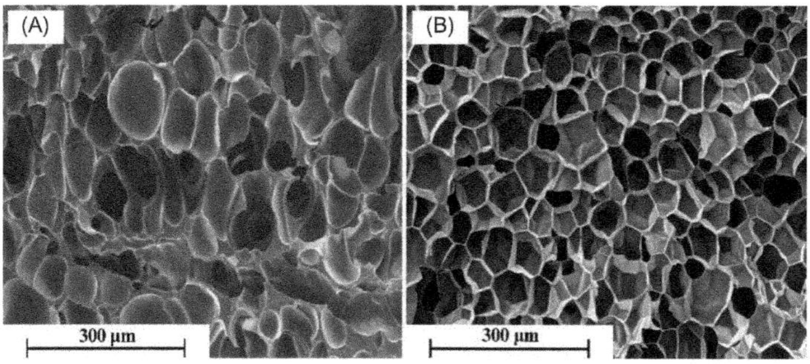

Figure 4.9 SEM micrographs of two foaming specimens of PP foams prepared at 152°C and 20 MPa: (A) non-cross-linked, (B) cross-linked (30 kGy).[34]

effectively prevent bubble collapse. The radiation effects on the $scCO_2$ foaming of atactic PP were also investigated.[43] The results showed that cross-linking can increase the number of nucleations and broaden the foaming temperature window.

The foaming behavior of radiation cross-linked isotactic PP/high-density PE blend using $scCO_2$ as the blowing agent was also investigated.[38] Tensile measurements showed that the elongation at break and tensile strength at break of the cross-linked PP/HDPE foams were higher than those of the noncross-linked ones. The most interesting observation was the increase of the foaming temperature window from 4°C for pristine PP to 8–12°C for the radiation cross-linked PP/HDPE blends. In this study, the gel content of PP/HDPE increased as the dose increased up to 60 kGy. The possible reaction and interactions of the irradiated PP and HDPE macromolecules likely occur in the PP-HDPE interface, and PP-graft-PE copolymers were also formed, which could greatly increase the gel content of the PP/HDPE blends. Micrographs of the gels were obtained by SEM. A SEM observation of the PP90/HDPE10 gel obtained after a dose of 20 kGy is shown in Fig. 4.10. It can be seen that the gel has a tangled and entwined mesh structure, and many macromolecular chains are intertwined together. The formation of the mesh structure is mainly caused by cross-linking, entanglement, and intertwining between macromolecules in the blend.

Cell morphology micrographs of the pristine PP and cross-linked PP90/HDPE10 foam are shown in Fig. 4.11. A well-defined

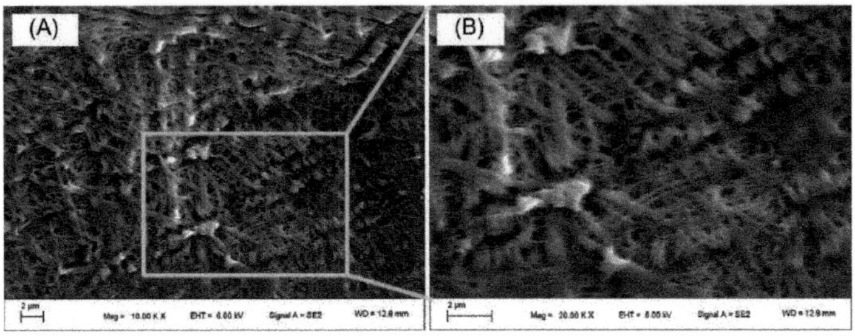

Figure 4.10 (A) Low- and (B) high-magnification micrographs of the gel structure of PP90/HDPE10 after irradiation at a absorbed dose of 20 kGy.[38]

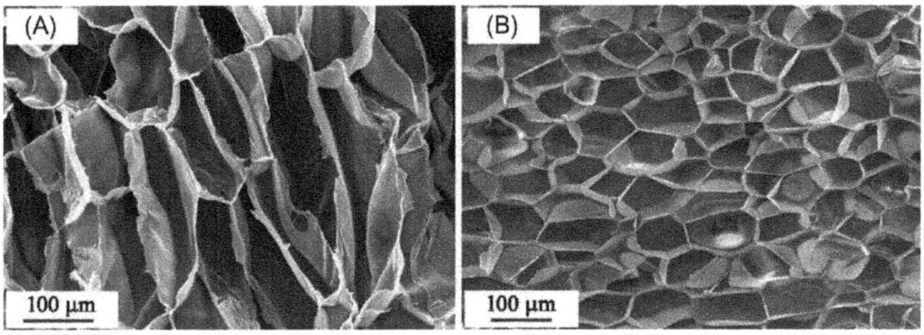

Figure 4.11 SEM micrographs of foamed PP prepared at 146°C and 20 MPa: (A) pristine PP, (B) PP90/HDPE10 (20 kGy).[38]

cellular structure of cross-linked PP was obtained and compared to the neat PP foam, which indicated the higher mechanical property of the PP/HDPE foam. The characteristic data of stress—strain are summarized in Table 4.4. The results show that the cross-linked PP foam with well-defined morphology possessed higher stress and strain.[38]

Wu et al.[44] prepared a closed-cell PP foam material with high-performance using $scCO_2$ foaming with the assistance of γ-ray radiation cross-linking (see Fig. 4.12). In this study, styrene-ethylene-butadiene-styrene (SEBS) copolymer was added to enhance radiation cross-linking and nucleation. The cell density of the cross-linked PP/SEBS foam greatly increased at a dose of 10 kGy and a high closed-cell ratio was obtained. The cell size distributions for the pristine PP foam and cross-linked PP98/SEBS2 foam are compared in Fig. 4.13. It can be clearly seen that the cross-linked PP foam has a more homogeneous cell size with a

Table 4.4 Tensile strength at break and elongation at break of PP and PP/HDPE foams[38]

Foam sample	Tensile strength at break (MPa)	Elongation at break (%)
Pristine PP	11.4 ± 1.1	65.3 ± 4.8
PP95/HDPE5	12.7 ± 0.9	75.1 ± 2.9
PP90/HDPE10	12.5 ± 1.2	89.2 ± 3.4
PP85/HDPE15	12.3 ± 1.0	98.4 ± 5.3
PP95/HDPE5-20 kGy	14.1 ± 1.3	187.1 ± 6.3
PP95/HDPE5-40 kGy	14.7 ± 1.1	171.2 ± 4.7
PP95/HDPE5-60 kGy	13.8 ± 1.6	95.6 ± 3.2
PP90/HDPE10-20 kGy	14.5 ± 1.0	134.3 ± 5.9
PP90/HDPE10-40 kGy	14.3 ± 1.3	131.2 ± 3.1
PP90/HDPE10-60 kGy	15.4 ± 1.5	81.1 ± 3.7
PP85/HDPE15-20 kGy	14.8 ± 0.9	114.3 ± 3.4
PP85/HDPE15-40 kGy	15.1 ± 1.2	87.3 ± 2.9
PP85/HDPE15-60 kGy	14.4 ± 1.0	78.6 ± 4.8

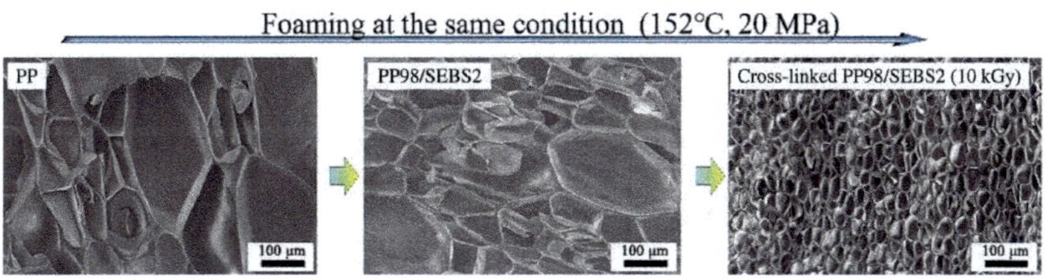

Figure 4.12 SEM micrographs of three different specimens of PP foam.[44]

higher closed-cell ratio. These results indicate that the addition of low amounts of SEBS further promotes the cross-linking and nucleation of the blend at a suitable dose. The tensile strength of the cross-linked PP/SEBS foams (10 kGy) improved from 14 to 20.7 MPa compared with the pristine PP foam (0 kGy). In addition, the cross-linked PP/SEBS blend exhibited a wider foaming temperature window (10°C) compared with the non-cross-linked blends (4°C).

It is believed that radiation cross-linking may be applicable for the supercritical foaming of PP in industrial production. How to

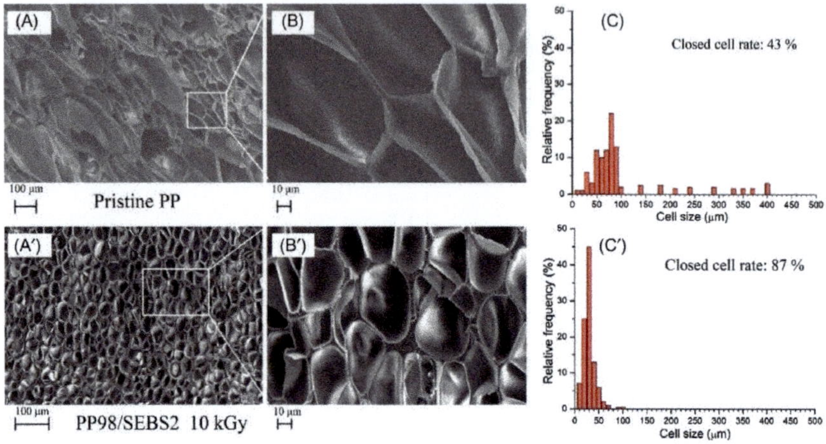

Figure 4.13 SEM (A) and (A′) low magnification and (B) and (B′) high magnification images of pristine PP foam and irradiated PP98/SEBS2 foam with (C) and (C′) cell size distributions.[44]

effectively control the microcellular PP foam structure is required in the experimental investigation.

REFERENCES

1. Martini JE. *The production and analysis of microcellular foam*. SM thesis. Dept. Mech. Eng, MIT; 1981.
2. Wypych G. *Handbook of foaming and blowing agents: introduction: lessons from living things. Self-healing materials*. Canada: Chem Tec Publishing; 2017. p. 1–5.
3. Hermann H, Virginia L, Timothy M, Heike K, Jackie S, Mark L. Impregnating solution for cosmetic cloths, patent, US; 2006.
4. Williams RS. Ethafoam and other polyethylene foams in conservation, patent, Canada; 1998.
5. Helen CM. Photographic film and a laminated protective coating therefor, patent, US; 1933.
6. Nelson JD, Steenstrup PV. Cellular phenolic resin, patent, US; 1948.
7. Fauser EE, Seeger NV. Polyisocyanates and mixtures thereof, patent, US; 1954.
8. Cheung P, Suwanda D, Balke ST. The reactive extrusion of polyethylene polypropylene blends. *Polym Eng Sci* 1990;**30**:1063–72.
9. Kumar V. Microcellular polymers-novel materials for the 21-century. *Cell Polym* 1993;**12**:207–23.
10. Collias DI, Baird DG, Borggreve RJM. Impact toughening of polycarbonate by microcellular foaming. *Polymer (Guildf)* 1994;**35**:3978–83.

11. Goel SK, Beckman EJ. Generation of microcellular polymeric foams using supercritical carbon-dioxide. 2. Cell-growth and skin formation. *Polym Eng Sci* 1994;**34**:1148–56.
12. Park CB, Baldwin DF, Suh NP. Effect of the pressure drip rate on cell nucleation in continuous processing of microcellular polymers. *Polym Eng Sci* 1995;**35**:432–40.
13. Cooper AI. Recent developments in materials synthesis and processing using supercritical CO_2. *Adv Mater* 2001;**13**:1111–14.
14. Nalawade SP, Picchioni F, Janssen LPBM. Supercritical carbon dioxide as a green solvent for processing polymer melts: processing aspects and applications. *Prog Polym Sci* 2006;**31**:19–43.
15. Fu SL, Zhou J, Huang MF, Ye Y, Li SD. Study on extraction of flavour from vanilla beans by supercritical CO_2 fluid. *Chem Res Appl* 2002;**14**(4):452–4 (in Chinese).
16. Parks KL, Beckman EJ. Generation of microcellular polyurethane foams via polymerization in carbon dioxide. I: Phase behavior of polyurethane precursors. *Polym Eng Sci* 1996;**36**:2404–16.
17. Kubo J, Otsuhata K, Ikeda S, Seguchi T. Electron-induced crosslinking of polypropylene with the addition of hydrogen-donating hydrocarbons. *J Appl Polym Sci* 1997;**64**:311–19.
18. Loy DA, Russick EM, Yamanaka SA, Baugher BM, Shea KJ. Direct formation of aerogels by sol – gel polymerizations of alkoxysilanes in supercritical carbon dioxide. *Chem Mater* 1997;**9**(11):2264–8.
19. Park CB, Cheung LK. A study of cell nucleation in the extrusion of polypropylene foams. *Polym Eng Sci* 1997;**37**:1–10.
20. Parks KL, Beckman EJ. Generation of microcellular polyurethane foams via polymerization in carbon dioxide. II: Foam formation and characterization. *Polym Eng Sci* 1996;**36**:2417–31.
21. Wood CD, Cooper AI. Synthesis of macroporous polymer beads by suspension polymerization using supercritical carbon dioxide as a pressure-adjustable porogen. *Macromolecules* 2001;**34**:5–8.
22. Leung SN, Wong A, Wang LC, Park CB. Mechanism of extensional stress-induced cell formation in polymeric foaming processes with the presence of nucleating agents. *J Supercrit Fluids* 2012;**63**:187–98.
23. Li S, Xiao M, Guan Y, Wei D, Xiao H, Zheng A. A novel strategy for the preparation of long chain branching polypropylene and the investigation on foamability and rheology. *Eur Polym J* 2012;**48**:362–71.
24. Ren Q, Wang J, Zhai W, Su S. Solid state foaming of poly(lactic acid) blown with compressed CO_2: influences of long chain branching and induced crystallization on foam expansion and cell morphology. *Ind Eng Chem Res* 2013;**52**:13411–21.
25. Zhai W, Leung SN, Wang L, Naguib HE, Park CB. Preparation of microcellular poly(ethylene-*co*-octene) rubber foam with supercritical carbon dioxide. *J Appl Polym Sci* 2010;**116**:1994–2004.
26. Jacobs MA, Kemmere MF, Keurentjes JT. Foam processing of poly (ethylene-*co*-vinyl acetate) rubber using supercritical carbon dioxide. *Polymer (Guildf)* 2004;**45**:7539–47.

27. Song L, Lu A, Feng P, Lu Z. Preparation of silicone rubber foam using supercritical carbon dioxide. *Mater Lett* 2014;**121**:126–8.
28. Jacobs L, Danen K, Kemmere M, Keurentjes J. A parametric study into the morphology of polystyrene-*co*-methyl methacrylate foams using supercritical carbon dioxide as a blowing agent. *Polymer (Guildf)* 2007;**48**:3771–80.
29. Vilaplana F, Morera-Escrich V, del Hierro-Navarro P, Monrabal B, Ribes-Greus A. Performance of crystallization analysis fractionation and preparative fractionation on the characterization of γ-irradiated low-density polyethylene. *J Appl Polym Sci* 2004;**94**:1803–14.
30. Lu HJ, Ma XY, Zhu GM, Liang GZ. Limited-oxygen irradiation cross-linking modification of HDPE. *Polym Mater Sci Eng* 2003;**19**(4):204–7 (in Chinese).
31. Xing Z, Wu GZ, Huang SR, Zeng HY. Preparation of microcellular cross-linking polyethylene foams using supercritical CO_2. *J Radiat Res Radiat Process* 2008;**26**(4):193–8 (in Chinese).
32. Zu JH, Liu XW, Zhou RM, Tu TC, Qiu SL. ESR study of radicals in irradiated LDPE and HDPE membrane. *J Radiat Res Radiat Process* 2005;**23** (3):174–8 (in Chinese).
33. Abou Zeid H, Ali Z, Maksoud A, Khafagy R. Structure–property behavior of polyethylene exposed to different types of radiation. *J Appl Polym Sci* 2000;**75**:179–200.
34. Yang C-G, Wang M-H, Zhang M-X, Li X-H, Wang H-L, Xing Z, et al. Supercritical CO_2 foaming of radiation cross-Linked isotactic polypropylene in the presence of TAIC. *Molecules* 2016;**21**:1660.
35. Xing Z, Wu G, Huang S, Chen S, Zeng H. Preparation of microcellular cross-linked polyethylene foams by a radiation and supercritical carbon dioxide approach. *J Supercrit Fluid* 2008;**47**:281–9.
36. Yang C, Xing Z, Wang M, Zhao Q, Wu G. Merits of the addition of PTFE micropowder in supercritical carbon dioxide foaming of polypropylene: ultrahigh cell density, high tensile strength, and good sound insulation. *Ind Eng Chem Res* 2018. Available from: https://doi.org/10.1021/acs.iecr.7b04644.
37. Wang MH, Xing Z, Liu WH, Zhang WF, Zhang WL, Wu GZ. Study on supercritical carbon dioxide foaming properties of radiation modified polypropylene. *J Radiat Res Radiat Process* 2014;**32**(2). 020303 (6). (in Chinese).
38. Yang C, Xing Z, Zhang M, Zhao Q, Wang M, Wu G. Supercritical CO_2 foaming of radiation cross-linked polypropylene/high-density polyethylene blend: cell structure and tensile property. *Radiat Phys Chem* 2017;**141**:276–83.
39. Liu TC, Zheng YS, Li LY, Zhang LY. Polypropylene foam sheet prepared by electron-beam irradiation. *Chin Plast* 2006;**20**(9):49–52 (in Chinese).
40. Sugimoto M, Tanaka T, Masubuchi Y, Takimoto JI, Koyama K. Effect of chain structure on the melt rheology of modified polypropylene. *J Appl Polym Sci* 1999;**73**:1493–500.
41. Lugão AB, Hutzler B, Ojeda T, Tokumoto S, Siemens R, Makuuchi K, et al. Reaction mechanism and rheological properties of polypropylene irradiated under various atmospheres. *Radiat Phys Chem* 2000;**57**:389–92.

42. Hong D, Yoon K, Baek W, Jung H, Lee J, Lee K, et al. Effects of irradiation cross-linking and molecular weight properties on cross-linked PP foaming process. *Polymer-Korea* 2002;**26**:508–15.
43. Yang C, Zhe X, Zhang M, Wang M, Wu G. Radiation effects on the foaming of atactic polypropylene with supercritical carbon dioxide. *Radiat Phys Chem* 2017;**131**:35–40.
44. Yang CG, Xing Z, Zhao Q, Wang MH, Wu GZ. A strategy for the preparation of closed-cell and cross-linked polypropylene foam by supercritical CO_2 foaming. *J Appl Polym Sci* 2018;**135**(6):45809.

CHAPTER FIVE

Radiation Degradation or Modification of Poly (tetrafluoroethylene) and Natural Polymers

Rong Li[1], Guozhong Wu[1], Yan Hao[2], Jing Peng[2] and Maolin Zhai[2]
[1]Shanghai Institute of Applied Physics, Chinese Academy of Sciences, Shanghai, P.R. China
[2]College of Chemistry and Molecular Engineering, Peking University, Beijing, P.R. China

5.1 RADIATION DEGRADATION, CROSS-LINKING, AND SURFACE MODIFICATION OF POLYTETRAFLUROETHYLENE

Polytetrafluroethylene (PTFE) is a self-lubricating polymeric material with high crystallinity and melting point (>320°C). In 1938, PTFE was discovered by Plunkett at the DuPont Company in the United States. PTFE can be synthesized by free radical polymerization (suspension polymerization or emulsion polymerization), which looks like a white waxy thermoplastic resin. The chemical formula of PTFE is shown in Fig. 5.1.

The ρ, glass transition temperature, and melting point of PTFE are 2.14–2.30 g cm^{-3}, ~127°C and ~327°C, respectively. PTFE is a kind of non-branched linear polymer, and its structure is completely symmetrical. The degree of crystallinity of PTFE is very high (92%–98%) due to its regular molecular structure. Since the radius (0.064 nm) of fluorine atoms is significantly larger than that of hydrogen atoms (0.028 nm), the plane and fully-stretched tortuous conformation of the C–C chain in PE is transformed into a helical conformation in PTFE. The bond energy of C–F (460.2 kJ mol^{-1}) is higher than that of C–H (410 kJ mol^{-1}) and C–C (372 kJ mol^{-1}), and the fluorine element outside the C–C

Radiation Technology for Advanced Materials.
DOI: https://doi.org/10.1016/B978-0-12-814017-8.00005-6

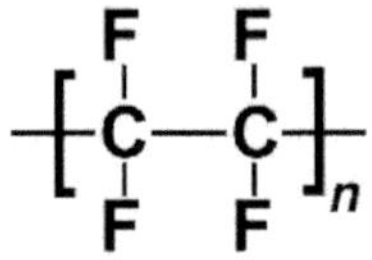

Figure 5.1 Chemical formula of PTFE.

chains, as the protective layer, enables the C—C backbone to be protected from any external reagent. Hence, PTFE has several unmatched properties, including solvent resistance, chemical stability, thermal stability, and low cohesive energy density. Due to the extremely high electro-negativity of fluorine atoms and the symmetry and regularity of PTFEs molecular structure, the low attraction between PTFE chains and low surface energy cause PTFE to have excellent lubricating properties and good ductility under low-temperature conditions. However, PTFE has some defects, including poor creep resistance, poor wear resistance, and it is prone to the cold flow phenomenon. In addition, PTFE is difficult to process using ordinary injection molding methods owing to its high crystallinity; it is usually processed by molding, sintering, extrusion, and machining methods.

In the past several decades, the radiation effects of PTFE have been extensively investigated. PTFE is a typical radiation-degradable polymer. Even a low absorbed dose can induce a significant reduction in the relative molecular mass of PTFE. The relative molecular weight of PTFE in emulsion can also be reduced by irradiation. Currently, PTFE ultrafine powder prepared by irradiation is widely used in industrial fields. In the radiation process, the high mobility of PTFE macromolecular chains in the molten state stimulates the coupling reaction of free radicals, resulting in the formation of cross-linked PTFE. Cross-linked PTFE exhibits excellent wear resistance, and can be used to prepare super wear-resistant commercial material, which has been commercialized in Japan. Surface modification of PTFE via radiation-induced grafting of polar or hydrophilic functional groups can effectively improve the compatibility or dispersibility of PTFE with other materials, which has great application value.

5.1.1 Structure and properties of polytetrafluroethylene

PTFE, known as the king of plastic, is one of the most important thermoplastic engineering plastics. PTFE has several advantages, including excellent thermal resistance, weather resistance, chemical stability, insulation, incombustibility, and good lubricity. PTFE has been used in the electronics, chemical, petroleum, textile, medical, food, machinery, environment, marine and other fields, owing to its superior performance. However, the application of PTFE is limited by its disadvantages, including its high coefficient of linear expansion, poor creep resistance, poor wear resistance, poor radiation resistance, low surface energy, and poor adhesion. Research on the improvement of PTFE's properties has been widely reported. γ-Ray or electron beam (EB) radiation cross-linking of PTFE has been employed to improve its resist creep, abrasion, and radiation resistance. Radiation- and plasma-induced surface modification and functionalization of PTFE have been employed to improve its surface property and adhesion, and ameliorate its hydrophilicity. Low-temperature plasma technology is particularly suitable for PTFE surface modification, because it does not result in the degradation of bulk PTFE.

5.1.2 Radiation degradation and cross-linking of polytetrafluroethylene

5.1.2.1 Radiation degradation of polytetrafluroethylene

PTFE is extremely sensitive to radiation, and prone to degradation by radiation, discovered by researchers at Oak Ridge National Laboratory, United States in the 1950s. Subsequently, a large number of studies have demonstrated that the absorbed dose of several kilogray can cause severe degradation of PTFE. As shown in Fig. 5.2, the degradation of PTFE increases with the absorbed dose, in combination with a decline in its mechanical properties (see Fig. 5.3).[1] In addition, PTFE is more prone to degradation by radiation under the conditions of high-temperature and high O_2 content, accompanied by the formation of toxic gases. It is universally accepted that radiation can cause the breakage of PTFE

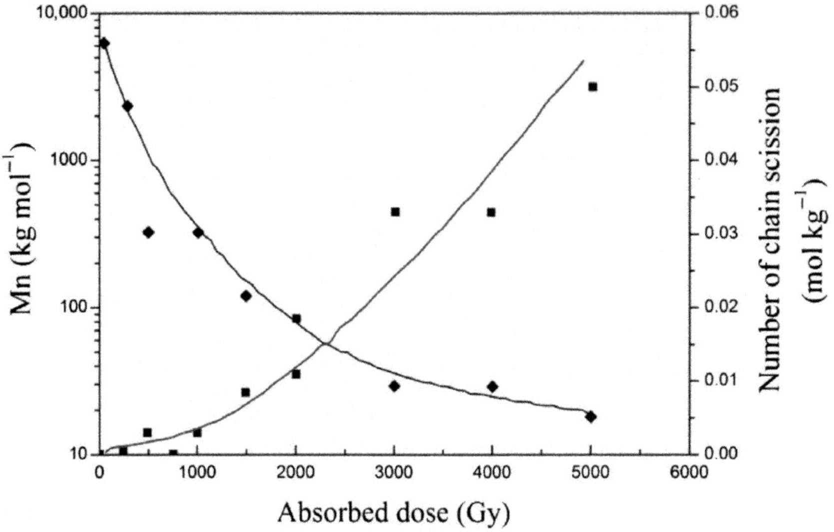

Figure 5.2 Average molar mass and number of chain scissions versus absorbed dose.[1]

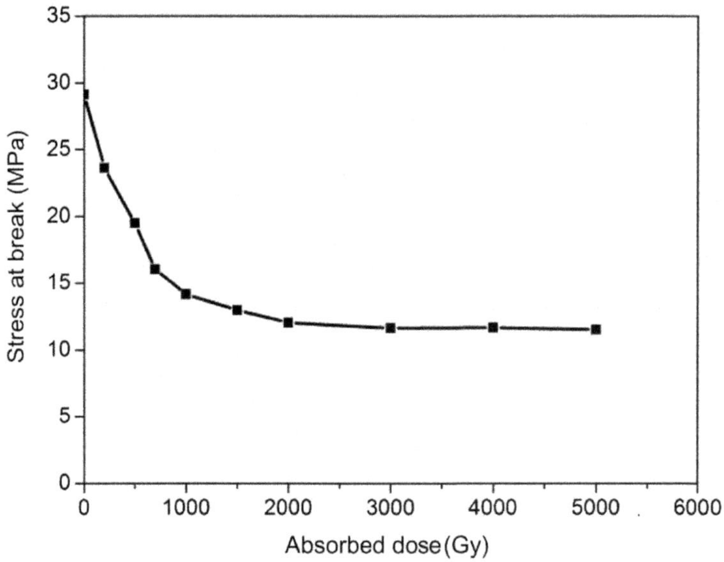

Figure 5.3 Stress at break versus absorbed dose.[1]

molecular chains at room temperature, thereby leading to a significant reduction in the molecular weight of PTFE. The C–C bonds for PTFE are more prone to degradation by radiation, because the bond energy of C—F is much stronger than that of C—C bond. However, the degradation of C—F bond does exist during the radiation process, because the side-production, corrosive fluorine gas, is detected.

PTFE ultrafine powder can be prepared at room temperature or below its melting point by radiation and grinding or air crushing, based on its radiation-degradable characteristics. PTFE ultrafine powder not only retains its own merits, including excellent chemical stability, low friction factor and good electrical properties, but also has good dispersion, and is easy to mix with other materials. PTFE ultrafine powder has been widely used to improve the lubricity, wear-resistance, and scratch-resistance of polymeric materials. In the current market, engineering plastics with good lubrication performance are almost all doped with PTFE superfine powder. For example, the wear-resistance and lubrication performance of nylon, polyformaldehyde (POM), polycarbonate, and other high-performance engineering resins can be significantly improved by adding PTFE ultrafine powder. Furthermore, these materials can be used to manufacture gears, bearings and pulleys, and can also be used to replace the metal parts in machines to reduce the weight and product cost. In addition, the addition of PTFE ultrafine powder in nylon can also significantly reduce its water absorption. PTFE superfine powder can be used as an ink additive to significantly improve the lubricity and gloss of ink, thereby making the surface smooth and improving the wear resistance of the printed product. For example, a black ballpoint pen can be used for writing smoothly in both winter and summer, since PTFE ultrafine powder was added to the ink. The antifriction property and lubrication performance under the conditions of high temperature and high pressure of silicone oil can be significantly improved by the addition of PTFE ultrafine powder. The highly viscous fluorine oil produced by the addition of PTFE ultrafine powder can be applied under the condition of high temperature, and this can also reduce the production costs. High-quality lubricants can be prepared by the addition of PTFE ultrafine powder to the grease, rosin and mineral oil used for ball bearings, wear-resistant bearings, lubrication rails, slides, open gears, chemical equipment valves, precision machining plane seal paste, etc. The viscosity, lubricity, abrasion resistance, and corrosion resistance of coatings can be improved by the addition of PTFE ultrafine powder to the coating, while reducing its hygroscopicity. In addition, the radiation-induced formation of hydroxyl and carboxyl groups in PTFE ultrafine powder can improve the compatibility of PTFE with matrix materials. PTFE waste and scrap are difficult to handle, and most were treated as industrial waste before the radiation-degradable property of PTFE was found. Today, PTFE waste degraded by radiation can be processed into ultrafine powder with unique performance.

5.1.2.2 Radiation cross-linking of polytetrafluroethylene

The cross-linking of PTFE was discovered in 1970 by Tsuchiya at the Tokyo Metropolitan Institute Isotope Research Center. Radiation cross-linking of PTFE was also proposed by a Chinese team in early times. From 1992, research on radiation cross-linking of PTFE was carried out by Professor Tabata et al., from the Takasaki Radiation Chemistry Institute, Japan. Radiation cross-linking of PTFE was then widely recognized. Radiation cross-linking of PTFE involves several steps: (1) the molten PTFE is irradiated by ^{60}Co γ-ray or EB; (2) free radicals are formed due to the pyrolysis of the PTFE molecular chains; (3) molecular chains are relatively free to flow, so that they can be close to each other to cross-link. As shown in Fig. 5.4, Y-type cross-linking for PTFE mainly occurs due to the steric effects. However, radiation cross-linking of PTFE should be carried out in a vacuum or inert atmosphere, because the presence of O_2 can react with the free radicals generated by radiation. In addition, the temperature range for cross-linking is narrow (330–340°C), that is, slightly higher than the melting point of PTFE. It is necessary to pay close attention to the high-temperature irradiation device and to controlling the temperature. In addition, close attention should be given to the poisonous gas formed in the radiation process. The cross-linking of PTFE prepared by EB irradiation is better than ^{60}Co γ-ray, because

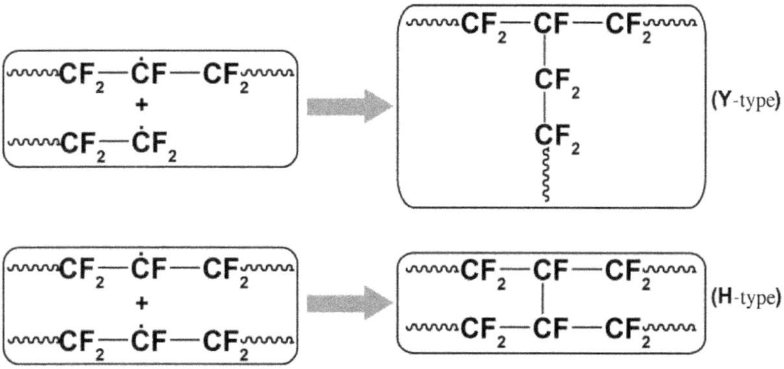

Figure 5.4 Classification of PTFE cross-linking.

Figure 5.5 Photographs of original PTFE and cross-linked PTFE (XPTFE) sheet.[2]

the dose rate of EB is much larger than that of ^{60}Co γ-ray. Cross-linked PTFE powder may be used as an additive in PTFE or other materials to enhance wear-resistance.

For PTFE films or sheets, the transparency is obviously increased after radiation cross-linking under high temperature.[2] As shown in Fig. 5.5, the transparency of PTFE films increases with the increase of absorbed dose from 500 to 3000 kGy. The change of transparency, as an intuitive indicator, can be used to evaluate whether the cross-linking of PTFE is successful. The transparency of cross-linked PTFE is improved due to the decrease of crystallinity. The temperature can significantly affect the tensile strength and elongation at break of PTFE. Below the melting point of PTFE, the tensile strength and elongation at break decrease with increase of temperature.[2] When the temperature is close to 340°C, the

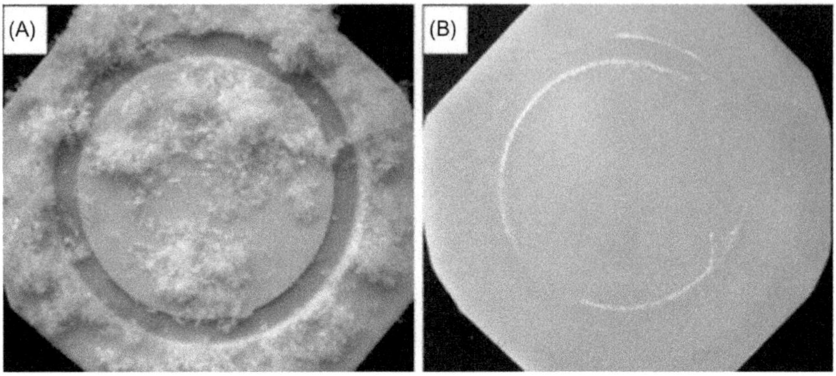

Figure 5.6 Photographs of abraded PTFE samples: (A) virgin PTFE, (B) cross-linked PTFE (150 kGy).[2]

elongation at break reaches 600%, much higher than that of conventional PTFE (300%).[2] The tensile strength is close to that of PTFE irradiated at ambient temperature. When the temperature is higher than 350°C, both tensile strength and elongation at break show a sharp decline.[2] After radiation cross-linking, the abrasion resistance of PTFE can also be significantly improved.[2] As portrayed in Fig. 5.6, the abrasion resistance of cross-linked PTFE sheet is much better than that of the un-cross-linked one.[2] The results show that the weight loss in the friction test decreases rapidly with the increase of absorbed dose in the range of 0–100 kGy. At a dose of 100 kGy, the percentage weight loss of PTFE is only 1‰. The wear-resistance performance of cross-linked PTFE is increased by three orders of magnitude (see Fig. 5.7).

In addition, the melting point of cross-linked PTFE also changes. As shown in Fig. 5.8, the melting point of cross-linked PTFE, tested by a differential scanning calorimeter, reduces gradually with the increase of absorbed dose or cross-linking degree, and may be reduced to ~280°C. At the same time, the crystal size of PTFE decreases with the increase of absorbed dose, based on the small angle X-ray scattering of the Shanghai Synchrotron Radiation Facility (see Fig. 5.9).[3]

Compared with non-cross-linked PTFE, cross-linked PTFE has many unique properties, which have not been fully explored. To

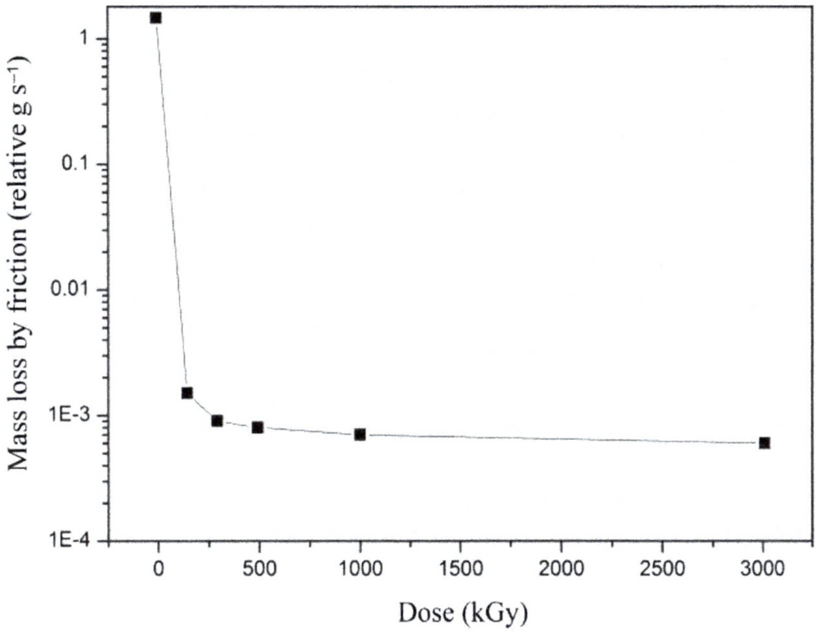

Figure 5.7 Effect of absorbed dose on mass loss of PTFE.[2]

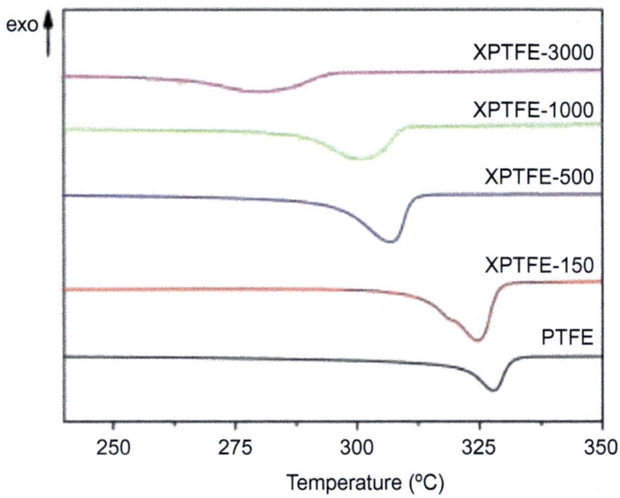

Figure 5.8 DSC spectra for PTFE and cross-linked PTFE with different absorbed doses.[3] DSC, differential scanning calorimeter.

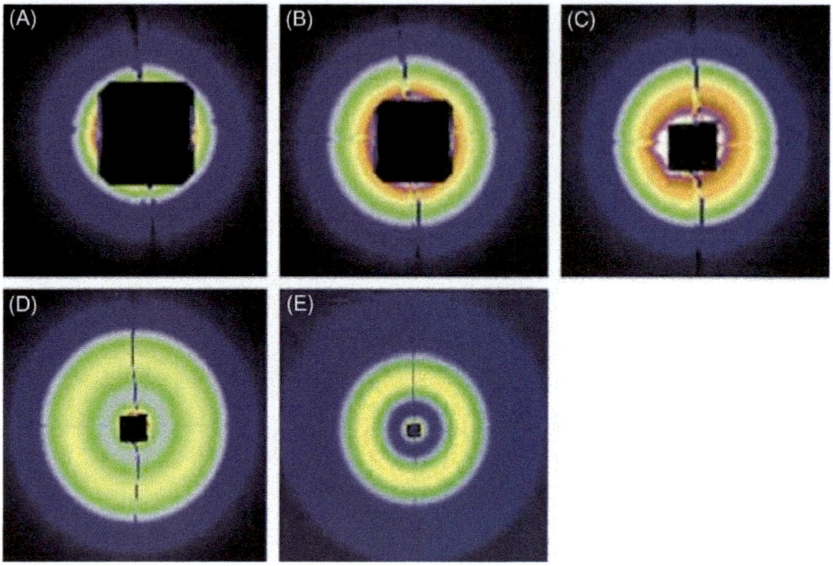

Figure 5.9 SAXS patterns of PTFE and cross-linked PTFE samples with different absorbed doses [(A) 0 kGy; (B) 150 kGy; (C) 500 kGy; (D) 1000 kGy; (E) 3000 kGy].[3] *SAXS*, small angle X-ray scattering.

date, more attention has been paid to the wear-resistance performance of cross-linked PTFE. A disadvantage of PTFE is its poor wear resistance, although its friction coefficient is the lowest of all polymer materials. Bronze powder, glass fiber, carbon fiber, etc., are usually added to PTFE to improve its wear resistance. The high abrasion resistance of cross-linked PTFE can compensate for this shortcoming by adding it to conventional PTFE to produce a bar, sheet, film, or self-lubricating material. At present, only the Japanese Hitachi Company has been using EB radiation technology to achieve large-scale production of cross-linked PTFE at high temperature.

In addition to preparing cross-linked PTFE materials by radiation at high temperatures, PTFE can be prepared by γ-ray irradiation-induced polymerization of tetrafluoroethylene, which is mixed with acetone in dry ice-cooled Dewar bottles. Such prepared small PTFE particles exhibit excellent wear resistance. This PTFE has good application prospects, because it is easy to suspend and disperse in organic solvents such as acetone. However, the radiation device should be close to the tetrafluoroethylene production line, and the entire production process presents many challenges.

5.1.3 Surface modification of polytetrafluroethylene

The material surface is the interface between the material and the external environment. Hence, the surface properties of the material often play an important role. However, the bond energy of C–F is high and the fluorine atom is closely distributed on the C–C main chain surface, regardless of the form of PTFE (powder, fiber, rod, film, microporous film, and sheet). Therefore, PTFE has high chemical inertness and low surface energy, which severely restrict its application. Surface modification and functionalization of PTFE can improve and optimize its surface performance while maintaining its advantages. Several methods have been trying to maintain the performance of PTFE substrate, and to achieve its surface functional modification for its diverse applications. The high-temperature melting modification method is used to embed the active material onto the surface of molten PTFE. However, toxic substances are released due to the degradation of PTFE at high temperature. The wet chemical modification method using sodium naphthalene or ammonia sodium solution, is the use of corrosive liquid to remove fluorine atoms from the PTFE surface. The formation of a carbonized layer, carboxyl groups, and other polar groups improves its surface activity and adhesion. Although this method is convenient, the surface color of the treated PTFE becomes dark gray, affecting the appearance of the material. In addition, this method is also not environmentally friendly, due to the production of a substantial amount of hazardous waste. Compared with the above-mentioned modification method, γ-ray/EB radiation modification and low-temperature plasma modification have the advantages of low energy consumption, ease of handling, and environmental friendliness, and are worthy of research.

5.1.3.1 *γ*-Ray/electron beam radiation grafting and its application

Electron spin resonance (ESR) studies have demonstrated that γ-ray/EB irradiation can induce the formation of a large number of stable free radicals or oxygen free radicals in PTFE main chains. As depicted in Fig. 5.10, these radicals are very stable at room temperature and even at higher temperatures, and can be used to trigger the graft polymerization of functional vinyl monomer onto PTFE.

Radiation grafting of PTFE has been widely investigated. Acrylic acid (AA) and vinyl pyridine has been grafted onto the PTFE surface by means of radiation grafting to prepare an ion-exchange membrane, and an alternating functional ion-exchange film.[4] Sulfonated PTFE membrane has been prepared by γ-ray/EB radiation grafting of styrene monomer and sulfonating treatment. The sulfonated PTFE membrane, to a certain extent, improved the mechanical properties, decreased the gas or methanol penetration, reduced the film thickness, and enhanced the performance of the battery.

Expanded PTFE has good physiological inertia and stability, and is an ideal biomedical material. Nevertheless, its poor biocompatibility should be improved. Chandler-Temple et al.[5] tried to graft 2-methyl 2-hydroxyethyl acrylate (HEA) phosphate monomer onto the surface of expanded PTFE using γ-ray simultaneous irradiation, in order to improve its in vitro biocompatibility and to explore its application in the field of medical implant materials.

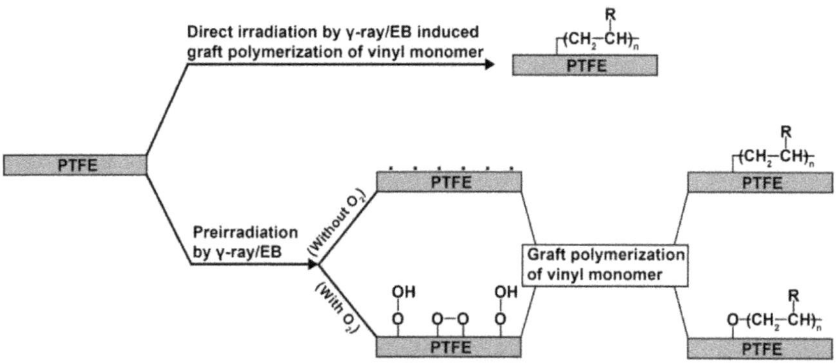

Figure 5.10 Diagrammatic sketch of graft polymerization of vinyl monomer onto PTFE via γ-ray/EB irradiation with/without O_2.

PTFE copper-clad laminate (CCL) is a high-frequency microwave dielectric material, owing to its low dielectric constant and dielectric loss factor. Ito et al.[6] prepared poly(AA) grafted PTFE film and perfluorinated ethylene propylene (PFEP) polymer by the EB pre-irradiation method. The results showed that the bonding strength between grafted PTFE/PFEP films and copper was significantly enhanced, while the dielectric properties of PTFE/PFEP films were kept constant.

The introduction of functional groups on the surface of PTFE ultrafine powder by radiation grafting can effectively improve its compatibility and dispersability. Klüpfel[7] significantly improved the compatibility between PTFE ultrafine powder and polyurea, rubber, etc., via introducing the hydroxyl and methacrylate groups on PTFE superfine powder by EB radiation and chemical modification. Wu et al.[8] successfully prepared poly(AA) and poly(Styrene) grafted PTFE powder, and the dispersibility of PTFE superfine powder was effectively improved in different solvents. Fig. 5.11 shows the effect of AA concentration and reaction time on the degree of grafting (DG) of PTFE powder (absorbed dose: 200 kGy; reaction temperature: 65°C). It was found that PTFE superfine powder grafted with a large amount of hydrophilic groups was stable enough to disperse in aqueous solution. Furthermore, the PTFE aqueous dispersion (diameter: ~300 nm) did not precipitate even at a centrifugation speed of 6000 round min^{-1} (see Fig. 5.12). This kind of PTFE aqueous dispersion is expected to be applied in water-based lubrication systems or water-based ink/coating, and it has great potential for application.

5.1.3.2 Low-temperature plasma-induced modification of polytetrafluroethylene

Compared with the degradation of bulk PTFE by γ-ray/EB irradiation, low-temperature plasma activates only the PTFE surface and improves its surface energy and surface wettability. Active sites induced by low-temperature plasma can be used to initiate graft polymerization of vinyl functional monomer to achieve permanent modification of the PTFE surface (Fig. 5.13). Hence,

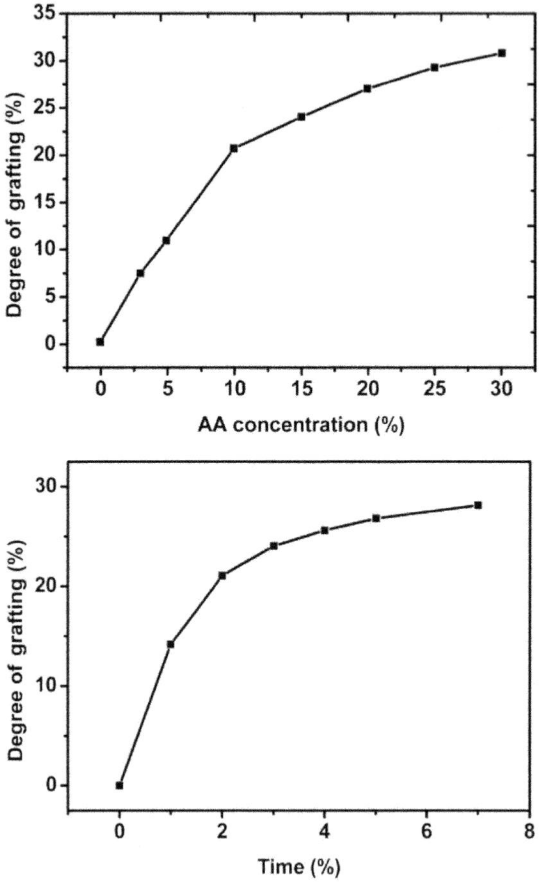

Figure 5.11 Degree of grafting as a function of AA concentration (65°C, 5 h) and time (65°C, 20 vol%-AA).[8]

low-temperature plasma is an ideal modification method for PTFE. It has been reported that the hydrophilicity of PTFE films increases with the prolongation of the treatment time of low-temperature plasma in air or N_2 atmosphere.[9] Cho et al.[9] improved the surface wettability of PTFE membrane by low-temperature plasma treatment in a hydrogen/Ar atmosphere, which induced the formation of carbonyl, hydroxyl, carboxyl and other active groups on its surface (see Fig. 5.14). Although low-temperature plasma treatment can enhance the surface wettability of PTFE, the surface will become hydrophobic over time. However, low-temperature plasma treatment-induced graft polymerization of functional monomers can give the PTFE surface a long-term stable property.

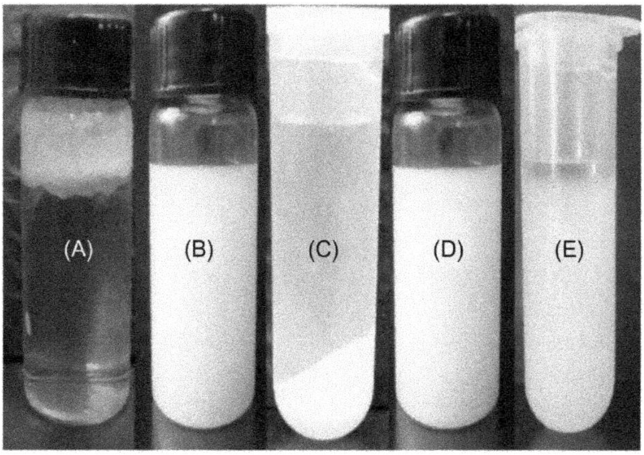

Figure 5.12 PTFE ultrafine powder dispersed in water [(A) pristine PTFE; (B) PTFE-*g*-PAA; (C) PTFE-*g*-PAA after centrifugation; (D) alkali treated PTFE-*g*-PAA; (E): alkali treated PTFE-*g*-PAA after centrifugation].[8]

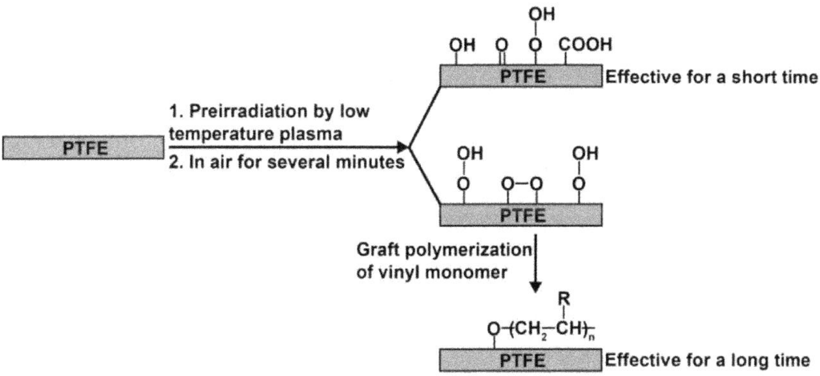

Figure 5.13 Sketch map of low-temperature plasma radiation-induced surface modification of PTFE.

In view of the low surface energy of PTFE and the poor adhesion between PTFE and metals/resins, Wu et al.[10] improved the adhesion between PTFE and copper via graft polymerization of glycidyl methacrylate (GMA), vinyl imidazole, HEA, AAm and other monomers on the surface of PTFE film by low-temperature plasma and ultraviolet (UV) light. In addition, the interface adhesion between modified PTFE and copper can be further enhanced ($\sim$19 N cm^{-1}) by treatment with grafted PTFE film using low-temperature oxygen plasma. Li et al.,[11] carried out research on the graft modification of PTFE membrane by low-temperature plasma-induced grafting of

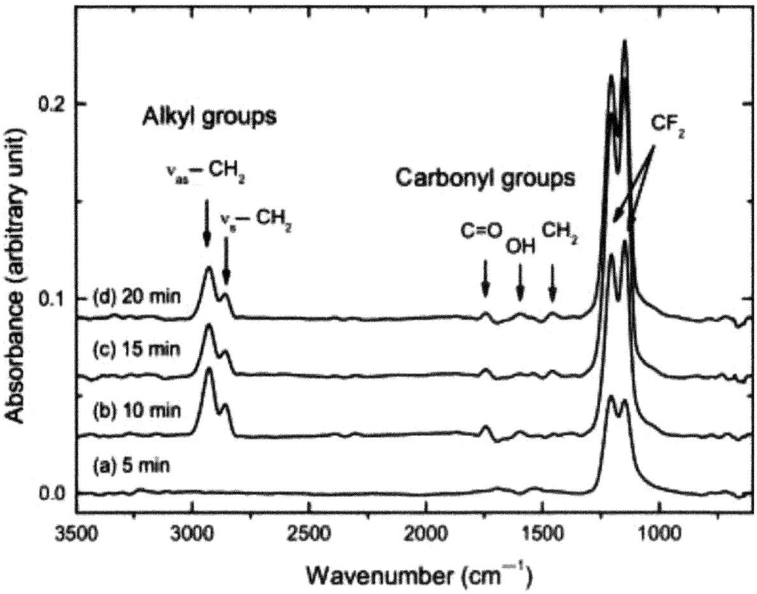

Figure 5.14 ATR-FTIR of PTFE treated with various hydrogen plasma treatment times: (a) 5 min; (b) 10 min; (c) 15 min; (d) 20 min.[9]

AA in order to improve its surface wettability and adhesive force (see the attenuated total reflectance Fourier transform infrared spectroscopy (ATR-FTIR) in Fig. 5.15). The results showed that the surface wettability of the grafted PTFE film was improved, and the adhesion between the PTFE film and the epoxy resin was remarkably improved, while its dielectric properties were unchanged.

The PTFE microporous membrane prepared by bidirectional stretching has become a very important separation membrane because of its high-temperature resistance, chemical resistance, and high porosity. However, surface hydrophobicity limits its application in water treatment and other fields. PTFE film was treated with low-temperature plasma in a hydrogen atmosphere and then treated with low-temperature plasma in a sulfur dioxide atmosphere or fuming sulfuric acid, and the surface hydrophilicity of the resulting PTFE film was significantly improved. The poly(2 acrylamido-2-methylpropanesulfonic acid) was grafted onto PTFE membrane using atmospheric low-temperature plasma and the photo-induced

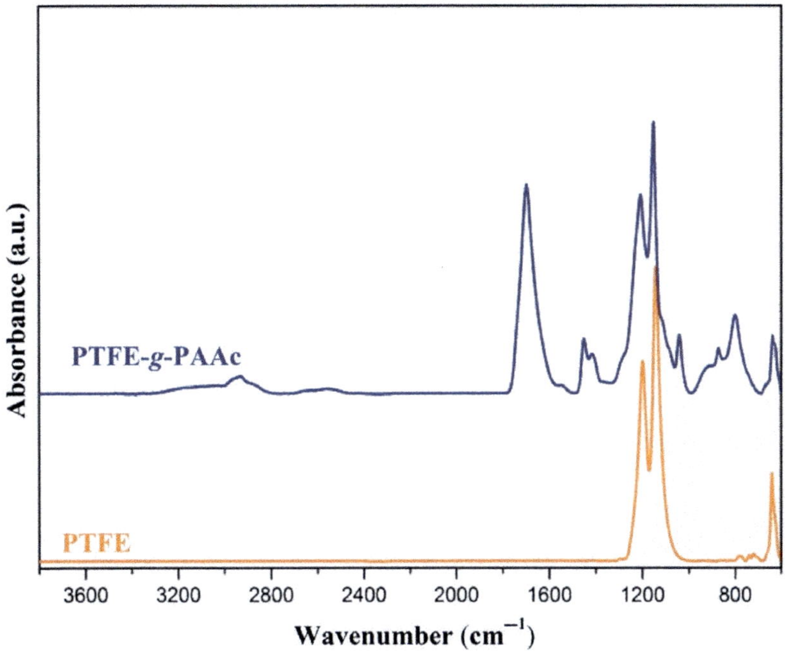

Figure 5.15 ATR-FTIR of PTFE before and after grafting of AA.[11]

grafting method. The oil repellency of the modified PTFE membrane was improved. Grafting of AA monomer onto PTFE membrane surface by low-temperature plasma treatment can significantly improve its hydrophilicity. Chuo et al.[12] prepared a new functional film with electrical resistance to biological contamination by low-temperature hydrogen plasma treatment-induced graft polymerization of GMA on the surface of PTFE membrane, coupled with further chemical modification. Venault et al.[13] immobilized polyethylene glycol methacrylate (PEGMA) monomer on the surface of an expanded PTFE membrane with atmospheric low-temperature plasma. The modified expanded PTFE film had good hydrophilicity and anti-biological contamination performance.

PTFE membrane is a good biomedical material (especially as a medical implant material such as artificial blood vessels, heart valves, etc.) due to its high O_2 permeability, biocompatibility, and lack of physical side-effects on the human body, etc. However, its long-term blood compatibility should be improved. Sun et al.[14]

deposited a fluorocarbon film on the PTFE surface by low-temperature plasma in an Ar atmosphere, followed by graft polymerization of GMA. The heparin was then immobilized onto the PTFE surface via the reaction between the carboxyl group and epoxy group. As shown in Table 5.1, the activated partial thromboplastin time and the thrombin time of the heparinized PTFE membrane were significantly elongated, and it exhibited good anticoagulant properties. Onder et al.[15] increased the polarity of the PTFE membrane surface by low-temperature hydrogen plasma, followed by grafting of AAm by low-temperature plasma treatment in an Ar atmosphere. Subsequently, the hirudin was immobilized onto the PTFE membrane surface. As portrayed in Fig. 5.16, the hirudin-modified PTFE membrane presented good blood compatibility and antithrombotic performance, and is expected to be used as artificial blood vessels. Chang et al.,[16] synthesized a poly (PEGMA) grafted PTFE membrane surface by low-temperature hydrogen plasma treatment. The results showed that the hydrophilicity and blood compatibility of the modified PTFE membrane significantly improved. Crombez et al.,[17] introduced vascular endothelial growth factor (VEGF) onto the surface of expanded PTFE artificial blood vessels by low-temperature plasma treatment in an ammonia atmosphere and a followed chemical reaction. The results showed that expanded PTFE artificial vascular surface modified with VEGF had a good effect on promoting the attachment and reproduction of endothelial cells.

Radiation cross-linking and degradation greatly enrich the types of PTFE products, and radiation grafting can significantly improve the surface properties of PTFE. Surface-modified PTFE products will be more widely applied in different areas in the future, such as

Table 5.1 APTT and TT of PTFE, fluorocarbon, and heparinized PTFE films

Clotting time	PTFE film	Fluorocarbon film	Heparinized PTFE film ($\mu g\ cm^{-2}$)				
			1.1	1.4	1.8	2.1	2.2
APTT (s)	18	21	38	42	43	44	44
TT (s)	14	17	29	32	35	36	37

APTT, activated partial thromboplastin time; *TT*, thrombin time.

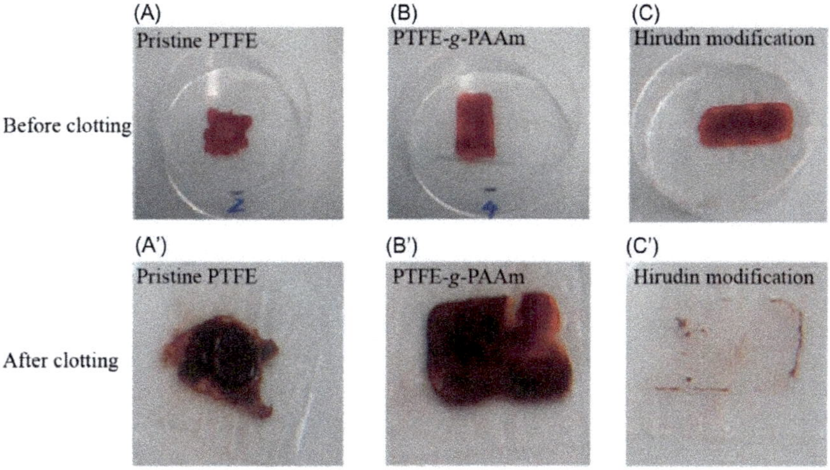

Figure 5.16 Antithrombotic performance of pristine and modified PTFE membrane: (A) before clotting for pristine PTFE; (B) before clotting for PTFE-g-PAAm; (C) before clotting for hirudin modified PTFE; (A') after clotting for pristine PTFE; (B') after clotting for PTFE-g-PAAm; (C') after clotting for hirudin modified PTFE.[17]

water purification filtration membranes, proton exchange membranes, hemodialysis membranes, artificial blood vessels, CCL, lubricants, seals, aerospace precision devices, and building materials.

5.2 POLYMER ULTRAFINE POWDER FABRICATION BY RADIATION

5.2.1 Polytetrafluroethylene ultrafine powder

PTFE ultrafine powder can be prepared by irradiation at room temperature, followed by ultrafine grinding. Fresh PTFE material, scrap or recycled PTFE waste, etc., can be used to fabricate ultrafine powder. The use of recycled PTFE or waste to produce ultrafine powder can reduce the cost and achieve the recycling of resources. PTFE ultrafine powder is widely used as a functional additive in the fields of engineering plastics, anticorrosion coatings, nonstick coatings, coil coatings, powder coatings, and inks. Shamrock Technologies (the United States), Solvay Company

(Italy), Lubrizol Corporation (Germany), Kitamura Company Ltd. (Japan) and many other companies produce PTFE ultrafine powder. In recent years, domestic Chinese enterprises have also made great progress in the technology of manufacturing PTFE ultrafine powder. Fig. 5.17A (average particle size: 2.43 μm) and Fig. 5.17C (average particle size: 4.26 μm) shows the appearance and particle size distribution of two kinds of PTFE ultrafine powder, prepared by Jiashan Senga Tech Co., Ltd., China.

The particle size of PTFE powder is usually tens to hundreds of microns, which is suitable for producing PTFE sheet and pipe. PTFE ultrafine powder can be made by polymerization, radiation degradation, and thermal cracking. The irradiated PTFE becomes very brittle, and can be further fabricated into ultrafine powder by grinding or air-jetting. The particle size is closely related to the absorbed dose. The higher the absorbed dose, the lower the relative molecular weight of PTFE, and the smaller the particle size of ultrafine powder. At present, PTFE ultrafine powder is mainly prepared by radiation degradation at home and abroad, since high

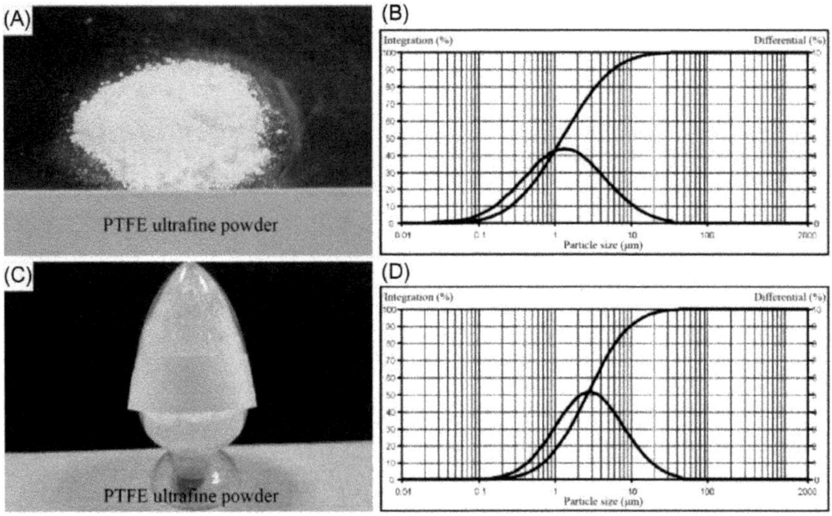

Figure 5.17 Photographs and particle sizes of two kinds of PTFE ultrafine powder: A, the appearance of PTFE ultrafine powder with the average particle size of 2.43 μm; B, the particle size distribution of sample A; C, the appearance of PTFE ultrafine powder with the average particle size of 4.26 μm; D, the particle size distribution of sample C.

energy beam can effectively break the molecular chains of PTFE at room temperature. For recycled PTFE, a very high absorbed dose is required for the fabrication of PTFE ultrafine powder. Hence, it is normally irradiated by EB accelerators. PTFE ultrafine powder is mainly used as an additive in lubricating oil and grease, ink, paint, engineering plastics, leather, rubber, etc., in order to improve anti-friction and scratch-resistance.

5.2.2 Other ultrafine powders

5.2.2.1 Chitin/chitosan ultrafine powder

Chitin is a natural polymer, and mainly exists in shrimp and crab shell together with calcium carbonate and protein. Chitin has huge reserves in nature, which is preceded only by cellulose. As shown in Fig. 5.18, chitin can be converted into chitosan by the removal of the acetyl group under conditions of strong alkali and heating. The positively-charged amino group in chitosan can attack negatively-charged harmful bacteria and viruses to make them dysfunctional. Therefore, chitosan is considered to have better antibacterial or antiviral functions, which is very different from other carbohydrate macromolecules, such as xylose. Chitosan has been widely applied in agriculture, biomaterials, medicine, health food, and cosmetics, because of its good biological activity. Research and applications related to chitosan are also very active.

Similar to PTFE, chitosan is prone to degradation after radiation. Chitosan is water insoluble, but can be dissolved in dilute

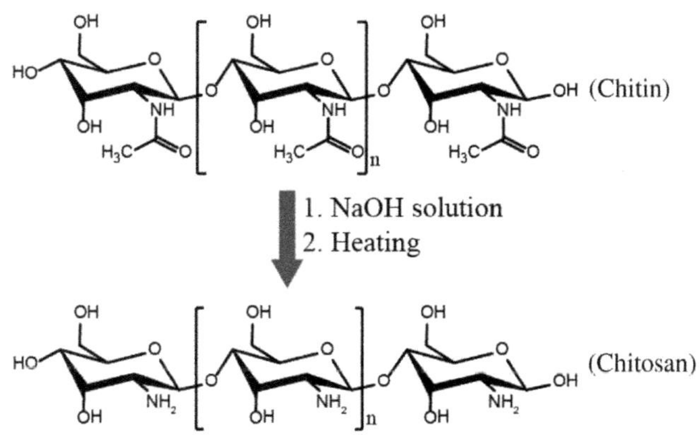

Figure 5.18 The conversion of chitin to chitosan.

hydrochloric acid or acetic acid solution. Chitosan in aqueous solution is very sensitive to degradation by radiation at a low absorbed dose (several kilograys). Radiation can significantly reduce its relative molecular mass. Chitosan can be fabricated into chito-oligosaccharides with a relative molecular mass of ~2000–3000 after irradiation to a high absorbed dose. The relative molecular mass of chitosan decreases rapidly with absorbed dose. However, when the absorbed dose reaches ~200 kGy, its relative molecular mass decreases slowly with absorbed dose. In order to reduce the absorbed dose and accelerate the degradation of chitosan, a small amount of hydrogen peroxide (H_2O_2) can be added to the aqueous solution of chitosan, which has a synergistic effect on degradation. This is because the decomposition of H_2O_2 produces hydroxyl radicals, and water radiolysis also produces hydroxyl radicals. Consequently, the hydrogen-abstraction reaction of hydroxyl radicals and chitosan is stimulated, thereby further reducing the relative molecular mass of chitosan. However, the residual H_2O_2 may be an issue for the application of chitosan.

Solid chitosan can also be degraded through irradiation, and its color turns yellow. However, the efficiency of degradation is lower than that of chitosan dissolved in solution. From another point of view, since the concentration of chitosan aqueous solution is usually only ~3%, the overall efficiency of radiation-induced degradation of solid chitosan is higher, and storage, transportation, and handling are more convenient in the production process. Wu et al.[18] produced chitosan ultrafine powder via irradiation of solid chitosan and jet milling. This technology can be used to produce chitosan ultrafine powder with different particle sizes, and the smallest particle size can reach 5–10 μm. The particle size is related to the absorbed dose and the crushing process. Chitosan ultrafine powder with its cleanliness and no impurities, has good and stable antibacterial properties. Chitosan can be used for feed, medical accessories, antibacterial film, and in other fields. The relative molecular mass of chitosan is often reduced through acid degradation, oxidative degradation, or enzymatic degradation. Compared with other techniques, the cost of radiation degradation of solid chitosan is relatively low, and it does not produce waste or waste water, and the amino group on the chitosan molecule is retained. Hence, this is an important practical value of radiation technology. Fig. 5.19A (average particle size: 8.78 μm) and Fig. 5.19C (average particle size: 38.53 μm) show the appearance and particle size distributions of two kinds of chitosan ultrafine powder, respectively. This ultrafine powder was produced by Jiashan Senga Tech Co., Ltd., China.

5.2.2.2 Polymethyl methacrylate ultrafine powder

Polymethyl methacrylate (PMMA), plexiglass, is a transparent polymer material, the molecular structure of which is shown in Fig. 5.20. PMMA is also prone to degradation by radiation. It is similar to PTFE, in that the solid PMMA can be fabricated to ultrafine powder by a combination of irradiation and grinding. When the absorbed dose is 300 kGy or more, PMMA ultrafine powder with an average particle size of ~10 μm can be fabricated. PMMA ultrafine powder has excellent gloss, but does not have good wear, lubrication, and temperature resistance. PMMA

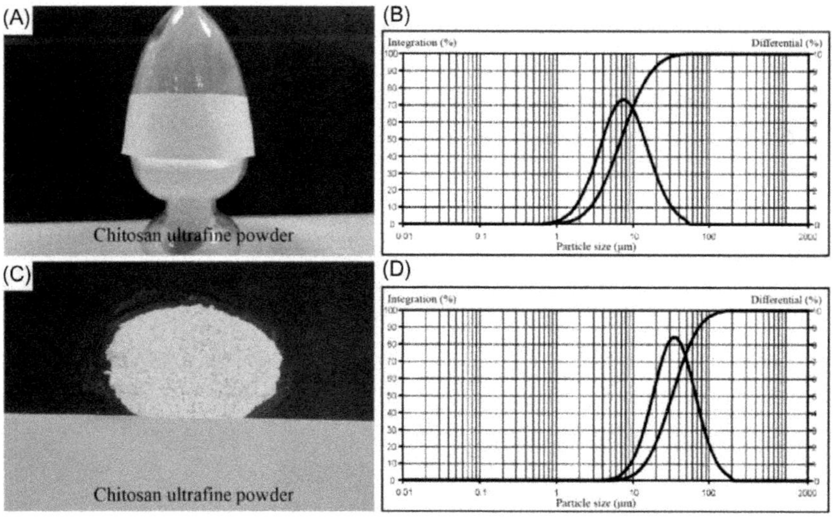

Figure 5.19 Photographs and particle sizes of two kinds of chitosan ultrafine powder: A, the appearance of chitosan ultrafine powder with the average particle size of 8.78 μm; B, the particle size distribution of sample A; C, the appearance of chitosan ultrafine powder with the average particle size of 38.53 μm; D, the particle size distribution of sample C.

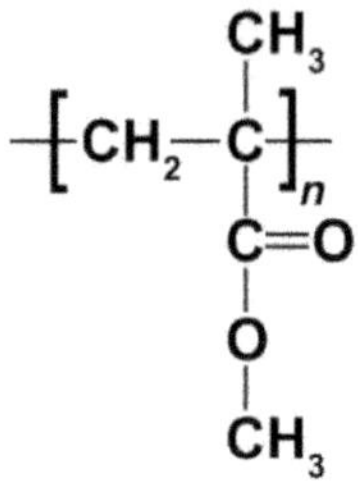

Figure 5.20 Chemical structure of PMMA. *PMMA*, polymethyl methacrylate.

ultrafine powder can be used together with PTFE ultrafine powder or other auxiliary materials to improve the gloss and auxiliary lubrication effect of paints and inks. In general, PMMA microspheres are prepared by dispersion polymerization, not by irradiation degradation. Mono-dispersed PMMA microspheres can be obtained by following a strict preparation process. PMMA microspheres are commercially used in cosmetics, medical materials, and chemical materials. The surface of PMMA ultrafine powder produced by radiation technology is relatively rough, and it is difficult to make uniform spherical particles. Nonetheless, this technology has a competitive advantage in the market, because of its simple production process and low cost. Fig. 5.21A (average particle size: 5.49 μm), and Fig. 5.21C (average particle size: 10.99 μm) depict the appearances and particle size distributions of PMMA ultrafine powder fabricated by Jiashan Senga Tech Co., Ltd., China.

5.2.2.3 Polyformaldehyde ultrafine powder

POM has relative thermal resistance and self-lubricating properties. The molecular structure of POM is shown in Fig. 5.22. As an important engineering plastic, POM is prone to degradation by radiation, and its relative molecular mass decreases. Consequently, POM ultrafine powder can be produced by irradiation. Compared with PTFE, POM is much easier to degrade by irradiation. In other words, the $G(s)$ value of POM is higher than that of PTFE under the same conditions. In air and at room temperature, the $G(s)$ value of PTFE is $\sim$4.4, while the $G(s)$ value of POM is $\sim$8.0. Jiashan Senga Technology Co., Ltd. manufactures POM ultrafine powder by radiation degradation. The fabricated POM

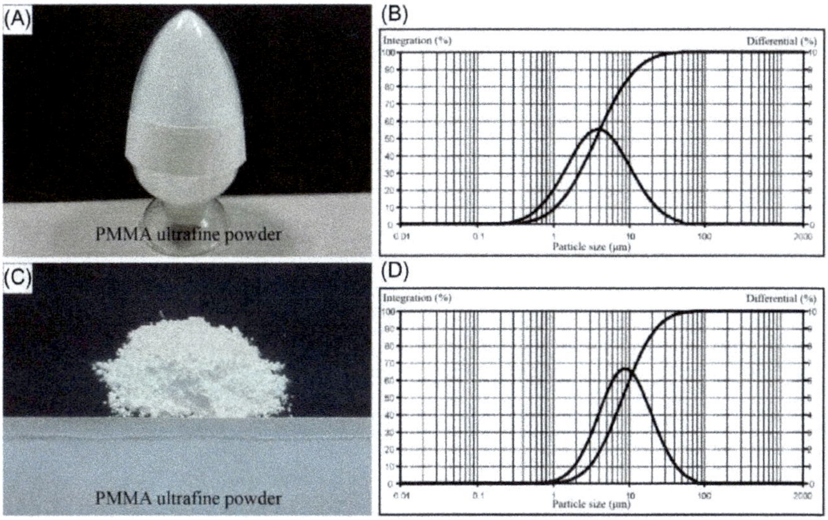

Figure 5.21 Photographs and particle size of two kinds of PMMA ultrafine powder: A, the appearance of PMMA ultrafine powder with the average particle size of 5.49 μm; B, the particle size distribution of sample A; C, the appearance of PMMA ultrafine powder with the average particle size of 10.99 μm; D, the particle size distribution of sample C. *PMMA*, polymethyl methacrylate.

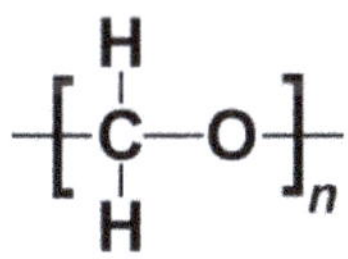

Figure 5.22 Chemical structure of POM. *POM*, polyformaldehyde.

ultrafine powder can partially replace PTFE ultrafine powder, since it also has self-lubricating and wear-resistance characteristics. However, the temperature resistance of POM is worse than that of PTFE. Hydrogen fluoride forms in the radiation degradation of PTFE and can be treated with alkaline substance. In contrast, the formaldehyde produced during the POM radiation degradation process is not easy to handle and control, and it is necessary to take strict measures to control it. This shortcoming, to a certain extent, limits the research and application of POM ultrafine powder. Fig. 5.23A (average particle size: 6.83 μm) and Fig. 5.23C (average particle size: 31.29 μm) show the appearance and particle size distributions of two different kinds of POM ultrafine powder.

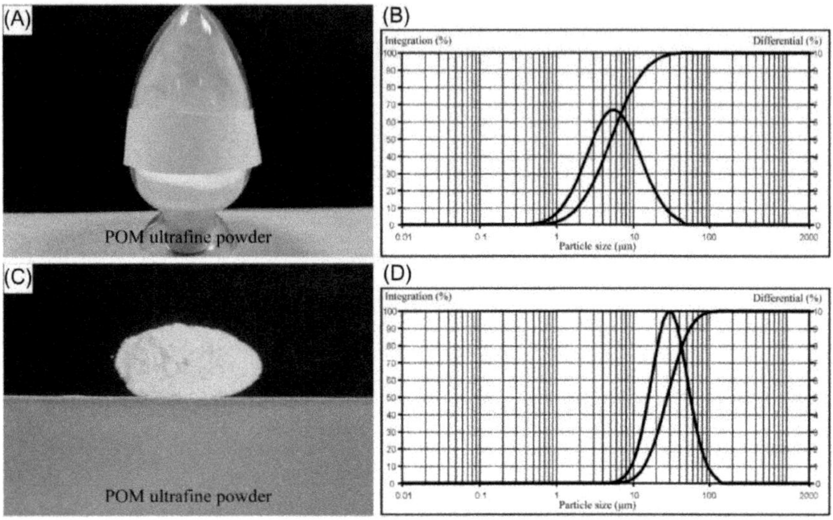

Figure 5.23 Photographs and particle sizes of two kinds of POM ultrafine powder: A, the appearance of POM ultrafine powder with the average particle size of 6.83 μm; B, the particle size distribution of sample A; C, the appearance of POM ultrafine powder with the average particle size of 31.29 μm; D, the particle size distribution of sample C. *POM*, polyformaldehyde.

5.3 RADIATION MODIFICATION OF POLYSACCHARIDES AND THEIR DERIVATIVES

Polysaccharides are polymeric carbohydrate molecules composed of long chains of monosaccharide units bound together by glycosidic linkages. Among polysaccharides, cellulose and chitosan are the two most important carbohydrates. Cellulose is a linear polymer of cellobiose consisting of two glucose sugar units that are linked by glucosidic linkages (C–O–C) at the C_1 and C_4 positions. Its structure is shown in Fig. 5.24. The repeat unit is comprised of two anhydroglucose rings linked together through an oxygen covalently bonded to C_1 of one glucose ring and C_4 of the adjoining ring (1 → 4 linkage), which is called the β(1 → 4) glucosidic bond. As an abundant biopolymer on the earth, cellulose is renewable and biodegradable.

Chitin is the second most abundant natural polymer on the earth next to cellulose. The chemical structure of chitin is similar to that

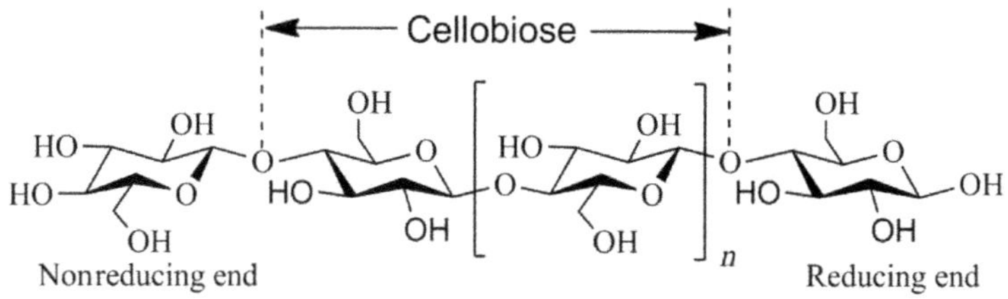

Figure 5.24 Structure of cellulose.

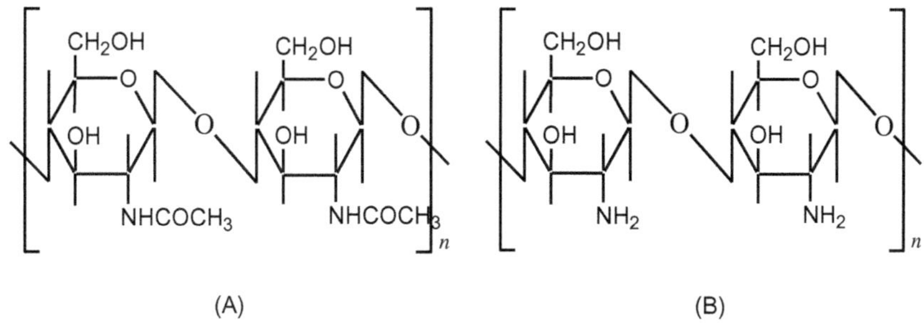

Figure 5.25 Structures of (A) chitin and (B) chitosan.

of cellulose with 2-acetamido-2-deoxy-*b*-D-glucose monomers attached via β(1→4) linkages, which are composed of 2-acetamido-2-deoxy-*b*-D-glucose (*N*-acetylglucosamine). Chitosan is the de-acetylated form of chitin, which is soluble in acidic solutions, and can be produced in numerous forms, such as powder, paste, film, fiber, and so on. Fig. 5.25 presents the structure of chitin and chitosan.

Over the last few years there has been a rapidly expanding interest in polysaccharides and their derivatives, which are expected to have a variety of applications in the industrial and biomedical fields, due to their low cost, nontoxicity, biodegradability, and biocompatibility. These applications are attracting much research interest in the development of functional polymers from polysaccharides. Radiation processing is a very convenient tool for the modification of polysaccharides by radiation cross-linking, degradation, and grafting. Herein we introduce the radiation grafting and cross-linking of polysaccharides (cellulose and chitosan) and their applications.

5.3.1 Radiation grafting and applications

5.3.1.1 Radiation grafting of cellulose and chitin/chitosan and their applications

As an abundant biopolymer on the earth, cellulose is renewable, biodegradable, and can be derivatized to produce various useful products. However, cellulose is difficult to dissolve in water and common organic solvents due to the formation of hydrogen-bonded supra-molecular structures in cellulose. In comparison with traditional methods using a high concentration of caustic reagents, such as acid and alkali, γ-ray irradiation is an environmentally-friendly technique. This technique can easily penetrate the crystalline regions in cellulose and increase the accessibility of these regions to solvents. Hence, radiation-induced modification of cellulose has been widely investigated.

Hao et al.[19] reported a homogeneous modification of microcrystalline cellulose (MCC) in ionic liquids via radiation-induced grafting. Thermo-sensitive poly(*N*-isopropylacrylamide) (PNIPAAm) was successfully grafted onto MCC in 1-butyl-3-methylimidazolium chloride ([Bmim]Cl) using γ-ray irradiation at room temperature, and the synthetic route is shown in Fig. 5.26. The cellulose grafted with PNIPAAm (cellulose-*g*-PNIPAAm) has higher thermal stability than that of ungrafted regenerated cellulose. The crystalline structure of the original MCC is destroyed during the dissolution process

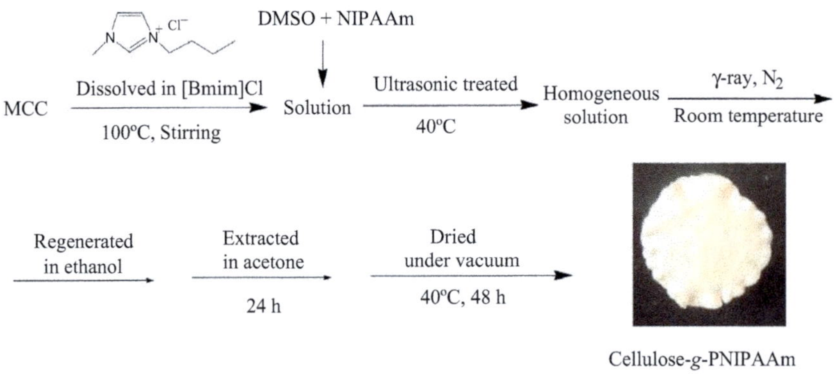

Figure 5.26 Procedure for preparation of cellulose-*g*-PNIPAAm in [Bmim]Cl by radiation-induced grafting copolymerization.[19]

according to the X-ray diffraction (XRD) profiles, and grafting of PNIPAAm onto cellulose further reduces the degree of crystallinity. Moreover, the resultant cellulose-*g*-PNIPAAm exhibits obvious thermal sensitivity with a lower critical solution temperature around 35°C. The cellulose-*g*-PNIPAAm could be expected to have a potential application in drug release systems, adsorption materials, and functional membrane materials.

Super-hydrophobic cotton fabrics can be prepared successfully through radiation-induced graft polymerization with a series of alkyl methacrylates as the functional monomers, such as methyl methacrylate (MMA), ethyl methacrylate, *n*-propyl methacrylate, *n*-butyl methacrylate, *n*-hexyl methacrylate, and *n*-lauryl methacrylate. Cotton fabric grafted with poly(alkyl methacrylate) with shorter alkyl chains requires higher grafting density to achieve full coverage of the fabric, whereas longer alkyl chain length results in lower critical grafting density. The resulting super-hydrophobic cotton fabrics exhibit considerable resistance to air permeability, whereas the water vapor transport is almost the same as that of pristine cotton. The combination of opposing properties decreases air permeability and ideal water vapor transmission rate, creates a breathable and comfortable fabric as clothing material.

Cellulose has been used to synthesize environmentally-friendly adsorbents. However, cellulose must be modified before it is used as an adsorbent due to its limited functional groups (only hydroxyl group) and poor adsorption capacity. Poly(vinylbenzyl trimethylammonium chloride) can be grafted successfully onto cotton by the mutual γ-irradiation grafting technique. The grafted cotton adsorbent shows an excellent adsorption of protein due to anion exchange adsorption mechanism. Similarly, Hesham et al. synthesized the ion-exchange adsorbents based on cellulose fabric wastes containing sulfonic acid and amine functional groups by radiation grafting of GMA with subsequent chemical modification of the epoxy groups of poly-GMA graft chains with sodium sulfite/H_2SO_4 and triethylamine, respectively. The synthetic adsorbents have a wide range of applications in the removal of different types of hazardous pollutants, such as acidic dye, cobalt, dichromate, and phenols from aqueous solutions.[20]

Cellulose can be fabricated in different forms, such as fabric, film, fiber, and sphere shapes. Cellulose-based microsphere materials, which can be used as column packing for separation, have potentially wide applications in industry, owing to their spherical morphology. Cellulose microsphere (CMS) adsorbent used for the removal of Cr(VI) is prepared by radiation-induced grafting of dimethylaminoethylmethacrylate (DMAEMA) onto CMS followed by a protonation process. The adsorption of Cr(VI) onto the resulting adsorbent is very fast, and equilibrium adsorption can be achieved within 15 min. The adsorption capacity strongly depends on the pH of the solution, which is attributed to the change of both the existing forms of Cr(VI) and the tertiary-ammonium group of PDMAEMA grafted CMS at different pH. A maximum Cr(VI) uptake ($\sim$78 mg g^{-1}) is obtained at the pH range of 3.0–6.0[21]. A novel cellulose-based adsorbent is prepared by radiation-induced grafting of styrene onto the surface of cellulose microspheres, followed by acetylation and amination processes.[22] The adsorption equilibrium can be achieved within 30 min due to the strong electrostatic interaction between the adsorbent and $HCrO_4^-$. The adsorption experimental data fitted well with the Langmuir isotherm model with an adsorption capacity of Cr(VI), 123.4 mg g^{-1} (Fig. 5.27). The optimum adsorption was observed at pH 3.1 for initial Cr(VI) concentration of 100 mg L^{-1}. Furthermore, even when the pH reaches 1.0 or 10.6, the anion exchange microspheres (GY = 100%) still exhibit a high Cr(VI) uptake of 60.7 and 44.1 mg g^{-1}, respectively. Therefore, this novel adsorbent can be applied in some extreme environments due to its high acidic and alkaline resistance.

Three *n*-(aminomethyl)pyridine (*n*-AMP, $n = 2, 3, 4$) functionalized adsorbents (*n*-AMPRs) for recovery of Au(III) were synthesized by radiation grafting of GMA onto MCC microsphere, followed by ring-opening processes with *n*-AMP.[23] The results indicated that *n*-AMPRs exhibited good adsorption abilities for Au (III) with adsorption capacities in the order of 2-AMPR > 3-AMPR > 4-AMPR. The dynamic experiment revealed that Au(III)

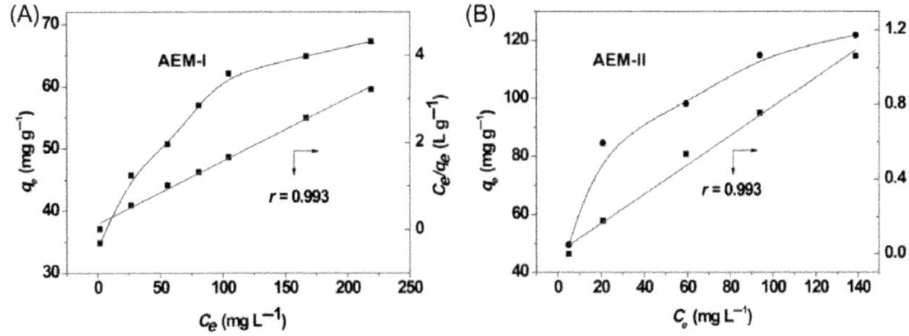

Figure 5.27 Experimental Cr(VI) adsorption and Langmuir linear plot on (A) AEM-I and (B) AEM-II (AEM-I: *GY* = 52%, absorbed dose: 25 kGy, N% = 2.6%; AEM-II: *GY* = 100%, absorbed dose: 50 kGy, N% = 3.7%).[22]

in HCl media can be efficiently adsorbed by 2-AMPR in the presence of equal amounts of Cu(II), Ni(II), and Cr(III), which demonstrates that 2-AMPR is a promising adsorbent for trace amounts of Au(III) ions from industrial effluents and wastes.

Recently, a cellulose-based microsphere containing glucamine groups, named cellulose grafted with 4-vinylbenzyl chloride and functionalized with N-methyl-D-glucamine (CVN), was successfully synthesized by radiation-induced graft polymerization of 4-vinylbenzyl chloride onto cellulose microspheres and subsequent functionalization with *N*-methyl-D-glucamine.[24] The synthetic route of CVN is shown in Fig. 5.28. The adsorption of CVN for boron (III) from aqueous solutions was evaluated systematically using a batch adsorption technique. The adsorption behavior fitted well with Langmuir models. The CVN adsorbents exhibited a high adsorption capacity up to 12.4 mg g^{-1} for boron (III) over the wide pH range of 5–8. With increasing ionic strength, the boron (III) uptake of CVN was enhanced, due to the compensation of surface charge generated by boron (III) adsorption to favor the adsorption. This work provides a new, sustainable, cost-effective material as a promising specific adsorbent for the removal of boron (III) from saline solutions.

5.3.1.2 Radiation-induced grafting polymerization of chitin/chitosan and their derivatives

In recent years, the active/controlled free radical polymerization method has been widely used in graft copolymerization of

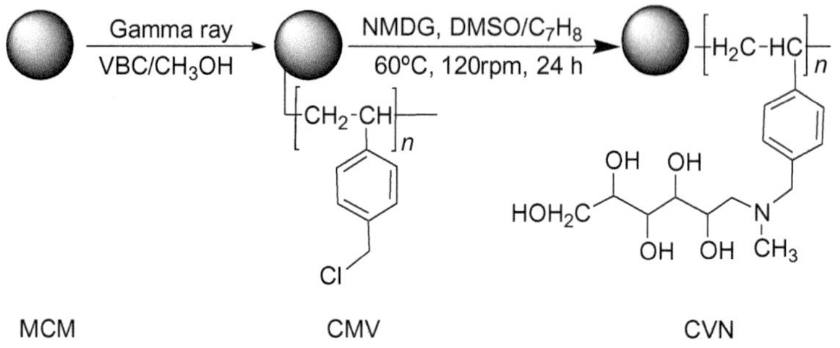

Figure 5.28 Synthetic route of CVN for adsorption of boron(III).[24]

polysaccharides and their derivatives, such as living free radical nitroxide-mediated polymerization (NMP), ATRP, and reversible addition-fragmentation chain transfer (RAFT). Hua et al.[25] investigated the modification of chitosan using the controlled free radical polymerization method. Temperature-responsive chitosan was synthesized by free radical polymerization of N-isopropyl acrylamide (NIPPAm) at 60°C in the presence of RAFT-chitosan agent. The chitosan was subsequently modified with phthalic anhydride and with *S*-1-dodecyl-*S*′-(α,α'-dimethyl-α''-acetic acid) trithiocarbonate to serve as RAFT agent. The temperature plays an important role in the self-assembly in H_2O dispersion and the morphologies of chitosan-*g*-PNIPPAm. The well-defined grafted poly(*N*-isopropyl acrylamide) and the marked thermally-responsive property enable chitosan to be selected for drug delivery applications. This is the first thermo-sensitive chitosan prepared by controlled graft modification of chitosan by RAFT polymerization. Another controlled graft modification of chitosan has been achieved first by NMP using a chitosan-2,2,6,6-Tetramethylpiperidine 1-oxyl (TEMPO) macro-initiator. The chitosan-TEMPO macro-initiator was obtained from the ^{60}Co γ-ray irradiation of *N*-phthaloylchitosan and 4-hydroxy-TEMPO in *N*,*N*-dimethyl formamide in an Ar atmosphere. The results indicate that the graft copolymers were successfully synthesized and the graft polymerization was well-controlled by the nitroxide-mediated process. The size distribution of chitosan-*g*-PS in benzene is very narrow, which may be associated with the well-defined PS grafted onto chitosan from NMP.[26]

Partially de-acetylated chitin with de-acetylation degree of ~40% was graft-copolymerized with acrylonitrile (AN) using a γ-ray pre-irradiation method.[27] The mixture ratio of 0.1 N $NH_2OH \cdot HCl$ to 0.1 N NaOH was selected to be 7:3 (V/V) for amidoxime conversion of cyano-groups on grafted chitin (Chi-*g*-AN). The adsorption equilibrium of As(III) onto Chi-*g*-AN converted amidoxime (Chi-*g*-AN-C) fitted with the Langmuir model and the maximal adsorption capacity was 19.7 $mg\,g^{-1}$. The content of As(III) in ground water was completely absorbed by Chi-*g*-AN-C packed in the column. These results demonstrate the feasibility of the use of novel materials in water treatment. The chitosan films were grafted with DMAEMA, 2-hydroxyethylmethacrylate (HEMA) and MMA using the ^{60}Co γ-ray irradiation. The solvent composition significantly affected the DG of chitosan films. Maximum DG was obtained in the water–methanol system. The controlled release of glucose as a model compound through the chitosan membranes grafted with HEMA was investigated, and it was found that the permeation of glucose through grafted membranes follows pore type of mechanism. The permeability coefficient of glucose through grafted membranes was in the order of 10^{-6}–10^{-7} $cm^2\,s^{-1}$.

Hien[28] reported the preparation of partially de-acetylated chitin (PD-chitin) by soaking chitin in NaOH solution with various concentrations from 10 to 50 wt% at room temperature for 4 days. The swelled PD-chitin sample was filtered and irradiated with ^{60}Co γ-ray at an absorbed dose of 4.8 kGy to induce the grafting of AA. The resulting product, called PD-chitin-*g*-PAA, was changed to the sodium form, PD-chitin-*g*-PANa, by treating it with NaOH and used as a metal ion adsorbent. The results showed that the adsorption capacities for Cu^{2+} were in the following order: chitin < PD-chitin < PD-chitin-*g*-PANa < chitosan (DG of 76%). The obtained product, PD-chitin-*g*-PANa, can be produced on a large scale at a competitive cost for use as a metal ion adsorbent for water purification, and for other purposes, such as the sorption of dyes and the immobilization of bioactive substances.[28]

5.3.2 Radiation cross-linking and application

5.3.2.1 Radiation effect on polysaccharide derivatives

Ramnani et al. investigated the synthesis of cross-linked chitosan using γ-irradiation in the presence of carbon tetrachloride as a sensitizer. The resultant cross-linked product has an intact $-NH_2$ group, as is evident from the results of ionic capacity determination.[29]

Huang et al. studied the radiation effects of γ-ray on carboxymethylated chitosan (CM-chitosan) in the solid state[30] and in the aqueous state,[31] respectively. Experimental results in solid-state condition indicated that random chain scissions occurred under γ-ray irradiation. The radiation chemical yield (Gd) of CM-chitosan in solid state with N_2-saturated was 0.49, indicating that CM-chitosan has high radiation stability. XRD patterns identified that the degradation of CM-chitosan occurred mostly in the amorphous region. Furthermore, aqueous solutions of CM-chitosan were irradiated with γ-ray under various conditions. The degradation of CM-chitosan was faster in the presence of nitrous oxide or hydrogen peroxide, but it was inhibited after adding isopropanol because of the changes of the concentration of hydroxyl radicals under the above different conditions. The intrinsic viscosity of CM-chitosan decreased faster than that of CM-chitin, indicating that amino groups may enhance the reactivity.

The CM-chitosan hydrogels were synthesized by γ-ray radiation-induced cross-linking in a paste-like state with the concentration of CM-chitosan from 12% to 22%. As shown in Fig. 5.29, the gel fraction first increased with adsorbed dose and leveled off at ~20 kGy. The diffusion behavior of water in CM-chitosan gels was Fickian diffusion, and the swelling of CM-chitosan gels was affected by the pH and ionic strength.

However, due to its tendency for radiation degradation, the mechanical strength of as-prepared CM-chitosan hydrogels is not high enough, which restricts their further application. Therefore, enhancement of cross-linking is important for the application of CM-chitosan hydrogels. Enhanced radiation cross-linking of CM-chitosan hydrogels was synthesized by γ-ray radiation-induced

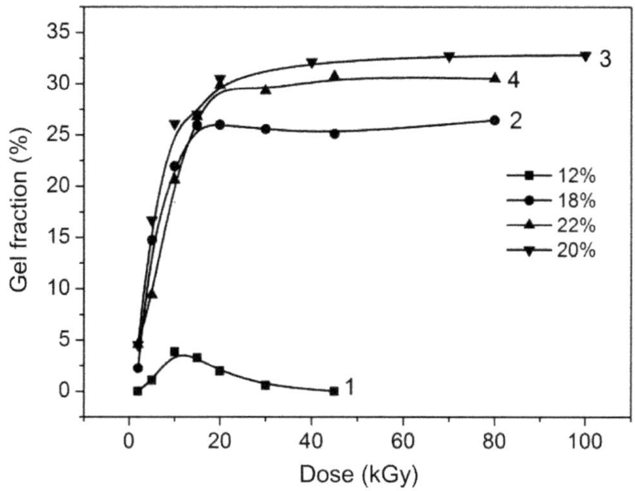

Figure 5.29 Gel fraction of CM-chitosan hydrogels [(1) 12% CM-chitosan; (2) 18% CM-chitosan; (3) 20% CM-chitosan; (4) 22% CM-chitosan].[30] *CM-chitosan*, carboxymethylated chitosan.

cross-linking in the presence of acids or polyfunctional monomers. Aqueous solutions of perchloric acid and acetic acid with different concentrations were used as solvents instead of deionized water to prepare the CM-chitosan hydrogels. Moderate acidification was able to enhance the cross-linking behavior of CM-chitosan so that the gel fraction increased after adding perchloric acid to paste-like CM-chitosan solutions, and it increased further by increasing the amount of acid. Similar results were found in the CM-chitosan aqueous systems with acetic acid as the solvent. In addition, *N,N′*-methylene bis-acrylamide and poly(ethylene glycol) dimethacrylate were also shown to be effective cross-linking reagents in the synthesis of hydrogels. The gel strength and gel fraction increased remarkably after adding the polyfunctional monomers.

5.3.2.2 Radiation cross-linking of polysaccharide derivatives for biomaterials

Radiation cross-linking is a green method to prepare biomedical materials. UV, EB, and γ-ray were frequently used to induce cross-linking of polymers. The energy of UV light is too weak to penetrate into deeper parts of materials. Therefore, cross-linking forms on the surface instead of the bulk of polymer materials and the extent of cross-linking is limited. However, both EB and γ-ray with high energy are effective for radiation cross-linking of polymer materials. Moreover, cross-linking and sterilization can be performed at the same time. Therefore, for the preparation of biomaterials, radiation cross-linking has great advantages over other methods.

Ionizing radiation is an efficient method to modify carboxymethyl cellulose (CMC), under which CMC cross-links and forms hydrogels with many excellent properties for the utilization of biomaterials, such as high water uptake, good biodegradability, and innocuity. CMC can be partially cross-linked to form hydrogel by radiation. The gel fraction increases with the absorbed dose. The cross-linking reaction of CMC is promoted in the presence of N_2 or N_2O due to the increase of free radicals on the CMC backbone, but the gel fraction of CMC hydrogel is not high. The experimental results indicated that strong acidity, strong basicity, a small amount of inorganic salts, and lower temperature can reduce the swelling ratio of CMC hydrogel. Because of the lack of toxicity and good biocompatibility of natural polymer, CMC-based hydrogel will have wide application in biomedical materials and agriculture. In order to improve the degree of substitution of CMC, sodium CMC with AAm groups (AAm-CMC-Na) was synthesized by grafting AAm onto CMC. Then, a novel modified CMC hydrogel based on AAm-CMC-Na aqueous solution was obtained using ^{60}Co γ-ray irradiation. Compared with CMC-Na hydrogel, the novel AAm-CMC-Na hydrogel can be produced with a lower absorbed dose and lower concentration. The swelling ratio of AAm-CMC-Na hydrogel in water was improved with the increase of temperature and had a maximum value in the range of $pH = 6-8$. As shown in Fig. 5.30, the resulting AAm-CMC-Na hydrogel showed good sensibility to temperature and pH value, implying this novel hydrogel is expected to be a water absorption material and water preserver.[32]

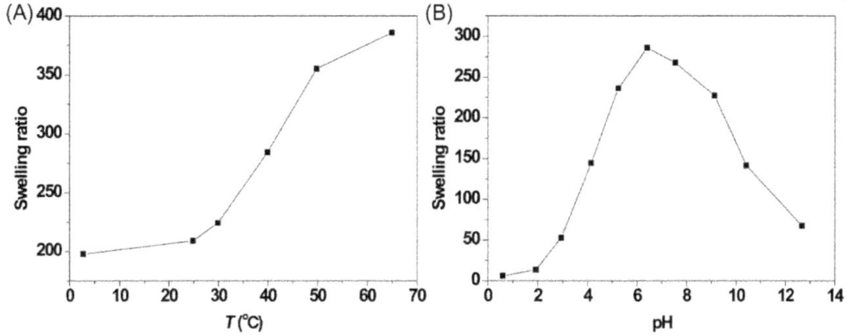

Figure 5.30 (A) Effect of swelling temperature and (B) pH on swelling ratio of AAm-CMC-Na hydrogel.[32]

Blending synthetic polymers with natural polysaccharides is an effective way to improve the gel properties of natural polymers, which could overcome the poor mechanical strength of single component hydrogels to extend the applications of hydrogels as a new kind of wound dressing material. A series of excellent hydrogels was prepared from poly(vinyl alcohol) (PVA) and CM-chitosan with EB at room temperature.[33] Electron spectroscopy analysis of the blend hydrogels revealed that good miscibility was sustained between CM-chitosan and PVA. The mechanical properties and equilibrium degree of swelling were clearly improved after adding CM-chitosan to PVA hydrogels, due to the grafting interaction between PVA and CM-chitosan molecules during irradiation. The blended hydrogels exhibited satisfying antibacterial activity against *Escherichia coli*, even when the CM-chitosan concentration was only 3 wt%.[33] Poly(*N*-vinyl pyrrolidone) (PVP) were blended with CMC to synthesize hydrogels by ^{60}Co γ-ray irradiation. Compared with pure PVP and CMC hydrogels, blended hydrogels possessed improved gel strength, flexibility, and transparency. The application of the wound dressing was performed by comparison with a commercial hydrogel wound dressing to find a similar moisture retention capability and improved swelling rate. The improved properties as well as the low cost of the materials suggest that PVP/CMC blended hydrogel may be a good candidate for wound dressing.

A series of biodegradable composite scaffolds was fabricated from an aqueous solution of gelatin, CM-chitosan, and β-tricalcium phosphate (β-TCP) by radiation-induced cross-linking at ambient temperature. Ultrasonic treatment on the polymer solutions significantly influenced the distribution of β-TCP particles. In vivo implantation of the scaffold into the mandible of beagle dog revealed that the scaffolds had excellent biocompatibility and the presence of β-TCP can accelerate bone regeneration. The comprehensive results of this study paved the way for the application of gelatin/CM-chitosan/β-TCP composite scaffolds as candidates for bone tissue engineering material.[34]

5.3.2.3 Radiation cross-linking of polysaccharide derivatives for metal adsorption

Owing to the high affinity of functional groups to metal ions, polysaccharides and their derivatives can be used as environmentally-friendly and economical materials in metal removal/recovery from dilute solutions with high efficiency. Using radiation cross-linking, hydrogel-based polysaccharide derivatives can be fabricated and used as adsorbents.

Wang et al. investigated the adsorption behavior of Sr^{2+}[35] and Fe^{3+}[36] ions using gel adsorbents based on three types of polysaccharide derivates: CMC, CM-chitosan, and hydroxypropyl methylcellulose phthalate (HPMCP). The adsorption kinetics study showed that the adsorption of Fe^{3+} ions into CM-chitosan gels was very fast ($\sim$20 min) due to the combination of Fe^{3+} ions with amino, hydroxyl, and carboxyl groups of CM-chitosan molecules. The maximum uptake of Fe^{3+} ions, based on the Langmuir equation, was determined to be 18.5 mg g^{-1}. For the adsorption of Sr^{2+}, the sorption rate is in the sequence of CM-chitosan > CMC > HPMCP, and the sequence of the adsorption capacity is CMC > CM-chitosan > HPMCP (Fig. 5.31). Desorption of Sr^{2+} ions was achieved by immersing the adsorbents in solutions with low pH or high ionic strength, or heating the adsorbents in a moist, hot environment.

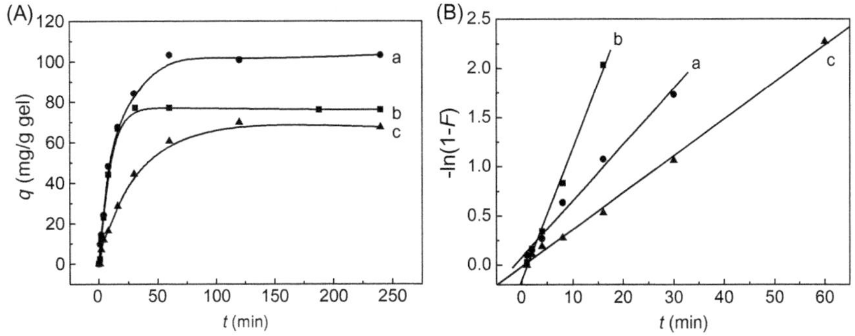

Figure 5.31 Adsorption kinetics of Sr (II) ions into CMC (a), CM-chitosan (b), and HPMCP (c) gels (A) and the linear curves of $-\ln(1-F)$ versus t (B).[36] *CM-chitosan*, carboxymethylated chitosan; *HPMCP*, hydroxypropyl methylcellulose phthalate.

A series of valuable results have shown that the radiation technique is a powerful tool for the modification of polysaccharides. The biomaterials prepared by radiation cross-linking are free of toxic additives, and can be sterilized during irradiation. However, since most of the modifications of cellulose and chitosan have been carried out in heterogeneous and nongreen solvent systems, it is necessary to further investigate the radiation-induced grafting of polysaccharides in green solvents for low-cost and large-scale production.

REFERENCES

1. Fayolle B, Audouin L, Verdu J. Radiation induced embrittlement of PTFE. *Polymer (Guildf)* 2003;**44**(9):2773–80.
2. Tang ZF, Wang MH, Tian F, Xu L, Wu GZ. Crystal size shrinking in radiation-induced crosslinking of polytetrafluoroethylene: synchrotron small angle X-ray scattering and scanning electron microscopy analysis. *Eur Polym J* 2014;**59**:156–60.
3. Tang ZF, Wang MH, Zhao YN, Wu GZ. Tribological properties of radiation cross-linked polytetrafluoroethylene sheets. *Wear* 2010;**269**(5–6):485–90.
4. Chapiro A. Préparation des copolymères greffés du polytetrafluoroéthylène (teflon) par voie radiochimique. *J Polym Sci* 1959;**34**(127):481–501.
5. Chandler-Temple A, Kingshott P, Wentrup-Byrne E, Cassady AI, Grondahl L. Surface chemistry of grafted expanded poly(tetrafluoroethylene) membranes modifies the in vitro proinflammatory response in macrophages. *J Biomed Mater Res Part A* 2013;**101A**(4):1047–58.

6. Ito N, Mase A, Seko N, Tamada M, Sakata E, Kogi Y. Surface treatment of poly(tetrafluoroethylene) and perfluoroethylene-propylene by radiation grafting. *Jpn J Appl Phys* 2006;**45**:9244–6.
7. Klüpfel B, Lehmann D. Functionalization of irradiated PTFE micropowder with methacryl- or hydroxy groups for chemical coupling of PTFE with different matrix polymers. *J Appl Polym Sci* 2006;**101**(5):2819–24.
8. Yang CQ, Xu L, Zeng HY, Tang ZF, Zhong L, Wu GZ. Water dispersible polytetrafluoroethylene microparticles prepared by grafting of poly(acrylic acid). *Radiat Phys Chem* 2014;**103**(5):103–7.
9. Cho YK, Park D, Kim H, Lee H, Park H, Kim HJ, et al. Bioactive surface modifications on inner walls of poly-tetra-fluoro-ethylene tubes using dielectric barrier discharge. *Appl Surf Sci* 2014;**296**(4):79–85.
10. Wu SY, Kang ET, Neoh KG. Surface modification of poly(tetrafluoroethylene) films by graft copolymerization for adhesion improvement with evaporated copper. *Macromolecules* 1999;**32**(25):186–93.
11. Li R, He XZ, Gao QH, Pang LJ, Wang HL, Hu JT, et al. Variations in surface and electrical properties of polytetrafluoroethylene film after plasma-induced grafting of acrylic acid. *Nucl Sci Tech* 2016;**27**(3):88–95.
12. Chuo TW, Wei TC, Chang Y, Liu YL. Electrically driven biofouling release of a poly(tetrafluoroethylene) membrane modified with an electrically induced reversibly cross-linked polymer. *ACS Appl Mater Interf* 2013;**5**(20):9918–25.
13. Venault A, Chang Y, Hsu HH, Jhong JF, Yang HS, Wei TC, et al. Biofouling-resistance control of expanded poly(tetrafluoroethylene) membrane via atmospheric plasma-induced surface PEGylation. *J Membr Sci* 2013;**439**(4):48–57.
14. Sun XQ, Li D, Liu B, Zhang Y, Ma Z. Preparation and anti-blood coagulation property of heparin/fluorocarbon composite film using radio frequency sputtering. *Surf Coat Technol* 2007;**201**(9–11):5659–63.
15. Onder S, Kazmanli K, Kok FN. Alteration of PTFE surface to increase its blood compatibility. *J Biomater Sci Polym Ed* 2011;**22**(11):1443–57.
16. Chang Y, Cheng TY, Shih YJ, Lee KR, Lai JY. Biofouling-resistance expanded poly(tetrafluoroethylene) membrane with a hydrogel-like layer of surface-immobilized poly(ethylene glycol) methacrylate for human plasma protein repulsions. *J Membr Sci* 2008;**323**(1):77–84.
17. Crombez M, Chevallier P, Gaudreault RC, Petitclerc E, Mantovani D, Laroche G. Improving arterial prosthesis neo-endothelialization: application of a proactive VEGF construct onto PTFE surfaces. *Biomaterials* 2005;**26**(35):7402–9.
18. Wu GZ, Long DW, Chen SP, Yao S. High efficient fabrication of chitosan micropowder by combination of gamma radiation and jet pulverization. *Carbohydr Polym* 2005;**60**(1):61–5.
19. Hao Y, Peng J, Li JQ, Zhai ML, Wei GS. An ionic liquid as reaction media for radiation-induced grafting of thermosensitive poly(*N*-isopropylacrylamide) onto microcrystalline cellulose. *Carbohydr Polym* 2009;**77**(4):779–84.
20. Sokker HH, Badawy SM, Zayed EM, Eldien FAN, Farag AM. Radiation-induced grafting of glycidyl methacrylate onto cotton fabric waste and its

modification for anchoring hazardous wastes from their solutions. *J Hazard Mater* 2009;**168**(1):137−44.

21. Li CC, Zhang YW, Peng J, Wu H, Li JQ, Zhai ML. Adsorption of Cr(VI) using cellulose microsphere-based adsorbent prepared by radiation-induced grafting. *Radiat Phys Chem* 2012;**81**(8):967−70.
22. Zhang Y, Xu L, Zhao L, Peng J, Li C, Li J, et al. Radiation synthesis and Cr (VI) removal of cellulose microsphere adsorbent. *Carbohydr Polym* 2012;**88**(3):931−8.
23. Dong Z, Liu J, Yuan W, Yi Y, Zhao L. Recovery of Au(III) by radiation synthesized aminomethyl pyridine functionalized adsorbents based on cellulose. *Chem Eng J* 2016;**283**:504−13.
24. Liu S, Xu M, Yu T, Han D, Peng J, Li J, et al. Radiation synthesis and performance of novel cellulose-based microsphere adsorbents for efficient removal of boron (III). *Carbohydr Polym* 2017;**174**:273−81.
25. Hua D, Deng W, Tang J, Cheng J, Zhu X. A new method of controlled grafting modification of chitosan via nitroxide-mediated polymerization using chitosan-TEMPO macroinitiator. *Int J Biol Macromol* 2008;**43**(1):43−7.
26. Hanh TT, Huy HT, Hien NQ. Pre-irradiation grafting of acrylonitrile onto chitin for adsorption of arsenic in water. *Radiat Phys Chem* 2015;**106**:235−41.
27. Hanh TT, Huy HT, Hien NQ. Pre-irradiation grafting of acrylonitrile onto chitin for adsorption of arsenic in water. *Radiat Phys Chem* 2015;**105**:235−41.
28. Hien NQ, Van PD, Duy NN, Huy HT. Radiation grafting of acrylic acid onto partially deacetylated chitin for metal ion adsorbent. *Nucl Instrum Methods B* 2005;**236**(1):606−10.
29. Ramnani SP, Chaudhari CV, Patil ND, Sabharwal S. Synthesis and characterization of crosslinked chitosan formed by gamma irradiation in the presence of carbontetrachloride as a sensitizer. *J Polym Sci Polym Chem* 2004;**42**(15):3897−909.
30. Huang L, Peng J, Zhai M, Li J, Wei G. Radiation-induced changes in carboxymethylated chitosan. *Radiat Phys Chem* 2007;**76**(11−12):1679−83.
31. Huang L, Zhai M, Peng J, Li J, Wei G. Radiation-induced degradation of carboxymethylated chitosan in aqueous solution. *Carbohydr Polym* 2007;**67**(3):305−12.
32. Liu P, Zhai M, Li J, Peng J, Wu J. Radiation preparation and swelling behavior of sodium carboxymethyl cellulose hydrogels. *Radiat Phys Chem* 2002;**63**(3):525−8.
33. Zhao L, Mitomo H, Zhai M, Yoshii F, Nagasawa N, Kume T. Synthesis of antibacterial PVA/CM-chitosan blend hydrogels with electron beam irradiation. *Carbohydr Polym* 2003;**53**(4):439−46.
34. Zhou Y, Xu L, Zhang X, Zhao Y, Wei S, Zhai M. Radiation synthesis of gelatin/CM-chitosan/β-tricalcium phosphate composite scaffold for bone tissue engineering. *Mater Sci Eng C* 2012;**32**(4):994−1000.

35. Wang M, Xu L, Peng J, Zhai M, Li J, Wei G. Adsorption and desorption of Sr (II) ions in the gels based on polysaccharide derivates. *J Hazard Mater* 2009;**171**(1):820–6.
36. Wang M, Xu L, Zhai M, Peng J, Li J, Wei G. γ-ray radiation-induced synthesis and Fe (III) ion adsorption of carboxymethylated chitosan hydrogels. *Carbohydr Polym* 2008;**74**(3):498–503.

CHAPTER SIX

Radiation Emulsion Polymerization

Mozhen Wang, Xuewu Ge and Zhicheng Zhang
Department of Polymer Science and Engineering, University of Science and Technology of China, Hefei, P.R. China

When emulsions composed of a dispersed monomer phase and a continuous water phase are exposed to ionizing radiation, polymerization reactions can be initiated directly by the H· and ·OH radicals produced by the radiolysis of water molecules, which is commonly termed radiation emulsion polymerization, to distinguish it from conventional emulsion polymerization that is initiated by the addition of chemical initiators. The distinctiveness of radiation emulsion polymerization is much more than that it requires no extra chemical initiators, and includes the following features: (1) polymerization reactions can be carried out at room temperature or even lower; (2) the change of absorbed dose rate can regulate the rate of generation of free radicals, thereby controlling the polymerization rate; (3) the size and relative molecular weight of the produced latex particles have narrow distributions; (4) the polymerization has little dependence on the pH of the emulsion due to the formation of free radicals by irradiation. Since the entire emulsion system is exposed in a high energy radiation environment, some side reactions, such as grafting of the emulsifier molecules on the polymer chains, are inevitably triggered during radiation emulsion polymerization, resulting in the formation of branched polymer products.

Radiation Technology for Advanced Materials.
DOI: https://doi.org/10.1016/B978-0-12-814017-8.00006-8

6.1 EMULSION POLYMERIZATION

As a unique chemical synthesis process, emulsion polymerization has been widely used in the areas of synthetic rubber, adhesives, thermoplastic resins, coatings, etc. It is essentially free-radical polymerization in a multicomponent complex system. The process involves two stages: first, the water-insoluble monomers are dispersed into water with the aid of emulsifier to form an oil-in-water (O/W) emulsion; next, the water-soluble or oil-soluble initiators decompose to initiate the polymerization of monomers. Typical monomers used in emulsion polymerization include acrylonitrile (AN), styrene (St), vinyl acetate (VAc), and acrylate. The stabilizer plays a critical role in an emulsion polymerization system. Since the latex particles grow with polymerization, resulting in the increase of the oil—water interfacial areas, an effective stabilizer is needed to prevent the aggregation of the growing latex particles and maintain the stability of the emulsion.

6.1.1 History and features of emulsion polymerization

The development of the emulsion polymerization technique has more than 100 years of history. It started in the early 20th century.[1] In 1909, Hoffman at Bayer (Germany) disclosed the polymerization of vinyl monomers carried out in the form of water emulsion, which can be considered as the embryo of the emulsion polymerization technique. Patents for emulsion polymerization with a formula similar to the current production appeared in the late 1920s. In the early 1930s, emulsion polymerization was applied in industrial production. Although large numbers of patents appeared before 1940, research papers on emulsion polymerization were seldom published during that period.

During World War II, new demands on the synthetic rubber industry in the countries of the world motivated research and development on the theories and techniques of the emulsion polymerization, and a great progress had been made. The most notable studies in 1940s were done by Harkins, Smith, and Ewart.

Harkins[2] first qualitatively demonstrated the mechanism and the physical concept of emulsion polymerization of water-insoluble monomers. Later, based on Harkins' theory, Smith and Ewart[3] determined the quantitative relationship between the number of latex particles and the concentration of the emulsifier and the initiator, respectively, and proposed the quantitative theory of emulsion polymerization. They arbitrarily divided the emulsion polymerization process into three stages (or intervals), that is, the particle nucleation stage (Interval I), the particle growth stage (Interval II), and the consumption stage of residual monomer (Interval III). The Smith–Ewart (S–E)–Harkins theory has been recognized as the classic mechanism of free-radical emulsion polymerization, and is still used to analyze emulsion polymerization today. Research on emulsion polymerization expanded following the establishment of the S–E–Harkins theory. The following research, including the nucleation mechanism of oligomers, the swelling thermodynamics of latex particles by monomers, the core-shell structure theory, as well as mathematic models for emulsion polymerization, in turn constantly amended and developed the classic S–E–Harkins theory. On the other hand, new emulsion polymerization processes also appeared continuously, including core/shell emulsion polymerization, soap-free emulsion polymerization, organic/inorganic complex emulsion polymerization, living emulsion polymerization, interpenetrating network polymerization, and microemulsion polymerization, and were used extensively in emulsion production. Today, emulsion polymerization processes have more and more significance for the industrial production of polymers, and have become one of the major synthetic methods in the production of numerous types of polymers, such as synthetic rubber, synthetic resin coatings, adhesives, flocculants, and impact copolymers. Myriads of polymeric products are produced by the emulsion polymerization technique every year around the world, and have great significance for the world economy.

The distinct advantages of emulsion polymerization compared with other polymerization processes include: (1) the heat of polymerization can be easily transmitted, since the good flowability of the reaction medium is retained during the polymerization process; (2) the polymerization reactions can be carried out at a relatively low temperature, and high-molecular-weight polymers can be made at a fast polymerization rate; (3) the polymerization processes are low-cost and environmentally-friendly due to the use of water as the medium, and operate flexibly using simple equipment; (4) the emulsion products (latex) can be directly used as waterborne coatings, adhesives, leather and paper, fabrics coating, and cement additives, with less flammable and environment pollution. All these features endow the emulsion polymerization technique with a powerful vitality.

6.1.2 Fundamental theory of emulsion polymerization

Emulsion polymerization is a type of heterogeneous polymerization of monomers dispersed in an emulsion formed with the help of emulsifiers, producing polymer latex product.[1] Emulsion polymerization involves high polymerization rate, steady polymerization process, and high-molecular-weight products, which are closely related to the presence of emulsifier.

An emulsion system is commonly composed of a continuous phase and a dispersed phase which is stabilized by the emulsifiers located at the phase interfaces. Conventional emulsions are commonly composed of water-insoluble monomer droplets and the continuous water phase, that is, an O/W emulsion (customarily referred to as a positive emulsion system). In contrast, water-soluble monomers are often dissolved in water solutions, and dispersed in a continuous oil phase to form a water-in-oil emulsion (customarily referred to as an inverse emulsion system). Obviously, emulsions are not thermodynamically stable systems. An appropriate emulsifier or combination of emulsifiers is critical to maintain the stability of an emulsion system. Taking positive emulsion as an example, the roles of emulsifiers in the emulsion polymerization can be stated as follows: (1) helping the oil-soluble monomers to be dispersed in water to form a stable emulsion. The lipophilic groups on the emulsifier molecules tend to be distributed in oil droplets or polymer latex, while the hydrophilic groups extend to the water phase so that an emulsifier molecular layer forms around the surface of droplets or latex to protect them from coalescence; (2) lowering the interfacial tension between the oil-soluble monomer and water to promote the dispersion of monomers in water; (3) enhancing the solubility of monomer in water through the solubilization of emulsifier micelles. Generally, emulsifiers in an aqueous solution exist mainly in the form of micelles, rarely in the form of a single molecule. When monomers are added under stirring, a part of the monomers enter the micelles, although most monomers are dispersed in the form of droplets stabilized by the emulsifier molecular layer. Only a few of the monomers are dissolved in water. There is a dynamic equilibrium between water, monomers, and the solubilizing micelles in a stable emulsion. A typical constitution of an emulsion system is illustrated in Fig. 6.1.

According to the classic S–E mechanism, free-radical emulsion polymerization passes through three stages. The first stage is the polymer latex particle formation period, named the nucleation period. Before emulsion polymerization, the diameters of micelles

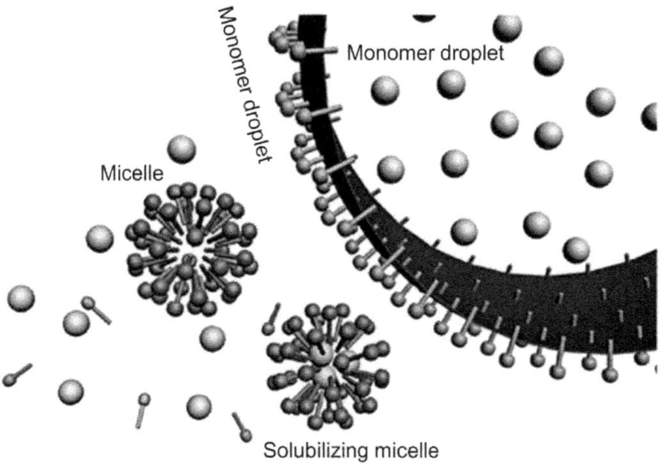

Figure 6.1 Emulsion system: micelle, solubilizing micelle, monomer droplet enwrapped by emulsifier molecules, monomer and emulsifier molecules dissolved in water.

and solubilizing micelles are $\sim 5.0 \times 10^{-9}$ and $\sim 1.0 \times 10^{-8}$ m, respectively. The number of micelles in 1 mL of an emulsion with a medium emulsifier concentration can reach up to 10^{18}. However, the diameter and the number of monomer droplets are 1–10 μm and 10^{10}, respectively, under the same conditions. Evidently, the micelles are far more numerous than the monomer droplets. This means the primary free radicals produced by the decomposition of the initiators in water have a much greater opportunity to enter the micelles than the monomer droplets. As a consequence, the polymerization reactions occur almost entirely in the solubilizing micelles at the initial stage of emulsion polymerization. Once the monomers in the solubilizing micelles are initiated, they will polymerize rapidly to form monomer-polymer latex particles. Polymerization is in a state of acceleration at this stage, and the monomers in monomer droplets and uninitiated solubilizing micelles diffuse into the reacted micelles. The number of latex particles is $\sim 10^{14}$ per milliliter. As polymerization proceeds, the monomer droplets with bigger volume are replaced by smaller latex particles, resulting in the increase of the total particle surface area in the system. This forces the free emulsifier molecules dissolved in water to diffuse to newly-formed interfaces until the concentration of free emulsifier molecules drops below the critical micelle concentration (CMC) and the micelles totally disappear. At this time, the monomer conversion is generally $\sim$15%. The number of latex particles reaches a maximum stable value with the disappearance of micelles, indicating the termination of the nucleation period.

Next, the monomers in droplets continuously diffuse into the latex particles to maintain a constant monomer concentration and a steady polymerization rate, known as the steady-state reaction stage. Zero-order reactions occur during this stage, which is considered to be a characteristic of emulsion polymerization. The latex particles gradually grow larger and larger.

When the monomer conversion is up to 50%–70%, the monomer droplets disappear, which breaks the balance between the supply of and demand for monomers, resulting in the decrease of the rate of polymerization. The polymerization reaction goes on in the latex particles until all the monomers are exhausted. The size of the latex particles is basically unchanged during this slowed-down reaction stage. The finally produced latex particles have a size of $\sim 0.1\ \mu m$, much smaller than the monomer droplets.

Based on the above S—E mechanism, the reaction sites in emulsion systems are isolated microparticles, in each of which only one living propagation chain can be accommodated, since the second free radical will induce the termination of chain propagation. In general, $\sim 10^{13}$ of free radicals will be created in 1 mL of emulsion per minute, while the number of latex particles in 1 mL of emulsion is 10^{14}. This means only one free radical can diffuse into a latex particle per 10 s. Therefore, every living propagating chain is terminated after 10 s. The propagation time of the living chains is much longer than those in solution polymerization and bulk polymerization. Consequently, the product in emulsion polymerization has a relatively high molecular weight. On the other hand, it can be inferred that only 1/2 of the latex particles are "living" during the steady-state reaction stage and the remaining 1/2 is "dead," but the particles alternate between "living" and "dead." Therefore, the concentration of living chains is half that of latex particles. If N is defined as the concentration of the latex particles, the living propagation chain concentration $[M\bullet]$ is $N/2$. Therefore, the polymerization rate R_p can be expressed in the following equation:

$$R_p = k_p[M\bullet][M] = \frac{1}{2}k_p[M]\bullet N \tag{6.1}$$

where k_p and $[M]$ are the chain propagation rate constant and the equilibrium monomer concentration in the latex particles. It should be noted that: (1) $[M]$ indicates the monomer concentration at the reaction site, that is, latex particle; (2) $[M\bullet]$ in emulsion polymerization is determined by N, which causes the factors affecting $[M\bullet]$ to differ from the general rules of polymerization, because N depends on the type and concentration of emulsifiers in addition to the general parameters such as the type and the dosage of monomer and initiator, temperature, etc.; (3) the effect of the high $[M\bullet]$ is not the same as the shielding effects that often occur in a common polymerization system (such as the gel effect). In the emulsion polymerization process, the living propagation chain is located in each isolated latex particle, that is, each latex particle is a reaction unit. Therefore, emulsion polymerization has a much higher reaction rate than other polymerization methods due to the large number of latex particles.

Generally, N is determined by the conditions of the initial polymerization stage, that is, the nucleation period, and has the following relationship:

$$N \propto [E]^{0.6} I^{0.4}$$

where $[E]$ is the concentration (dose) of the emulsifier, and I is the generation rate of free radicals, which is related to the initiator concentration and temperature. Therefore,

$$R_p = k[E]^{0.6} I^{0.4} [M] \tag{6.2}$$

In the emulsion polymerization process, N influences not only R_p, but also the degree of polymerization. The larger the number of N, the longer the interval time for two free radicals entering the same latex particle in succession, resulting in a longer lifetime of living propagation chains and the greater polymerization degree. In the case of the polymerization reaction with unnoticeable chain transfer (such as the emulsion polymerization of St), the chain growth is mainly terminated in a mode of combination termination, and the average polymerization degree can therefore be expressed as:

$$\overline{X}_n = 2\nu = 2\frac{R_p}{\rho} = \frac{k_p[M]N}{\rho} \tag{6.3}$$

where ρ is the initiation rate of the free radicals entering the micelles. The following formula can be derived as follows:

$$\overline{X}_n = k_n[E]^{0.6} I^{-0.6} [M] \tag{6.4}$$

Based from the above discussion, it can be concluded that both R_p and X_n increase with the dosage of the emulsifier within a certain range, especially for emulsifier with a small CMC. It is worth noting that the above descriptions are only applicable to the emulsion polymerization of monomers with little water solubility [such as St, butyl methacrylate (BMA)]. It is not always the case for monomers with high water solubility (such as methyl acrylate, VAc, etc.).

Polarity is also an important factor influencing emulsion polymerization since it is the most important parameter determining the phase interfacial energy. The polarity of the polymer in the dispersed phase strongly affects the adsorption of the emulsifiers, the formation mechanism of latex particles, the aggregation of particles, and the structure of the formed polymer.

6.1.3 Radiation emulsion polymerization

Radiation emulsion polymerization is a polymerization method in which the polymerization reaction is initiated by the free radicals produced by the radiolysis of water when the emulsion is exposed to high energy ionizing radiation.[4] No extra chemical initiators and heat sources are needed to trigger and control the polymerization process. Ionizing radiation induces the dissociation of the medium molecules to generate active free radicals; for example, γ-ray can break the water molecules in the continuous phase into H• and •OH, according to the following processes:

$$H_2O \xrightarrow{\gamma} H_2O^+ + e^- \tag{6.5}$$

$$H_2O \xrightarrow{\gamma} H_2O* \tag{6.6}$$

$$H_2O^+ + H_2O \rightarrow H_3O^+ + \bullet OH \tag{6.7}$$

$$H_2O* \rightarrow H\bullet + \bullet OH \tag{6.8}$$

$$e^- + nH_2O \rightarrow e_{aq}^- \tag{6.9}$$

$$H_3O^+ + e_{aq}^- \rightarrow H\bullet + H_2O \tag{6.10}$$

The molecules of monomer and emulsifier also have the possibility to be decomposed into free radicals. However, in general, the free radicals from the radiolysis of water play the major role in an emulsion system. These free radicals can not only diffuse into the solubilizing micelles or latex particles to initiate the polymerization of monomers in oil phases, but can also directly initiate the polymerization of the monomers dissolved in water. The latter is a very important process in the initial nucleation period for emulsion systems containing monomers with high water solubility.

To date, radiation emulsion polymerization has no special theoretical system of its own, and is analyzed basically according to the conventional emulsion polymerization theory. As a matter of fact, the S–E mechanism can be applied equally to the radiation emulsion polymerization process in emulsions containing monomers with little water solubility. Following the same analysis for chemically-initiated polymerization, R_p and X_n of the radiation emulsion polymerization are expressed as follows:

$$R_p = k_p[M\bullet][M] = \frac{1}{2}k_p[M]\bullet N \tag{6.11}$$

$$R_p = k[E]^{0.6}\dot{D}^{0.4}[M] \tag{6.12}$$

$$\overline{X}_n = k_n[E]^{0.6}\dot{D}^{-0.6}[M] \tag{6.13}$$

where $\dot{D}$ is the absorbed dose rate.

In other words, according to the classic emulsion polymerization theory, the polymerization rate and the relative molecular weight of the polymers formed in radiation emulsion polymerization are mainly affected by three factors: the emulsifier, dose rate, and monomer concentration.

Despite the lack of accurate theoretical guidance, domestic and foreign researchers have carried out extensive practical work on radiation emulsion polymerization using different monomers, radiation sources, and other conditions. Radiation emulsion polymerization is a practical and effective emulsion polymerization method that is worthy of commercial production.

6.2 TYPICAL APPLICATION EXAMPLES OF RADIATION EMULSION POLYMERIZATION

6.2.1 Radiation emulsion polymerization of vinyl acetate

Polyvinyl alcohol (PVA) is a water-soluble polymer that has unique strong cohesiveness and film toughness, as well as other features, including good smoothness, oil and solvent resistance, air barrier properties, abrasion resistance, and the property of protective colloid. After being specially treated, it can have good water resistance. PVA is widely used in many areas including textiles, architecture, paper-making, packaging, medicine, agriculture, and polymer chemical engineering. With the growing significance of high-strength and high-modulus fiber in current fiber materials science, PVA has also been rediscovered as a base material for high-performance fibers. Unlike most vinyl polymers, PVA cannot be prepared directly by the polymerization of the corresponding monomer, vinyl alcohol. It is prepared from the partial or total hydrolysis of polyvinyl acetate.

The emulsion polymerization of VAc monomers can be carried out at a relatively low temperature if it is initiated by the radiation of high energy ray, which can greatly reduce the activation energy of the polymerization reaction. This is beneficial for heat dissipation, avoiding explosive polymerization, and decreasing the possibility of free radicals transferring to other species. Therefore, the relative molecular weight of the produced polymer is relatively high and can be easily controlled by regulating the temperature. In addition, the distribution of the latex particle sizes and the relative molecular weight of polymers produced by radiation emulsion polymerization are narrow.

Li et al.[5] studied the kinetics and processing conditions of the γ-ray radiation emulsion polymerization of VAc. The results showed that the nucleation period ended at a low monomer conversion of 10%, followed by a rather long period of constant polymerization rate. The diameter of the latex particles was

~0.5–3.0 μm, larger than that of the particles produced by conventional emulsion polymerization. The data indicated that three nucleation mechanisms, that is, micelle nucleation, homogeneous nucleation, and droplet nucleation, may coexist in this system. Homogeneous nucleation means the monomers dissolved in water are initiated to polymerize into water-insoluble polymeric microparticles, which precipitate out of water, and are quickly enwrapped by emulsifiers, becoming the seed for further polymerization reactions. The water solubility of monomers is the major factor affecting the probability of homogeneous nucleation.

6.2.2 Preparation of polymer composite by radiation emulsion polymerization

Dai et al.[6] studied the seeded emulsion copolymerization of butyl acrylate (BA) and methyl methacrylate (MMA) initiated by γ-ray irradiation, and prepared core-shell structured poly(BA-*co*-MMA) latex particles, making use of the excellent performance of these two polymers. The obtained poly(BA-*co*-MMA) latex had good film-forming ability, and the composite film exhibited improved water and high-temperature resistance properties.

Wang et al.[7] prepared a novel core-shell structured complex with poly(BA)-intercalated or exfoliated organophilic montmorillonite as the core and polystyrene (PS) as the shell, using radiation-seeded emulsion polymerization. The prepared complex combines the advantages of the high strength of silicate layers and the high toughness of the flexible polymer. When it was blended in a pure rigid PS matrix, the impact strength of the blend was dramatically improved with little loss of the tensile strength.

Xu et al.[8] studied the γ-ray radiation microemulsion copolymerization of St and two kinds of acrylate monomers with different water-solubilities, BA and MMA, respectively, using 12-butinoyloxy-9-octadecenoica acid as the emulsifier. It was found that during the early stage of polymerization, the reaction loci were in the emulsifier layer around the small micelles swollen by monomers when initiated by γ-ray irradiation, rather than the core of large latex particles when initiated with an oil-soluble chemical initiator such as benzoyl peroxide. The kinetics of the radiation microemulsion copolymerization of St and BA or St and MMA are similar to those of bulk copolymerization, and finally cause the formation of ultrafine copolymeric latex particles.

Jia et al.[9] prepared an low-density polyethylene (LDPE)-based composite material uniformly filled with carbon black particles, taking advantage of the radiation emulsion polymerization initiated on the surface of the carbon black particles. The results indicated that the absorbed dose remarkably affected the positive temperature coefficient and negative temperature coefficient of the composite material.

6.2.3 Radiation emulsion copolymerization

Wu et al.[10] prepared a series of poly(butyl acrylate-*co*-methyl methacrylate)/poly(ethyl acrylate-*co*-acrylic acid) [poly(BA-*co*-MMA)/poly(EA-*co*-AA)] interpenetrating polymer networks in a latex form. The morphologies of the latex particles were studied using a transmission electron microscope, indicating the existence of three kinds of particulate morphologies: a gradient shell structure, a cellular structure, and a core-shell structure. The reason for the formation of different particulate morphologies may be related to the special radiation initiation mode and the different compositions of the core latex particles. The produced core-shell structured poly (BA-*co*-MMA)/poly(EA-*co*-AA) can be used as waterproof fabric coating, and has much higher water resistance than one-component coatings. In addition, the radiation emulsion polymerization process is much simpler and easier.

Wang et al.[11] synthesized a series of random copolymers, poly [(*n*-BMA)-*co*-(*tert*-BMA)], with different molecular weights and compositions via reversible addition-fragmentation chain transfer (RAFT) living free-radical polymerization initiated by γ-ray irradiation. Alkali-soluble polymeric surfactants (ASPS) were then derived from the copolymers to be used for the emulsion polymerization of St. Compared with the emulsion system using the common emulsifier, the polymerization rate can be accelerated by increasing the dose of ASPS. The more the carboxyl groups in

ASPS, the smaller the PS latex particles produced, indicating that surfactant fabricated from the radiation RAFT living polymerization is more effective in the preparation of PS latex particles with small size and narrow size distributions than surfactant fabricated via normal radiation polymerization methods.

Egusa et al.[12] also prepared copolymer lattices of *n*-BMA, 2-hydroxyethyl methacrylate, and AA with ^{60}Co γ-ray as initiator and sodium dodecylsulfate as emulsifier. A redox initiation system of $(NH_4)_2S_2O_8/NaHSO_3$ was also adopted as a comparison. The viscosity of the latex produced by radiation emulsion polymerization was higher than that produced by chemically-initiated emulsion polymerization. The latex particle size increased with the polymerization temperature during radiation emulsion polymerization, while chemically-initiated emulsion polymerization had the opposite event. The reason is that in the case of chemical initiation, as the decomposition rate of the initiators increases with the temperature, so does the nucleation rate. As a result, the more numerous nuclei, the smaller the size of the final latex particles. On the other hand, in the case of radiation initiation, the initiation rate had little dependence on the temperature. However, the solubility of monomer in water is increased with rising temperature, resulting in an increased swelling rate of the initiated seed nuclei, and fewer numbers and larger sizes of the formed latex particles. Therefore, the low temperature at the initial polymerization stage is the unique advantage of radiation emulsion polymerization.

Ge et al.[13] investigated the inverse emulsion copolymerization of (2-methacryloyloxyethyl) trimethylammonium chloride with AAm initiated by γ-ray irradiation. An aqueous solution of water-soluble monomers was dispersed in kerosene with the aid of a mixed surfactant system (Span 80 and OP10). The relative molecular weight of the produced polymers increased with monomer conversion, and a distinct gel effect was proved by the relationship between the monomer conversion and the radiation time. Based on the influence of dose rate, composition and concentration of monomers, dosage of emulsifier, and reaction temperature on monomer conversion and the inherent viscosity of the polymers, it can be concluded that the relative molecular weight of the formed polymers will increase with the absorbed dose and monomer concentration.

Zhang et al.[14,15] prepared PS latex by γ-ray radiation emulsion polymerization with polyurethane (PU) as seed particles that were synthesized in the water phase. The relative molecular weight and CMC of the PU were characterized. The conversion-time curves under different concentrations of the stabilizer and dose rate were investigated. Component analysis demonstrated grafting between PS and PU. The work showed that γ-ray initiation is the more effective way to achieve the grafting of PS and PU, compared with chemical initiation.

6.2.4 Radiation emulsifier-free emulsion polymerization

Wang et al.[16] prepared monodisperse PS latex particles by ^{60}Co γ-ray radiation-induced emulsifier-free emulsion polymerization in the presence of a surfactant monomer, 10(9)-hydroxyl-9(10)-allyl ether octadecanoic acid at room temperature. The surfactant monomer is not only a comonomer that participates in the polymerization reaction, but also becomes the stabilizing unit on the copolymer chains to achieve the stabilization of the latex particles. Since there is no individual surfactant in the system, the product has good purity. The results of the kinetics research showed that polymerization rate was not steady during the process of the radiation surfactant-free emulsion polymerization so that the nucleation mechanism was more likely to be droplet nucleation. This study indicates that radiation-initiated surfactant-free emulsion polymerization may be an effective way to fabricate polymer nanoparticles.

6.2.5 Other progress in radiation emulsion polymerization

Zhang et al.[17] gave a brief introduction to the technological process for industrialization of radiation-induced emulsion polymerization in a paper. They adopted a batch process rather than a continuous process in the industrial-scale production. The advantages of radiation-induced emulsion polymerization were also described in comparison with the chemically-initiated process in the literature.

Jiang et al.[18] studied the kinetics of ^{60}Co γ-ray initiated inverse emulsion polymerization of sodium acrylate using kerosene as the medium and Span 80 as the emulsifier. Based on the monomer droplet nucleation mechanism, the expression for the relationship between the monomer conversion and reaction time was proposed. The detailed kinetic analysis suggested that during a dose-rate-independent polymerization process, there is around 0.5 active radical per droplet (nucleus) on average, that is, in the droplet containing one active growing chain, the polymer radical is not terminated until another radical produced from the radiolysis enters. On the other hand, for the dormant droplets, once a new radical enters, it initiates the growth of a new polymer chain immediately. That is, at any given time, either one growing chain or no growing chains exists in a droplet (nucleus).

In addition to the conventional products of solid polymer latex particles, the preparation of various hollow or porous polymer-based latex particles has been developed using radiation emulsion polymerization in recent decades. For example, Wang et al.[19] proposed the one-step synthesis of MnO_2/PS nanocomposite foam in a high internal-phase emulsion (HIPE) under ^{60}Co γ-ray irradiation. Although HIPE is classified as an inverse emulsion system, HIPE emulsion polymerization has a distinct difference from classic inverse emulsion polymerization. In this HIPE, St monomer was the continuous phase, while the aqueous solution of $KMnO_4$ was the dispersed phase. Therefore, under γ-ray irradiation, the active free radicals produced from the radiolysis of water in the dispersed phase initiated the polymerization of St in the continuous phase. At the same time, $KMnO_4$ was reduced to form MnO_2 nanoparticles by the reductive active species (H•) produced from water radiolysis. The authors also investigated the catalytic activity of loaded MnO_2 nanoparticles in wastewater treatment. The results indicated that loaded MnO_2 nanoparticles are more effective in the decomposition of methylene blue than free MnO_2 nanoparticles.

6.3 PIGMENT PRINTING BINDER PREPARED BY RADIATION EMULSION POLYMERIZATION

6.3.1 Representative basic formula for pigment printing binder

Soft feel is a critical feature for a binder to be used in pigment printing. Therefore, a good balance of "hard" and "soft" monomers should be taken into consideration in the basic formula for the binder. Here, "hard" monomer refers to those monomers that homopolymer has a relatively high glass transition temperature (T_g), such as AN, MMA, and St. In contrast "soft" monomer refers to those monomers that homopolymer has a relatively low T_g, such as BA and EA. If only soft monomers are used in the formula, the synthesized binder will be a little sticky, resulting in an uncomfortable hand feeling, although it has enough softness. Therefore the combination of a few hard monomers in the formula is crucial for a successful binder product. There are countless open formulas for pigment printing binder (PPB). Here, taking a product with a solid content of 40% as an example, a typical formula (in terms of mass fraction) is provided for reference, BA: 30%; AN: 7%; *N*-methylol acrylamide: 0.4%; AA: 1.0%; emulsifier: 2%; peregal (O-25): 2%; H_2O: 57.6%.[20]

Here, AA is added in two batches. The second addition can be operated after the polymerization reaction starts and the system temperature nears the maximum temperature during polymerization, in order to allow AA to attach to the surface of the latex particles to stabilize the system and especially improve the shear stability to a great extent.[20]

6.3.2 Processing

First, all of the materials are emulsified in the emulgator for ~0.5 h under a N_2 atmosphere. Next, the emulsion is transferred to a special reactor for radiation polymerization, and irradiated by ^{60}Co γ-ray at a low-speed material flow. The average absorbed dose rate is 1–2 kGy h^{-1}. The total absorbed dose should be maintained at 5–8 kGy. There is an induction period of 0.5 h. The system temperature rises to a maximum value after ~1 h. Then, AA is added to the system after being diluted with a certain amount of water. The heat of the reaction can be dissipated by cooling water. The extent of the reaction can be judged by the change of the system temperature. Generally, the discharged temperature should be a little higher than room temperature. The typical process of radiation emulsion polymerization is illustrated in Fig. 6.2.

The reactors should be designed to fit the detailed ^{60}Co device. For example, a cylindroid reactor and double stirrer are suitable for the plate source. Conventional emulsion polymerization is commonly carried out under continuous stirring. At the emulsifying dispersion stage, monomers must be highly dispersed in water with

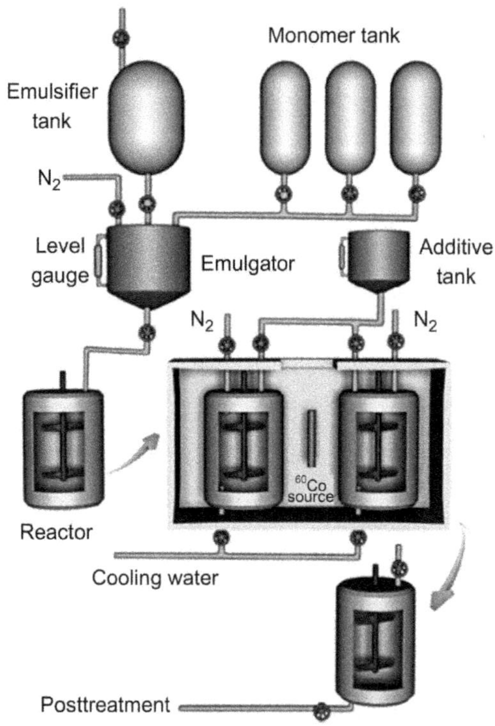

Figure 6.2 Radiation emulsion polymerization processing.

the aid of emulsifiers to form numerous fine solubilizing micelles and droplets. A low stirring intensity at this stage may cause the formation of large monomer droplets, and even the creaming of the emulsion. However, the stirring intensity at the steady polymerization period should be lowered, since excessive stirring readily triggers the aggregation of the reacted latex particles, resulting in local coalescence and the formation of "clumps," as well as slowing down the polymerization reaction. On the other hand, appropriate stirring is necessary to prevent the monomers from being separated by upward drifting, to cause the whole system to be irradiated homogeneously, and to dissipate the heat of reaction, so that the polymerization reaction can be carried out steadily.

Posttreatment, mainly involving the adjustment of the pH of the latex, is also crucial after the discharge of the latex. Posttreated products have a certain consistency and better stability for storage. Some materials such as silicon oil can be added on the demand of users in posttreatment processing.

6.4 POLYACRYLATE THICKENERS PREPARED BY RADIATION EMULSION POLYMERIZATION

Polyacrylate emulsion thickeners primarily involve partly cross-linked ammonium polyacrylate and sodium polyacrylate thickeners, as well as a copolymeric thickener composed of acrylate and acrylamide (AAm). The product is commonly in the form of an inverse emulsion. Due to their low-cost, energy-saving, and environmentally-friendly efficacy, ammonium polyacrylate thickeners are mainly used in fabric pigment printing pastes to replace the pastes for old processing such as pigment paste A. Sodium polyacrylate thickeners are mainly used for the preparation of economical dye printing pastes to replace part or all of some conventional pastes produced based on alginate thickeners.[21]

6.4.1 Typical basic formula

1. Aqueous solution of polyacrylate with high concentration: 76 wt%;
2. White oil or other solvent oil: 21 wt%;
3. Emulsifiers with low hydrophilic lipophilic balance value (such as Span 80): 3 wt%;
4. Tiny amount of cross-linking monomer (80 g of *N,N′*-methylene-bisacrylamide in 1 t of materials);

It should be noted that the processing stability and the properties of the product (such as thickening effect, antielectrolyte, etc.) may be affected by the concentration and pH of the aqueous solution of polyacrylate as well as the presence of tiny amounts of copolymer monomers.

6.4.2 Operation flow

Since the thickener is prepared by inverse emulsion polymerization which proceeds mainly according to the droplet nucleation mechanism shown in Fig. 6.3, it is critical for the major aqueous phase to be well-dispersed in the minor oil phase during the emulsification stage.

The entire process can be briefly described as follows:

1. Neutralization: The aqueous solution of AA is neutralized to a pH of 6.5 with ammonium water (for the preparation of ammonium polyacrylate thickener) or sodium hydroxide solution (for the preparation of sodium polyacrylate thickener). Caution: Pay attention to the cooling of the system in this process. Finally, the volume is made up by the addition of water.

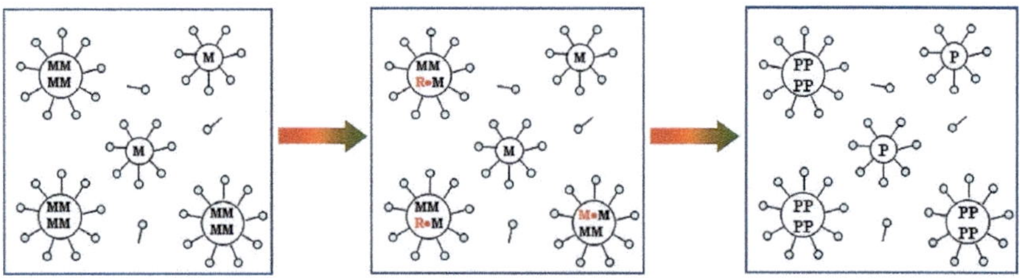

Figure 6.3 Inductive inverse emulsion polymerization process: microdroplet nucleation mechanism.

2. Emulsification: Put oil, emulsifier, and part of the acrylate solution in the emulgator. The system is emulsified under high-speed agitation, and the rest of the acrylate solution is added. The high-speed agitation is kept for 0.5–1.0 h to obtain a sticky emulsion.
3. The as-prepared sticky emulsion is poured into the reactor for γ-ray irradiation. The stirring intensity should be well-controlled. Too low and too high strength stirring is not suggested. Since explosive polymerization readily occurs during this polymerization process, a cooling system is vital to take away the heat of reaction promptly. Experience indicates that products produced in summer usually have better quality than those produced in winter, which confirms the importance of temperature control for an emulsion polymerization system.
4. The control of irradiation time (at a fixed relatively low dose rate) is very important for product quality. Insufficient exposure in the radiation field will cause the product have an unsatisfactory thickening effect, while excessive exposure will reduce the antielectrolyte properties of the product and influence the color when the product is used in pigment printing. Therefore, an optimum irradiation time should be selected to determine the optimum absorbed dose, which is commonly ~5 kGy.

6.5 ADVANTAGES AND DISADVANTAGES OF RADIATION EMULSION POLYMERIZATION TECHNIQUE

As a new technique, the preparation of fabric PPB via radiation emulsion polymerization was industrialized in the early 1990s in China. This unique technique was developed by the University of Science and Technology of China. To date, five companies are engaged in production based on the radiation emulsion polymerization technique, seven ^{60}Co facilities are in operation, and there is an annual output of up to 200 million China Yuan. The great vitality of this technique is based on the following advantages:

1. No need for initiators. This not only enables the production of high-purity products, but also excludes chain transference which causes a remarkable reduction in the relative molecular weight in most cases. For example, the viscosity of the product prepared by the radiation emulsion copolymerization of St and methyl styrene is 2–2.5 times as high as that of products produced by chemically-initiated copolymerization.
2. The radiation initiation rate is independent of temperature. It provides the possibility to carry out the reaction at low temperature so as to avoid some undesirable side reactions (such as chain transference), which can improve both the structural regularity and the relative molecular weight of the formed polymers.
3. The polymerization process is readily controllable. The complete conversion of monomers can be achieved by the suitable extension of irradiation time.
4. The product has good reproducibility since there are fewer interfering factors.

However, the obvious disadvantages of radiation emulsion polymerization, include the large capital investment required and nuclear security regulations. Safe production can be achieved on the premise of strict management and a certain production scale. Moreover, the unit production cost does not differ greatly from that of chemically-initiated production methods.

REFERENCES

1. Bassett DR, Hamielec AE. *Emulsion polymers and emulsion polymerization. ACS symposium series*. Washington, DC: American Chemical Society; 1981.
2. Harkins WD. A general theory of the mechanism of emulsion polymerization. *J Am Chem Soc* 1947;**69**(6):1428–44.
3. Smith WV, Ewart RH. Kinetics of emulsion polymerization. *J Chem Phys* 1948;**16**(6):592–9.
4. Zhang ZC, Ge XW, Zhang MW. *Radiation polymerization and radiation chemistry of polymers (in Chinese)*. Hefei, China: Press of University of Science and Technology of China; 2000.
5. Li JZ, Wang Q, Zhang ZC, Zhang MW. The study of kinetics and processing condition of γ-radiation emulsion polymerization of vinyl acetate. *J Radiat Res Radiat Process* 1998;**16**(1):10–14 (in Chinese).

6. Dai Q, Zhang ZC, Ge XW, Liu J, Ye Q. Preparation of composite latex consisted of polysiloxane (PSI)/polyacrylate by seeded emulsion polymerization initiated by ^{60}Co γ-rays irradiation. *Polym Mater Sci Eng* 2004;**20**(1):72–4 (in Chinese).
7. Wang T, Wang MZ, Zhang ZC, Ge XW, Fang YE. Preparation of core (PBA/layered silicate)–shell (PS) structured complex via γ-ray radiation seeded emulsion polymerization. *Mater Lett* 2006;**60**:2544–8.
8. Xu XL, Zhang ZC, Ge XW, Zhang MW. Microemulsion copolymerization of styrene with acrylates. *Polym Mater Sci Eng* 2000;**16**(3):70–3 (in Chinese).
9. Jia SJ, Jiang PK, Zhang ZC, Wang ZG. Effect of carbon-black treatment by radiation emulsion polymerization on temperature dependence of resistivity of carbon-black-filled polymer blends. *Radiat Phys Chem* 2006;**75**(4):524–31.
10. Wu MH, Shen WP, Ma Z. Synthesis of polyacrylate core-shell structure latex by radiation techniques. *Radiat Phys Chem* 1993;**42**(1–3):171–4.
11. Wang SJ, Ye Q, Zhang ZC, Wang XB. Emulsion polymerization stabilized by polymeric surfactant prepared by gamma-ray radiation. *Colloids Surf, A* 2006;**281**(1–3):156–62.
12. Egusa S, Makuuchi K. Characterization of carboxylated latices prepared by radiation-initiated emulsion polymerization. Comparison with chemically prepared latices. *J Colloid Interface Sci* 1981;**79**(2):350–64.
13. Ge XW, Ye W, Xu ML, Zhang ZC. Radiation copolymerization of acrylamide and cationic monomer in an inverse emulsion. *Polymer* 1998;**39**(10):1917–20.
14. Zhang GX, Zhang ZC, Hu ZQ, Xi HX. Seeded-emulsion polymerization of styrene with waterborne polyurethane stabilizer via ^{60}Co gamma ray. *Colloids Surf, A* 2005;**264**(1–3):37–42.
15. Zhang GX, Zhang ZC. The ^{60}Co-γ ray-initiated seeded-emulsion polymerization of methyl methacrylate in the presence of waterborne polyurethane seeds. *Radiat Phys Chem* 2004;**71**(1–2):273–6.
16. Wang XB, Zhang ZC. Preparation of polystyrene latex particles by γ-rays-induced emulsifier-free emulsion polymerization. *Radiat Phys Chem* 2006;**75**:1001–5.
17. Zhang ZC, Zhang MW. Industrialization of radiation-induced emulsion polymerization—technological process and its advantages. *Radiat Phys Chem* 1993;**42**(1–3):175–8.
18. Jiang PY, Zhang ZC, Zhang MW. Kinetic analysis of the ^{60}Co γ ray-initiated inverse emulsion polymerization of sodium acrylate solutions. *Macromolecules* 1996;**29**(21):6710–16.
19. Wang SF, Zhang ZC, Liu HR, Zhang W, Qian Z, Wang BB. One-step synthesis of manganese dioxide/polystyrene nanocomposite foams via high internal phase emulsion and study of their catalytic activity. *Colloid Polym Sci* 2010;**288**(9):1031–9.
20. Zhang MW, Zhang ZC, Wu CP, Feng XL, Chen DA. Patent CN86108672A; 1988.
21. Zhang ZC, Wu CP, Jiang PY, Zhang MW, Feng XL. Patent 88104060.6; 1992.

CHAPTER SEVEN

Radiation-Grafted Membranes for Applications in Renewable Energy Technology

Jun Ma, Jing Peng and Maolin Zhai
College of Chemistry and Molecular Engineering, Peking University, Beijing, P.R. China

This chapter shows the unique advantages of the radiation-induced grafting technique as applied to the synthesis of advanced functional membranes. As the key component, a radiation-grafted membrane with designed strategies offers an attractive option to meet the requirements of vanadium redox batteries and fuel cells.

7.1 ION-EXCHANGE MEMBRANE AND RADIATION-INDUCED GRAFTING COPOLYMERIZATION

In the first section, a brief outline of the basic fundamentals of membranes, grafting methods, and polymeric matrices is presented.

7.1.1 Ion-exchange membrane

Ion-exchange membrane is as a functional polymer material bearing ionic groups that allows the selective transportation of dissolved ions across the membrane.[1] The membrane typically consists of three necessary components: polymer matrix, ion-exchange group and mobile ion attached to the group. Depending on the charge state of the ionic group, ion-exchange membranes can be classified into three categories: cationic, anionic, and amphoteric (including bipolar). Negatively charged functional groups can provide cation-exchange capacity, that is, sulfonic group ($-SO_3^-$), phosphate group

Radiation Technology for Advanced Materials.
DOI: https://doi.org/10.1016/B978-0-12-814017-8.00007-X

($-PO_3H^-$), and carboxylic acid base ($-COO^-$), etc.; on the other hand, anion-exchange membranes comprising primary, secondary, tertiary, and quaternary amino ($-NH_3^+$, $-RNH_2^+$, $-R_2NH^+$, $-R_3N^+$) groups or vinyl pyridine rings, are capable of exchanging anions. The selectivity of the membrane therefore correlates with the number of ionic groups attached (ion-exchange capacity) at different charged states.

Ions that have the opposite charge on the membrane are defined as counter ions, while those with the same charge are named fixed ions. The electrostatic effect of the same ion is described as the Donnan exclusion. The Donnan equilibrium and Donnan exclusion between the bulk membrane phase and the electrolyte solution are considered to be the main driving forces for ion migration and retention across the membrane. For example, a cationic membrane carrying a fixed anion allows the transmission of the cations in the solution through the membrane. In contrast, owning to coulombic exclusion, the anions dissolved in solution are suppressed to crossover. As a result, the selectivity and effectiveness of a membrane is in relation to the overall effect of the equilibrium of adsorption and the exclusion of ions.

An ideal manufactured ion-exchange membrane (IEM) should offer superior permeability of specific ions, suitable ion-exchange capacity, and excellent swelling stability, while being cost-effective. The properties are affected and manipulated by several important factors that is, the synthesis process, the structure of the polymer matrix, the type and number of introduced ionic groups, as well as the specific operating chemical environment. In the past few decades, IEMs have played a significant role in the separation industry, renewable energy technology, biomedical engineering, and many other fields. For this reason, the diverse membranes with a variety of specific electrochemical properties are demanded. With the aim of tailoring ion-exchange membrane materials with desirable molecular architectures and physic-chemical properties, interfacial polymerization, in situ polymerization and graft copolymerization have been developed to a significant extent. Compared with these techniques, radiation-grafting copolymerization has emerged as a useful

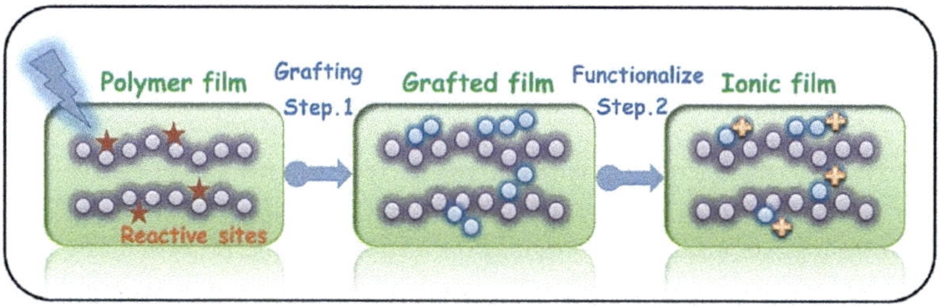

Figure 7.1 Schematic description of ion-exchange membrane synthesized via radiation-induced grafting polymerization.

technique for advanced functional membranes that offers unique advantages as it provides a system that does not require any chemical initiators and makes grafting reaction conditions feasible.

Over many decades, radiation-induced grafting has been demonstrated to be one of the most versatile and fascinating ways to prepare a range of membranes with different functionalities, so that they can be adapted to a variety of industrial applications with slight modification.[2] The general principle, as shown in Fig. 7.1, is the incorporation of the ion functional groups in low-cost polymer film matrix via free-radical copolymerization initiated by various types of commercially available high energy radiation sources, such as ^{60}Co, electron beam (EB), X-ray, and ion beam. Since grafting copolymerization is the key step for the synthesis of radiation-grafted membranes, the following part of this chapter provides detailed illustration of the classification of grafting methods that have been applied; it includes simultaneous grafting, preirradiation grafting, and control/living radiation-induced grafting, as outlined in Fig. 7.2.

7.1.2 Methods of radiation-induced copolymerization

7.1.2.1 Simultaneous radiation grafting

Simultaneous grafting is the simplest method of radiation-induced copolymerization grafting. In this process, the polymer substrate and the monomer are simultaneously irradiated, and the monomer could either be a gas, a solution, or a pure monomer liquid. Irradiation is usually carried out under the protection of an inert gas (N_2 or Ar), and highly reactive free radicals are generated in

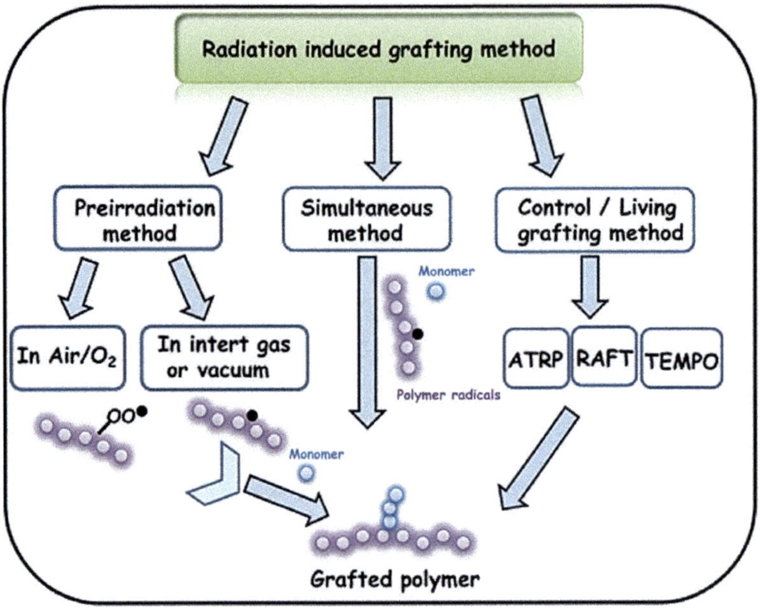

Figure 7.2 Outline of methods of radiation-induced grafting.

both the polymer backbone and the monomer. When the radiolytic yield of active sites produced in a polymer substance (G value, mol J^{-1}) is much higher than that in the monomer, a graft reaction is likely to occur. If it is the opposite case, the homopolymerization reaction of monomers tends to be the dominant process.

The operation of simultaneous grafting is relatively simple, but the monomer during the irradiation easily forms a homopolymer, leading to a dramatic decrease of grafting efficiency. In practical cases, to suppress homopolymerization, the degree of grafting (DG) of grafted polymer can be increased by optimizing the reaction conditions, for instance, by adding inhibitors (Cu^{2+} or Fe^{3+}) or trace amounts of inorganic acids, selecting suitable solvents, operating at a lower dose rate and so on.

7.1.2.2 Preirradiation grafting

The method of preirradiation grafting is relatively complicated and it involves two-step grafting. First, the polymer substrate is irradiated to produce active radicals and then stored at low temperature to prevent free-radical quenching; a monomer solution is subsequently injected into the irradiated polymer film to make the grafting reaction occur under appropriate conditions. If the irradiation is carried out in the air or in an O_2 atmosphere, the polymer radicals generated on the substrate first react with O_2 to form peroxide or hydroperoxide, and the grafting reaction is initiated by peroxide thermal decomposition. The pregraft reaction technique is frequently used in practical situations because this process produces fewer homopolymers and it is more flexible as the grafting step can be carried out at any time after irradiation as long as the radicals are not completely quenched.

7.1.2.3 Control/living radiation grafting

Controlled-radiation graft copolymerization is a new type of radiation-grafting method along with the advancement of controlled living radical polymerization. It can be divided into three categories according to the reversible-deactivation of free radicals: atom-transfer radical-polymerization (ATRP), reversible addition–fragmentation chain-transfer polymerization (RAFT), NMP (nitroxide-mediated radical polymerization)-controlled radical polymerization in the presence of nitroxide. Radiation grafting has several advantages for industrial manufacture; the majority of monomers are theoretically available for free-radical grafting polymerization, and the reaction conditions are mild and easily controlled. Unfortunately, in conventional-free-radical graft polymerization, the free radicals of the chains are prone to bimolecular coupling or disproportionation termination, resulting in limitations in precisely controlling the molecular weight, molecular weight distribution, chain sequence, and terminating group of grafted polymers on the substrate, etc. Therefore, the combination of controllable reactive radical polymerization with radiation-induced initiation not only enriches the synthesis method of IEMs, achieving a specifically designed polymeric microscopic architecture, but also optimizes the physical and chemical properties of membrane material to meet the requirements of particular applications. In the following section, we briefly describe the synthesis and application of IEM using the control-living radiation-grafting process.

7.1.2.3.1 ATRP-mediated radiation

ATRP-mediated radiation graft copolymerization is a two-step grafting method. First, the vinyl monomer containing halogen elements (chlorine and bromine) is introduced into the grafted substrate by the traditional radiation-grafting method, as shown in Fig. 7.3. Next, ATRP-mediated grafting is controlled by equilibrium between propagating radicals and dormant species, predominantly in the form of initiating alkyl halides/macromolecular species. The most commonly used grafted monomers are styrene and methyl acrylate.

7.1.2.3.2 NMP-mediated radiation

NMP-mediated radiation graft copolymerization is a graft copolymerization method using nitrogen oxide compounds 2,2,6,6-tetramethylpiperidine-1-oxyl (TEMPO) to trap and deactivate reactive polymer radicals produced by ionizing radiation, and subsequently proceed to the living polymerization with vinyl monomers in a controlled manner (see Fig. 7.4). In contrast to the traditional

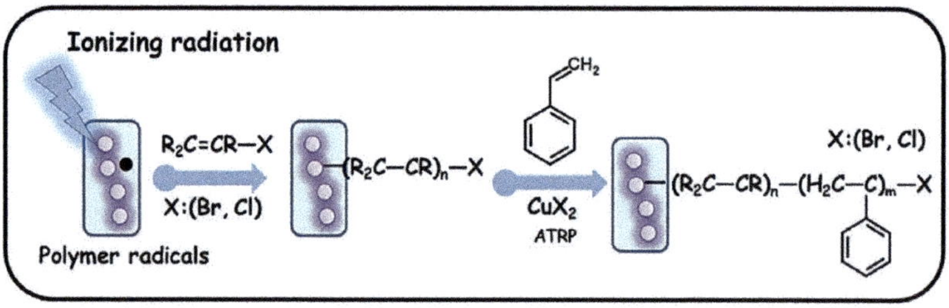

Figure 7.3 ATRP-mediated radiation grafting.

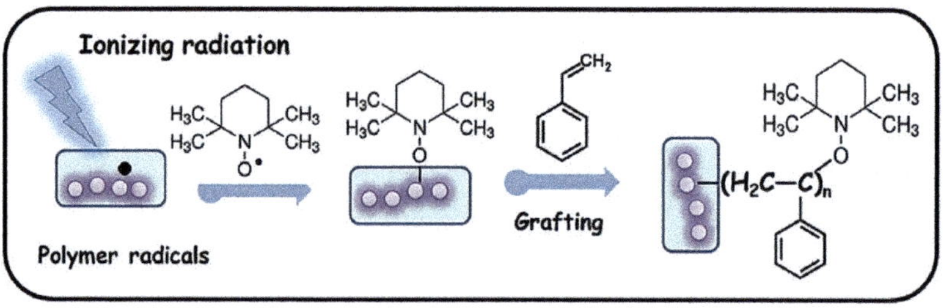

Figure 7.4 NMP-mediated radiation grafting. *NMP*, nitroxide-mediated radical polymerization.

grafting method, the grafting rate obtained by this method is obviously low. However, Miwa et al.[3] have shown that the homopolymers of graft chains have almost identical molecular weight and molecular weight distribution, and the dispersity index or formerly the polydispersity index (PDI) is ~1.2, which is one of the features of living polymerization.

7.1.2.3.3 RAFT-mediated radiation grafting

In RAFT grafting, as shown in Fig. 7.5, a dithioester derivative namely dithiobenzoic acid or dithiocarbamate, etc. is usually present in the grafting system as a chain transfer reagent to the chain of free radicals and to form a dormant intermediate. The irreversible double radical termination reaction between the growths of chain free radicals allows the polymerization reaction to be controlled. A large number of monomers, such as styrene, methyl acrylate, butyl acrylate, methyl methacrylate (MAA), maleic anhydride and acrylic acid (AA), have been successfully polymerized by RAFT polymerization initiated under mild conditions (at room temperature). A few monomers, such as styrene and AA, have also been successfully introduced into polymers such as polypropylene, polyvinylidene fluoride (PVDF), and cellulose using radiation-induced RAFT graft polymerization.[4] Similar to the NMP graft polymerization initiated by radiation, Barner et al. showed that the molecular weight and PDI of the homopolymers in bulk solution and the grafted polymer were not significantly different after grafting of styrene on the surface of cellulose by γ-ray irradiation in the presence of RAFT agent.[5] These results verify the control of γ-initiated free-radical grafting via the RAFT technique without prior modification of the surfaces.

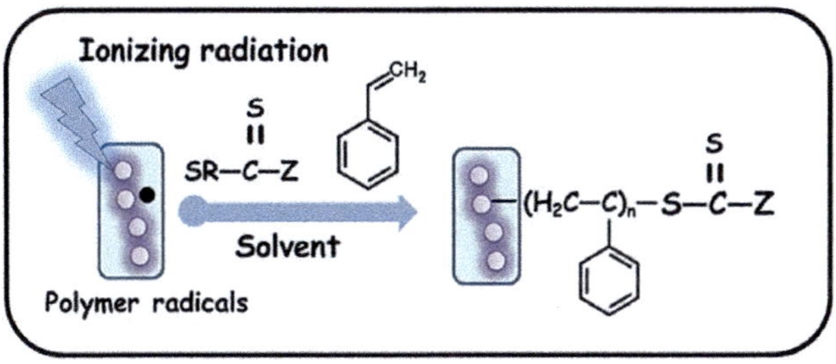

Figure 7.5 RAFT-mediated radiation grafting.

7.2 NATURE OF POLYMER MATRIX

The chemical structure, crystallinity, radiation effect (cross-linking and degradation), and the lifetime of polymer radicals are critical for the properties of IEMs. The polymer substrate film reported to date for the successful radiation synthesis of IEM is shown in Table 7.1. It can be simply classified by type of fluoropolymer and fluorine-free polymer. Fluoropolymer is a polymer containing one or more fluorine atoms with multiple strong carbon–fluorine bonds. The most prominent feature of fluoropolymers is their excellent chemical and thermodynamic stability. Therefore, fluoropolymer film has been widely used in harsh chemical or thermodynamic environments.

Previous studies have shown that the chemical properties of fluoropolymers are quite different to those of hydrocarbon polymers although they have similar structures.[6] For example, the

Table 7.1 Polymer films for radiation-grafted membranes

Polymer film	Abbreviation	Repeating units
Hydrocarbon membrane		
Polyethylene	PE	$-CH_2-CH_2-$
Polypropylene	PP	$-CH_2-CH(CH_3)-$
Polyvinyl chloride	PVC	$-CH_2-CHCl-$
Fluoride film		
Polyvinyl Fluoride	PVF	$-CH_2-CHF-$
Polyvinylidene fluoride	PVDF	$-CH_2-CF_2-$
Polytetrafluoroethylene	PTFE	$-CF_2-CF_2-$
Perfluoroethylene propylene copolymer	FEP	$-CF_2-CF_2-CF_2-CF(CF_3)-$
Perfluoroalkoxy	PFA	$-CF_2-CF_2-CF_2-CF(O-C_3F_7)-$
Ethylene tetrafluoroethylene	ETFE	$-CH_2-CH_2-CF_2-CF_2-$
Polyether ether ketone	PEEK	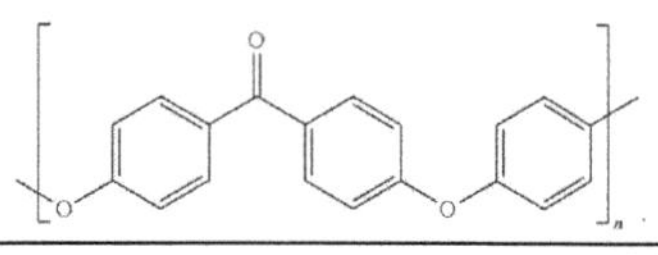

disproportionation of carbon free radicals formed by the radiolysis of fluoropolymers is often absent. This difference mainly corresponds to the unique C–F bond: in all natural elements, fluorine and carbon form a bond with the highest energy up to 544 kJ mol^{-1}, and the electronegativity (3.98) of element fluorine is the highest, which makes C–F bonds very stable. In the irradiation of fluorine compounds, the fluorine atom transfer reaction is also not likely to occur. Polytetrafluoroethylene (PTFE) has long been used as a substrate for radiation grafting as a fluorocarbon polymer. However, PTFE has very high radiation sensitivity; even at lower absorbed doses of γ-ray irradiation, PTFE undergoes rapid degradation.[6] The study also found that the degradation of PTFE mainly takes place in the backbone fracture. Under vacuum conditions, the degree of degradation is relatively lower than in the air. This radiation sensitivity greatly limits the application of PTFE in radiation grafting.[7] Recent studies have found that PTFE can be cross-linked at the melting point of PTFE while the radiation resistance of cross-linked PTFE is significantly enhanced.[8]

Unlike PTFE, perfluoroethylene propylene copolymer (FEP) has high radiation stability and can form stable free radicals. It can also achieve high graft efficiency with many monomers.[9] This allows a higher graft yield to be obtained without reducing the mechanical strength of the membrane. Other fluorocarbon polymers such as perfluoroalkoxy (PFA) copolymer[10] are also used as polymer substrates, but the scope of application is not as extensive as that of FEP.

Ethylene tetrafluoroethylene (ETFE), due to its alternating structure of PTFE and polyethylene (PE), has combined functional properties from both fluorocarbon and hydrocarbon polymers.[11] For example, ETFE is one of the most lightweight and transparent cladding materials, with the advantages of better elasticity than other perfluorinated polymers such as PTFE, PFA, and FEP. In addition, ETFE has excellent thermal stability and good stability for common chemical solvents, and greater resistance to radiation ageing. Owing to these superior properties, ETFE is a good option for the preparation of membrane materials, polymer electrolytes, ion-exchange resins and so on.

PVDF is also a very attractive substrate, mainly due to its unique electrical properties, high durability, biocompatibility, and high resistance to radiation. Therefore, PVDF is widely used as substrate in the preparation of pristine membranes and functionalized membranes. In the preparation of functional membranes, a variety of monomers have been incorporated by radiation-induced grafting.[12] In addition, PVDF powder can be used to graft with monomers via irradiation and subsequently cast into film via the solution inversion method. This new process is more flexible and suitable for industrial manufacture, and it also solves the long-term problem of the inhomogeneous distribution of graft chains in the film.[13]

Polyvinyl fluoride (PVF)[11] is an uncommon fluorine-containing substrate film because PVF is prone to oxidative degradation during the sulfonation process, resulting in its inability to be used in the preparation process involving sulfonation. However, PVF will be a promising fluoropolymer substrate if styrene sulfonate can be directly grafted onto PVF without the oxidization process.

Polyetheretherketone (PEEK)[14] is a nonfluorine ultra-high performance engineering plastic, with high thermal stability and chemical resistance. The low crystallinity of PEEK can also be applied to directly graft with styrene sulfonate to avoid the subsequent sulfonation process.

7.3 RADIATION-GRAFTED MEMBRANE

The properties of the IEM depend on the nature of the functional monomer and the DG.[2] Table 7.2 shows that the monomers for radiation-grafting copolymerization can be classified into two types: functional monomers such as AA, MAA; and nonfunctional monomers such as styrene, *N*-vinylpyridine and chloromethylstyrene. Grafting functional monomers can directly provide ion-exchange capacity to polymer substrates, whereas grafting of nonfunctional monomers such as styrene requires further chemical modifications to obtain ion-exchange capacity. The most commonly used processes are sulfonation and quaternization. The former introduces a canionic group to obtain a cation-exchange membrane, while the latter introduces an anionic group to obtain an anion-exchange membrane (AEM). Therefore, the type of grafting monomer and subsequent chemical treatment determine the physical and chemical characteristics of the membrane. The following examples illustrate the progress of radiation synthesis of these membranes.

Table 7.2 Monomers for radiation-grafted membrane

Monomer	Abbreviation	Molecular structure
Acrylic monomer		
Acrylic acid	AA	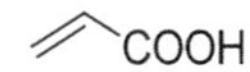
Methyl acrylate	MA	
Acrylamide	AAm	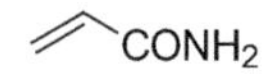
Acrylonitrile	AN	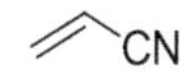
Methacrylate	MAA	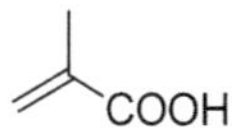
Methyl methacrylate	MMA	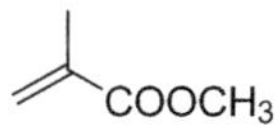
Vinyl acetate	VAc	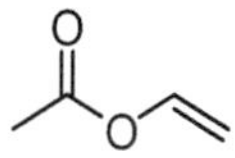
Glycidyl methacrylate	GMA	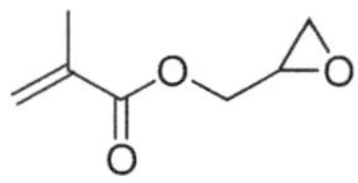
Vinyl monomer		
Styrene	St	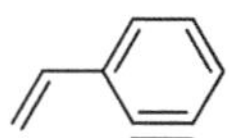
4-Vinylpyridine	4-VP	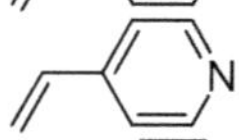
2-Vinylpyridine	2-VP	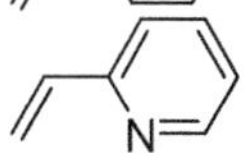
N-Vinylpyrrolidone	NVP	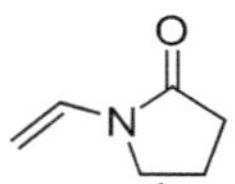
N-vinylimidazole		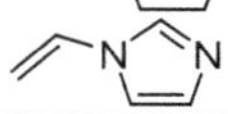

7.3.1 Cation-exchange membrane

Chen et al.[15] reported for the first time the study of radiation grafting of styrene (St) on a PE film to prepare a cation-exchange membrane. Since then, a variety of radiation-grafting processes and functional monomers have been applied to prepare a wide range of IEMs. In the cation-exchange membrane, the most commonly used ion-exchange groups are carboxylic acid groups and sulfonic acid groups, the former being weak acid, while the latter are strongly acidic. Sulfonates are currently ideal proton-conducting groups because of the value of $pK_a \leq 1$, which is stronger than the acidity of other cation-exchange groups such as carboxylic acid groups ($pK_a = 2-3$). There are two methods for the preparation of cationic exchange membranes based on carboxylic acid groups: direct grafting of allyl monomers such as AA or MAA, or cografting grafted epoxy acrylate monomers such as glycidyl acrylate and then converting the epoxy groups into carboxylic acid groups.

Strongly acidic cation-exchange membranes are typically prepared by grafting St and then sulfonating on a polymer, that is, PTFE film (see Fig. 7.6A). The use of this indirect grafting method instead of directly grafting functional monomers containing sulfonic acid group is because the sulfonic acid group is hydrophilic, and it forms a hydration layer to prevent monomer diffusion to the interior of the polymer matrix. For this reason, it inhibits the graft reaction, leading to a very low yield in the direct grafting of styrene.[14] The sulfonation process can be operated by chlorosulfonic acid, sulfonyl chloride, or concentrated sulfuric acid diluted in inert solvents such as dichloromethane, tetrachloroethane, or carbon tetrachloride. The selection of diluent depends on the swelling ratio of the polymer film. The degree of sulfonation varies with factors including the concentration of sulfonating reagent and reaction time, as well as the temperature of the sulfonation. Alternatively, sulfonic acid group-containing cation-exchange membrane can also be prepared by grafting glycidyl acrylate or GMA monomers. The grafted film is treated with sodium sulfite to introduce epoxy ring-opening into the sulfonic acid group, and eventually to yield to the

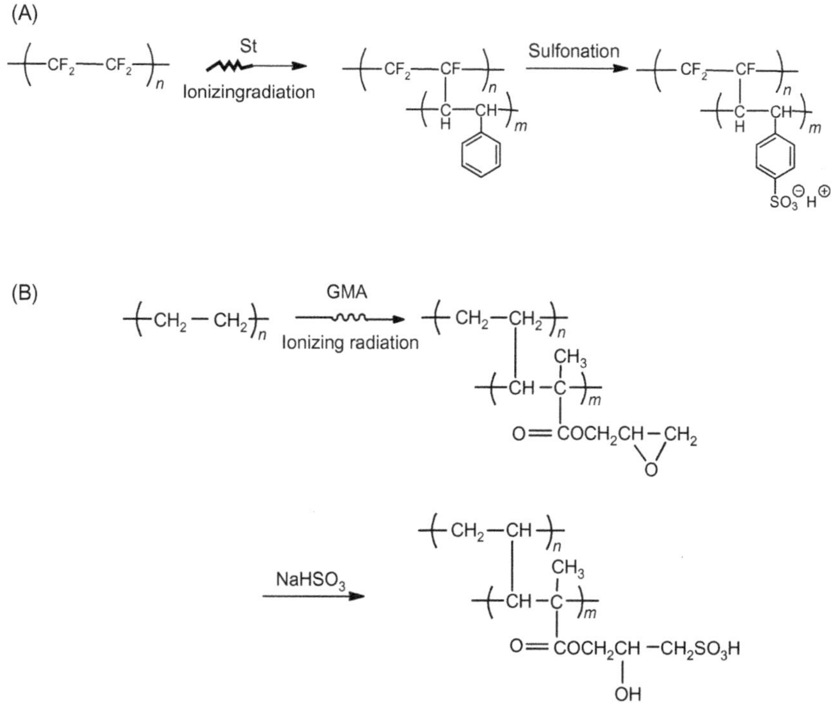

Figure 7.6 Schematic description of cation-exchange membranes with sulfonic acid group by radiation grafting: (A) introducing the cation-exchange group via sulfonation[14]; and (B) introducing the cation-exchange group via epoxy ring-opening.[16]

sulfonic acid group-containing cation-exchange membrane[16] (see Fig. 7.6B). Phosphate-type cation-exchange membrane can also be synthesized by treating the grafted film with phosphoric acid.[17]

Zhai et al.[18] recently developed a fluoropolymer film based on radiation cografting of the binary monomer for the preparation of cation-exchange membrane (Fig. 7.7). This work showed that St and MAn (maleic anhydride) have a synergistic effect in grafting degree by adding appropriate amounts of MAn (see Fig. 7.8). Based on the reactivity ratio of monomer copolymerization ($r_1 = 0.60$, $r_2 = 0.032$),[19] it is suggested that copolymerization tends to form an alternating copolymer. In addition, the hydrolysis of MAn introduces the two carboxyl groups as ion-exchange groups, resulting in the high ion-exchange capacity of the grafted film.

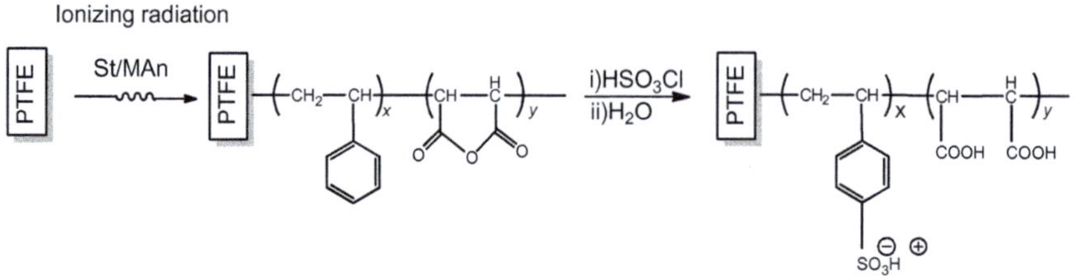

Figure 7.7 Typical synthesis process of cation-exchange membranes by radiation grafting.[18]

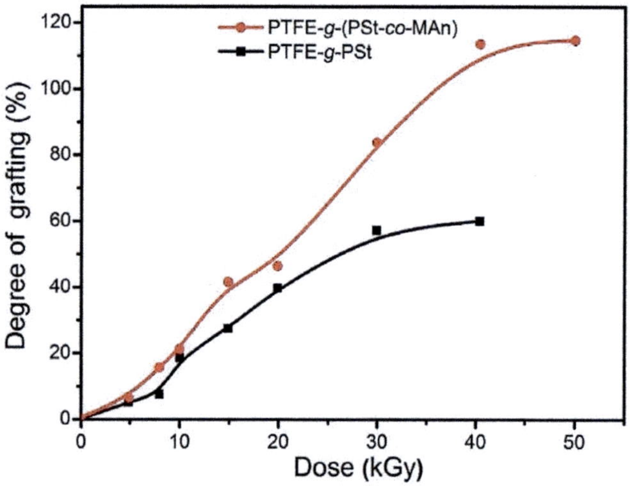

Figure 7.8 Absorbed dose effect on DG of St and St/MAn on PTFE membrane.[18] *PTFE*, polytetrafluoroethylene.

Conductivity test analysis showed that the conductivity of a binary cation-exchange membrane with 20% graft degree can reach a level similar to that of commercial Nafion117 membrane. This process demonstrates that a cationic exchange membrane with high conductivity can be prepared at low absorbed dose by introducing MAn.

The grafting chain is not homogeneously distributed on the IEM prepared by the conventional radiation-grafting technique. Holmberg et al.[20] used radiation grafting of St on a PVDF film via a combination of radiation grafting and NMP-mediated living free-radical polymerization, followed by sulfonation of poly (styrene) to sulfonic acid groups. The results showed this strategy allows the grafting reaction to occur through the polymer substrate at a low grafting degree (10%–20%). The graft chain has a uniform distribution in the IEM, displaying improved conductivity. Instead of film based grafting, Li et al.[13] investigated on radiation grafting radiation grafting of monomer on PVDF and PE powders, respectively. The grafted film was subsequently prepared by solution phase casting or melting-casting to obtain a cation-exchange membrane. The results showed that compared with membranes prepared by film-based grafting, polymer powder-based grafting followed by solution phase inversion also displayed homogeneous graft distribution and higher conductivity at the same level of DG.

IEM-based electrochemical devices such as fuel cells or redox flow batteries are usually operated in corrosive chemical environments, which require an IEM with good chemical stability and durability. It has been reported that the following methods are used to improve the chemical stability of the IEM: (1) the use of corrosion-resistant fluoropolymers as the substrate,[2] (2) the use of cross-linking agent on the membrane for graft chain,[21] (3) grafted fluorinated monomers such as α,β,β-trifluorostyrene and perfluoroalkyl vinyl ether.[22,23] This is because the bond energy of the C–F bond is much higher than that of the C–H bond, which inhibits the degradation of the graft chain. Zhai et al.[24] recently conducted a study on the fluoropolymer IEM in cooperation with the Japan Atomic Energy Research Institute. Perfluoroalkyl vinyl ether monomer was grafted onto fluoropolymer film and further prepared by ATRP. This cation-exchange membrane shows high conductivity and good chemical stability. The synthetic route is present in Fig. 7.9.[24] The results show that the conductivity of the cation-exchange membrane with DG of 15% can reach the level of

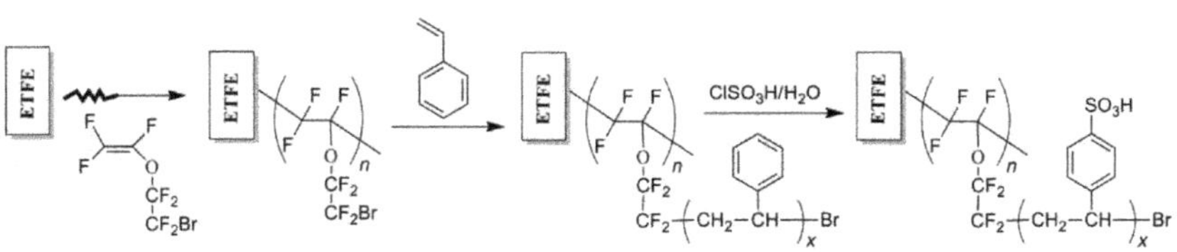

Figure 7.9 Preparation of fluorinated cation-exchange membrane by radiation grafting and atom transfer-free polymerization.[24]

DuPont Nafion117 membrane. This may be due to the fact that the polymer chains grafted with ATRP are more regular. The ion-exchange membrane also displayed a higher chemical stability due to the incorporation of the perfluoroalkyl vinyl ether as the graft chain compared with conventional radiation-grafted styrene (St) polymers.

7.3.2 Anion-exchange membrane

The radiation synthesis of AEMs generally involves the grafting of vinyl monomers such as 4-vinylpyridine (4-VP), 2-vinylpyridine (2-VP) or chloromethylstyrene (VBC). The ion-exchange groups of the AEMs may be strongly basic, such as quaternary ammonium salts, or weakly basic, such as primary, secondary, and tertiary amines. In PE, PTFE, ETFE, or PVDF and other polymeric substrates, grafting of 4-VP and quaternization of halogenated hydrocarbon to synthesis AEM have been previously reported.[25,26] Fig. 7.10A shows a typical synthetic route for the preparation of an AEM by grafting of 4-VP onto PTFE.[25] *N*-Vinylpyridine, 2-vinylpyridine, 2-methyl-5-vinylpyridine have also been used to prepare AEMs. In the work on radiation grafting of AEMs, vinyl pyridine is the most widely used monomer.

Another way to prepare AEMs is to graft chloromethylstyrene (VBC) and then react with trimethylamine.[27] Fig. 7.10B shows the

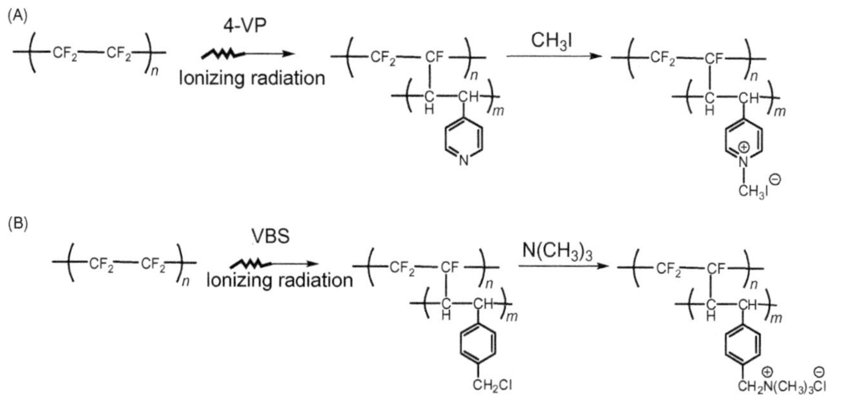

Figure 7.10 Radiation synthesis of quaternary ammonium salt anion-exchange membrane: (A) vinyl pyridine quaternization method[25] and (B) chloromethylstyrene (VBC) quaternization method.[27]

route of grafting VBC onto PTFE to prepare an AEM by quaternization. AEMs can also be prepared by irradiating the film onto the membrane, followed by chloromethylation and quaternization.[28] Similar to the preparation of the cation-exchange membrane, the chemical stability of the membrane can be effectively improved by the use of preferred film substrate, monomer or cross-linking agent in the process of radiation grafting.

Direct grafting of amino-containing monomers is a simple method for preparing AEMs. Based on the requirements of the vanadium flow battery, Qiu et al.[29] grafted N,N-dimethylaminoethyl methacrylate (DMAEMA) on a fluoropolymer matrix, and further synthesized the IEM by a simple protonation process. An alkaline AEM (Fig. 7.11) has been successfully synthesized.[29] RAFT living graft polymerization has also been successfully used to synthesize this type of AEM. Similar to the ATRP-mediated radiation-grafting method, the graft chain of the membrane synthesized by the RAFT method is homogenously distributed, forming a continuous ion channel for effective ion transport.

7.3.3 Amphoteric ion-exchange membrane

The amphoteric IEM, first proposed by Sollner[30] refers to an IEM with both cationic and anion-exchange groups. The bipolar membrane can be considered as a special case of an amphoteric IEM, usually referring to the membrane on both sides containing

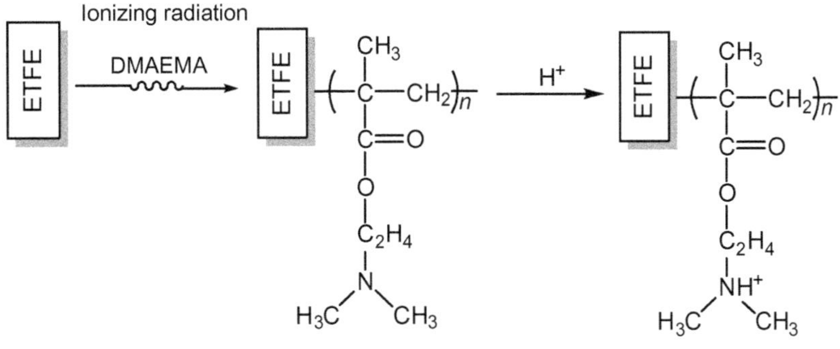

Figure 7.11 Typical synthesis process of anion-exchange membranes by radiation grafting.[29]

cationic and anionic groups of the film. The amphoteric IEM is able to exchange both anions and cations, and its performance in specific environments is mainly determined by the factors of pH, temperature, ion concentration and so on. These properties of membranes can be adjusted in accordance with the performance requirements, making amphoteric IEM an excellent candidate for a new generation of IEM. At present, the application of amphoteric IEM has attracted the attention of many researchers in sewage treatment and desalination of seawater.

From a molecular point of view, the architecture of amphoteric IEM mainly is correlated with the type and structure of the functional monomer introduced in the membrane substrate. Therefore, the monomer used in the grafting process determines the corresponding properties of the amphoteric IEM. The types of cationic and anionic groups selected by amphoteric IEMs are similar to those in the preparation of the cation-exchange membranes and AEMs described above, that is, the anion-exchange groups are predominantly carboxyl or sulfonic acid groups and the cation-exchange groups are predominantly amine or quaternary ammonium salt groups.[31,32] The methods for preparing amphoteric IEM are as follows: (1) the one-step method is simultaneously grafting two different monomers and then converting them into cationic and cationic groups; (2) the two-step grafting method is grafting two different monomers in a different order into graft chains and then converting them into cationic and anion-exchange groups, respectively; (3) grafting a monomer, such as styrene, and then using sulfonation followed by introducing an anion-exchange group-containing graft chain via chloromethylation and quaternization.

Zhai et al.[32] recently used the binary monomer radiation cografting method (Fig. 7.12) and two-step radiation grafting (Fig. 7.13). Amphoteric IEM was synthesized by radiation grafting of St and DMAEMA monomer on the fluoropolymer film, followed by sulfonation and protonation reaction. The results show that the sulfonation reaction of polystyrene (PS) graft causes a certain degree of damage to the grafted grafts of poly N,N-dimethylaminoethyl methacrylate (PDMAEMA) during the preparation of the amphoteric IEM by the binary monomer coradiation grafting method. The

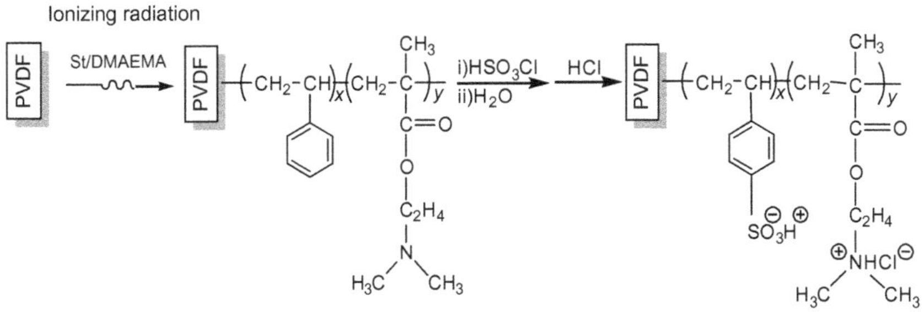

Figure 7.12 Typical synthesis process of amphoteric ion-exchange membrane by radiation grafting.[32]

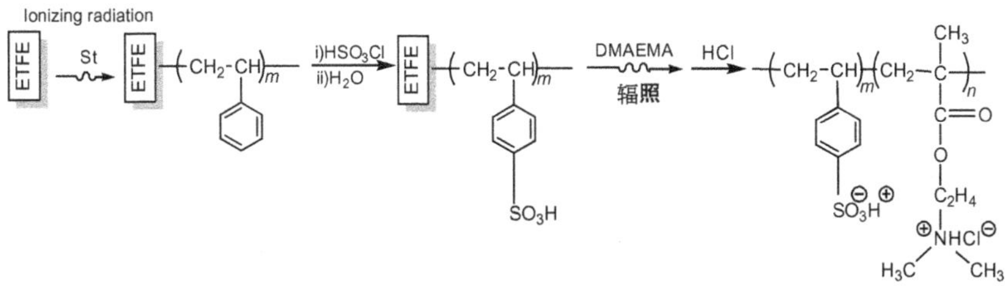

Figure 7.13 Two-step radiation synthesis of amphoteric ion-exchange membrane.[32]

undesirable effect of sulfonation on the grafting of PDMAEMA was clear and two-step grafting is proposed to further optimize the properties of the membranes. The first step in the two-step grafting process is to irradiate the grafted monomer St onto the ETFE film and the sulfonation of sulfonic acid groups. The second step is the irradiation of the grafted ETFE-*g*-PSSA cation-exchange membrane, and finally the protonation of PDMAEMA grafting leads to the introduction of tertiary amine groups.

7.4 APPLICATIONS IN RENEWABLE ENERGY TECHNOLOGY

Radiation-grafting synthesis of IEM has been developed more than 40 years. In the 1970s, Japan and other countries successfully realized the industrial production of radiation-grafted PE film to produce silver-zinc button batteries. In recent years, radiation-grafted membranes have been promising in the applications of vanadium redox batteries and fuel cells.

7.4.1 Radiation-grafted membrane for vanadium redox battery

The vanadium redox battery (VRB), developed by Skyllas-Kazacos et al.[33] in the 1980s, is a promising energy storage system that has emerged in response to the increasing global implementation of renewable energy technologies. It is a type of rechargeable redox flow battery that employs vanadium ions in four different oxidation states as the electrolytes to store the chemical energy. The aqueous electrolyte in the positive half-cells contains VO^{2+} and VO_2^+ ions, and the electrolyte in the negative half-cells are V^{3+} and V^{2+} ions. The two positively charged electrolytes are separated by a proton exchange membrane in the assembly of a power cell to complete the charging and discharging processes, as shown in Fig. 7.14. In past decades, VRB has attracted more and more attention from numerous companies and organizations due to its excellent features, including its flexible design. The IEM is one of the key components in VRBs and is used to prevent the cross-over of vanadium ions, and allows the transport of ions to complete the conducting

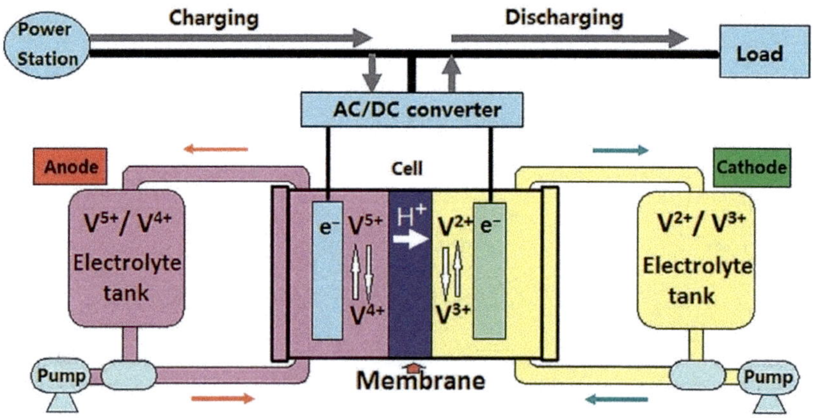

Figure 7.14 System of vanadium redox battery.

circuit. The ideal membrane has the following properties: low permeation rates of vanadium ions to minimize self-discharging, high conductivity to improve columbic efficiency, good chemical stability for long-term running, and cost-competitive. Unfortunately, the currently available commercial membranes cannot meet all of the above requirements. For instance, the Selemion CMV membrane (Asahi Glass Co., Japan) exhibits very poor chemical stability in VRB electrolyte solution due to the oxidation decomposition induced by V(V) ions. The perfluorinated Nafion (DuPont, United States) membranes with high conductivity and excellent chemical stability suffer from the cross-over of vanadium ions as well as high cost.[34] Consequently, studies of novel IEMs for VRB with higher electrochemical performance and lower costs are of great importance for the successful large-scale commercial development of VRBs.

In view of the challenges of vanadium batteries, in recent years, Qiu et al. developed a series of IEM materials for vanadium batteries using radiation-grafting technology. The amphoteric IEM described in Section 3.3 is the most promising,[32] which contains not only sulfonic acid cation-exchange groups to ensure higher proton conductivity but also protonated anion-exchange groups to inhibit the cross-over of vanadium ions on both sides of the membrane by Donnan exclusion. This membrane is produced by a two-step grafting method and has a high conductivity and a very low vanadium ion permeability coefficient; its permeability coefficient is only ~1/200 that of the Nafion117 film. The open circuit voltage can be maintained for 300 h, and its current efficiency and energy efficiency reach as high as 95.6% and 75.1%, respectively, which are superior to the Nafion117 membrane efficiency (87.9% and 72.6%). In addition, after 40 cycles of battery charging and discharge, it does not show any significant efficiency reduction (Fig. 7.15).

In order to meet the needs of industrial production, these amphoteric IEMs have been further optimized in terms of chemical stability and preparation routines.

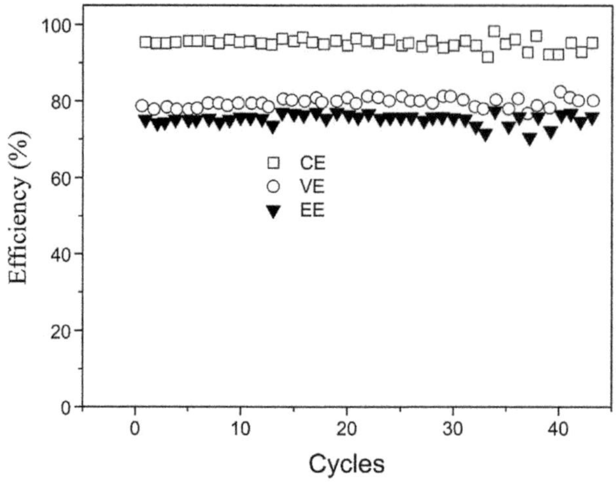

Figure 7.15 Cycle performance of VRB with amphoteric ion exchange membrane (AIEM).[32]

7.4.1.1 Optimization of chemical stability of membranes

The chemical stability of the membrane is an important property affecting the large-scale manufacture of vanadium batteries. In the process of the operation of the vanadium battery, the hydrophilic graft chain in the IEM is susceptible to oxidative degradation, due to the V(V) ions and the strong oxidizing properties of the sulfuric acid in the positive electrode electrolyte, resulting in a reduction of the performance of the battery. When AEMs grafted with DMAEMA are put in an oxidizing environment, for example in an aqueous solution 3% H_2O_2 (60°C) containing V(V) ions, compared to sulfonic acid group-containing cation-exchange membranes and some other commercial films, they show good chemical stability and durability.[35] Protonated PDMAEMA is an anion-exchange group that can prevent the migration of high valent vanadium ions from the membrane. In contrast, the chemical stability of grafted styrene sulfonic acid film is not sufficient. When monomers of alpha-methylstyrene and DMAEMA are mixed in the appropriate proportions, the two monomers can be successfully grafted onto the PVDF membrane in the presence of catalyst $AlCl_3$.[35] As shown in Fig. 7.16, the addition of α-methylstyrene in place of styrene and DMAEMA improves the chemical stability of the amphoteric

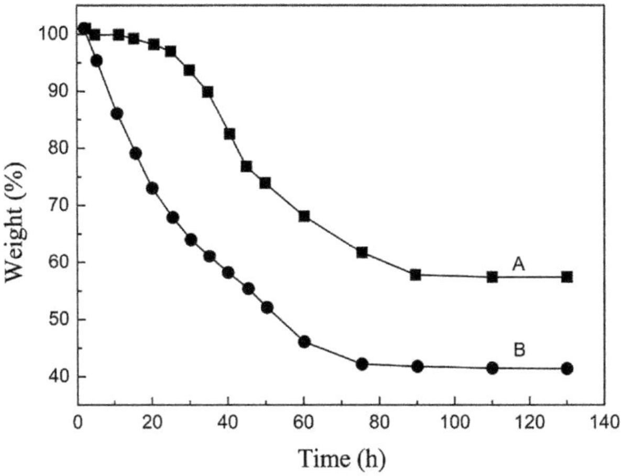

Figure 7.16 (A) Chemical stability test of styrene/DMAEMA amphoteric ion-exchange membranes evaluated in a 6% H_2O_2 aqueous solution at 60°C; (B) chemical stability test of the grafted α-methylstyrene/DMAEMA amphoteric ion-exchange membrane.[35]

IEM. For example, AMS/DMAEMA amphoteric IEMs with a graft yield of 40% contain 0.72 mol g^{-1} of cation-exchange capacity after 100 h of accelerated oxidation in hydrogen peroxide solution, whereas the ion-exchange groups of the traditional amphoteric membrane completely degrade.[35]

Perfluorosulfonic acid IEMs, that is, DuPont Nafion membrane, are a highly chemically stable cation-exchange membrane. However, vanadium ions on both sides of the membrane easily undergo cross-over, resulting in serious reduction of current efficiency. In order to avoid the osmotic effect, the perfluorosulfonic acid film is usually blended with weak interactions (intermolecular force, hydrogen bonds) with other functional or nonfunctional polymers and inorganic particles of different sizes (SiO_2, TiO_2). Alternatively, another strategy is proposed as shown in Fig. 7.17: the use of radiation technology. The oxidative-resistant DMAEMA monomer is directly grafted on the perfluorosulfonic acid proton exchange membrane, leading to a type of amphoteric IEM incorporated with a positive layer. This designed structure not only significantly improves chemical stability, but also inhibits vanadium

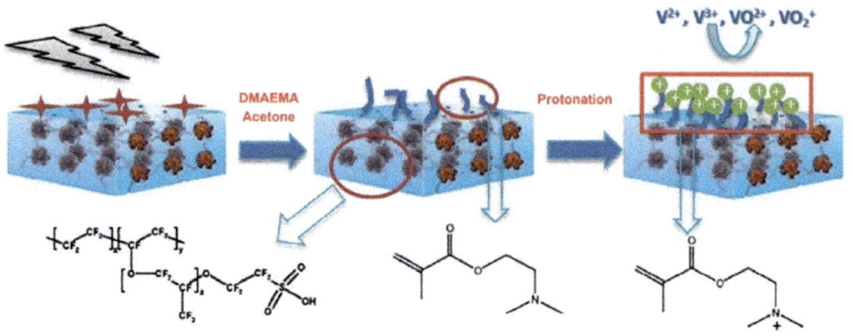

Figure 7.17 Nafion-*g*-poly(DMAEMA) amphoteric ion-exchange membrane.[36]

ion penetration[36] (see Fig. 7.17). The results also show that the vanadium permeability coefficient of the new IEM is 1.55 S min cm^{-3} at a grafting rate of 13.5%, which is one fifth that of the Nafion membrane and hence the efficiency of the inhibition of vanadium cross-over is improved.[36]

7.4.1.2 Optimization of membrane synthesis process

IEMs synthesized by the conventional radiation-grafting process suffer from the inhomogeneous distribution of graft chains; in other words, the graft chains are mainly enriched on the membrane surface. This heterogeneity leads to the lack of ion-exchange groups in the internal structure of the IEM at low DG ($<$20%), so that hydrophilic proton transfer channels are not formed. On the other hand, the mechanical strength of the IEM at a high DG is remarkably low, even leading to difficulty of assembling the membranes for battery performance testing. Therefore, under the condition of low DG, the problem of the heterogeneity distribution of the internal structure of the graft chain is a shortcoming that limits the industrial application of radiation-grafted film.

In order to solve the above-mentioned problems, Ma et al.[37] designed a new process in which PVDF powder instead of film is used as the substrate. The grafted powders are cast into different sizes of membrane materials, and finally amphoteric IEMs are obtained by sulfonation and protonation. The preparation procedure is shown in Fig. 7.18.

SEM-EDX analysis showed that the grafted membrane (A) grafted and ion-exchange groups were not enriched on the surface compared with conventional film-based grafting, forming a uniform ion channel (see Fig. 7.19). Conductivity testing of the film further indicates that this uniform structure facilitates the preservation of water molecules inside the membrane, facilitating proton transfer.

In the overall process of ion-exchange membrane production, the sulfonation process is a key step in the introduction of the sulfonic acid group, but it is also a severe oxidative reaction and causes environmental concerns. Although it has been reported that monomeric sodium styrene sulfonate (SSS) can be grafted directly onto the PVDF membrane, it requires an EB as a radiation source and the presence of dimethylformamide (DMF) as well as a high concentration of sulfuric acid solution. However, when DMAEMA is mixed with SSS at the same concentration, the graft reaction can be successfully achieved with higher DG, especially in the presence of RAFT chain transfer reagents, to meet the needs of vanadium batteries for IEMs.[38] Fig. 7.20 is a schematic representation of the

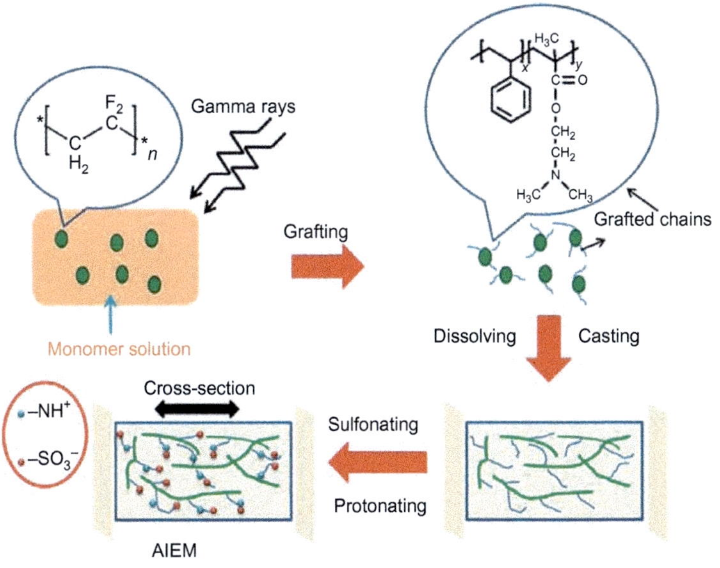

Figure 7.18 A novel process to prepare amphoteric ion exchange membranes (AIEMs) using a combination of radiation-grafted techniques and solution casting methods.[37]

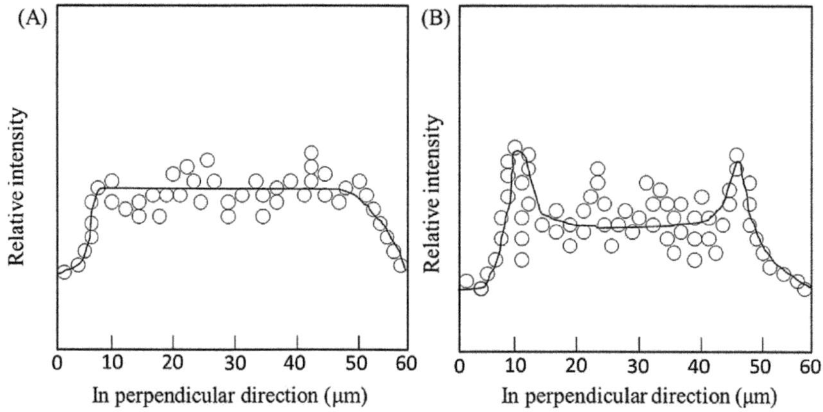

Figure 7.19 Scanning electron microscopy with energy dispersive X-ray spectroscopy (SEM-EDX) cross-section image of AIEM-based PVDF powder grafting with GY of 6.7% (A) and AIEM-based PVDF film grafting with GY of 11.4% (B), respectively.[37] *PVDF*, polyvinylidene fluoride.

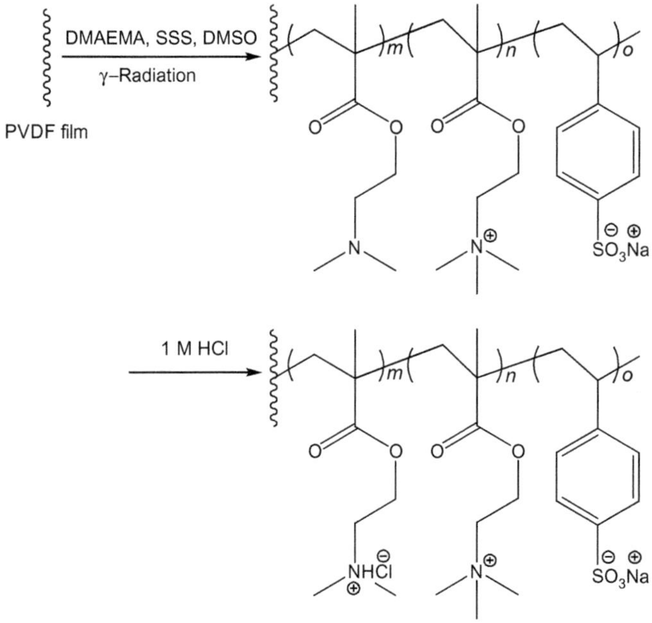

Figure 7.20 Schematic description of PVDF film simultaneous irradiation induced grafting DMAEMA/SSS binary monomer to synthesis amphoteric ion-exchange membrane.[38] *PVDF*, polyvinylidene fluoride; *SSS*, sodium styrene sulfonate.

synthetic route of the method. Compared with Nafion117 membrane, the amphoteric IEM with the DG of 43% has a lower vanadium ion permeability coefficient and similar conductivity. The open circuit voltage of VRB assembled with the new type of amphoteric IEM can be maintained above 1.4 V for 85 h, much higher than that of Nafion 117.[38]

Unlike the proton exchange membrane in a fuel cell, the IEM in a vanadium flow cell needs to be a barrier to prevent the transportation of each vanadium electrolyte. The amphoteric IEM with fluoropolymer as substrate has good chemical stability and proton transfer and hinders vanadium ion permeation. In the single cell test, the electrochemical properties of the membrane exhibited by the film assembly are superior to those of the commercial perfluorosulfonic acid film, and therefore has good application prospects. However, current research and development of radiation-grafted membrane still remain in the laboratory stage. Many challenges exist, such as efficiency, scale-up process, and cost reduction to meet the needs of vanadium flow batteries.

7.4.2 Radiation-grafted membrane for fuel cell

A fuel cell is an environmentally friendly power generation device that converts energy and electricity in a fuel and oxidant directly into electricity.[39] The common fuels are hydrogen and methanol. Fuel cells can be divided into proton exchange membrane fuel cells (PEMFCs), solid oxide fuel cells, molten carbonate fuel cells, phosphoric acid fuel cells, and alkaline fuel cells (AFCs). They are based on a similar working framework, consisting mainly of three adjacent sections: the anode, the electrolyte, and the cathode. Two chemical reactions occur between the interfaces of the three different sections. The final result of the reaction is the consumption of fuel (hydrogen, methanol) and oxidant, producing water or CO_2.

The PEMFC using the proton exchange membrane as the electrolyte contains hydrogen as the fuel, and the direct methanol fuel cell (DMFC) uses methanol as the fuel (as shown in Fig. 7.21). The PEMFC is typically operated at temperatures between 60 and 80°C

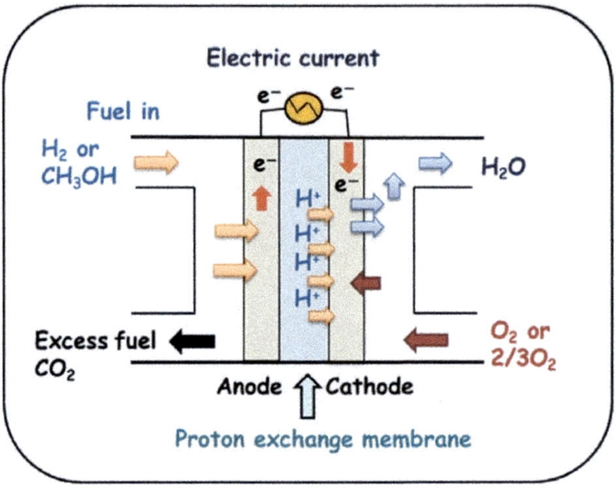

Figure 7.21 Structure and working principle of proton exchange membrane fuel cell.[39]

under 1–3 atmospheres. Considering that the evaluated temperature of the battery not only accelerates the electrochemical kinetics of the electrode reaction, and simplifies the water and heat management system, but it also improves the efficiency of the catalyst to prevent side reactions from impurities such as CO or sulfur. The PEMFC has gradually been developed into a high-temperature proton fuel cell; that is to say, the battery runs at temperatures greater than 100°C. Unlike the film used in the liquid cell, the proton exchange membrane in the fuel cell not only separates the reactants (hydrogen or methanol) from the products (water or CO_2), but also acts as a battery electrolyte in the form of a swollen polymer for proton or hydroxide ion transport. Therefore, the requirements of the fuel cell-based IEM are divisive and high. It should effectively hinder the penetration of fuel or O_2, with high ion conductivity and chemical stability, and have a good affinity with electrode material. In addition, at different temperatures, it also requires a suitable degree of swelling and sufficient mechanical properties to facilitate the assembly of the battery, and the cost should be as low as possible to permit commercial development. At present, Nafion series membrane proton exchange membrane-based fuel cells are commonly used. The perfluorosulfonic acid

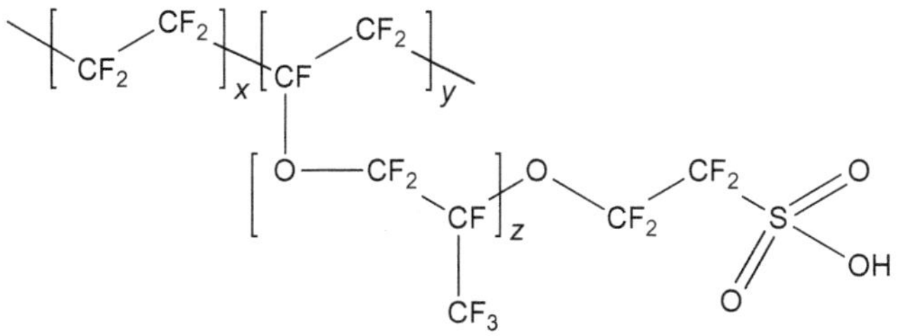

Figure 7.22 Chemical structure of perfluorosulfonic acid membrane.[39]

membrane contains the main chain of Teflon, and electron-rich fluorine atoms are tightly wrapped around the carbon–carbon backbone to protect the carbon skeleton from the oxidation of free-radical intermediates generated by the electrochemical reactions (its chemical structure is shown in Fig. 7.22). The Nafion membrane has good ion conductivity and electrochemical stability, and the running time of the fuel cell is up to 60,000 h at 80°C.[39] However, it is not cost-effective at \$800–\$1000 per square meter. In addition, its water retention rate decreases especially in high-temperature (>100°C) running conditions, so the proton conductivity reduces. Furthermore, the cross-over of methanol in DMFC also becomes serious. These shortcomings strongly hinder the commercialization of fuel cells. Therefore, the development of a new low-cost, better performing IEM is the key to the large-scale manufacture of these batteries.

7.4.3 Application of radiation-grafted film in proton exchange membrane fuel cells

The application of radiation-grafted membranes to fuel cells began in the 1990s. In recent years, a large number of researchers have studied the use of radiation technology to meet the needs of the IEM for PEMFC application. By varying the synthesis conditions (absorbed dose, dose rate, and radiation source), the type of substrate film, and the type of monomer and so on, the correlation between the chemical structure and the properties of the membrane has been comprehensively studied and many important results have been obtained.[40]

The radiation-grafted membrane of PE-*g*-PSSA was initially developed for fuel cells and based on PE as the base film and styrene (St) as the monomer. The further sulfonation reaction introduces a sulfonic acid cation-exchange group. The ion-exchange capacity of this membrane can reach 0.9 mmol g^{-1} when the DG is more than 18% and its conductivity is 75 mS cm^{-1}, which is better than that of Nafion membrane.[41–42] However, the performance of the film with low-temperature (60–80°C) oxyhydrogen or methanol fuel cells is not ideal. Although the maximum output power can reach 45 mW cm^{-2}, which is higher than that of the assembled Nafion 117 membrane (33 mW cm^{-2}); the duration of the battery with this membrane in working condition is not satisfactory (less than 100 h), due to the chemical stability of the membrane.[43] Researchers soon found that the use of fluorine-containing substrate film improves the chemical stability of the IEM to a significant extent. Fluorinated polymers, such as PVDF, FEP, ETFE, PTFE, and PFA, and radiation cross-linked PTFE have been applied in the radiation synthesis of IEMs.[2,12,43] Perfluoro FEP and radiation cross-linked PTFE membranes show excellent chemical stability.[43] At the same time, the appropriate amount of cross-linking agent (mass fraction 2%–10%) is widely added in the radiation-grafting process.[21–22,44] This cross-linked polymer film not only protects the membranes from fast degradation in the cell, but also prevents the diffusion of fuel. The most commonly used cross-linking agents are divinylbenzene, bovinylphenylethane, TAC (triallylcyanuric trioxide), and methylene bisacrylamide, and their chemical structures are shown in Fig. 7.23.

The above-mentioned IEMs are generally considered to be the first generation of radiation-grafted film materials for fuel cells, which use a fluorine-containing polymer chain as the main chain and polystyrene sulfonic acid (PSSA) as the graft chain. The macromolecular chains are linked to each other by cross-linking, and the chemical structures of the cross-linkers are presented in Fig. 7.24. In terms of durability, the fluorine-containing graft film has been significantly improved. For example, the current density of the PVDF-*g*-PSSA film with a graft ratio of 50% in the methanol fuel

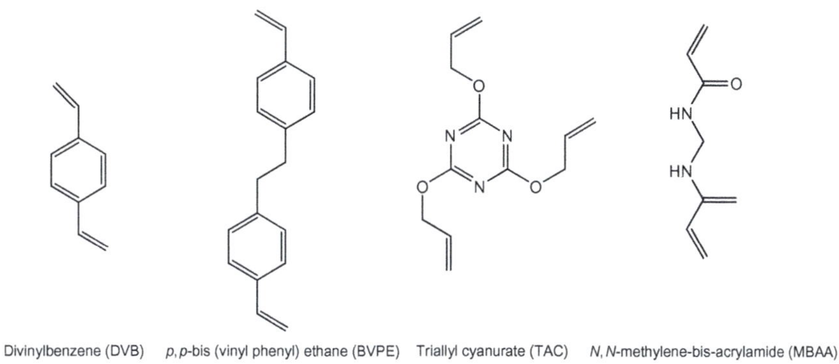

Figure 7.23 Chemical structure of cross-linker used in radiation-grafted membrane.

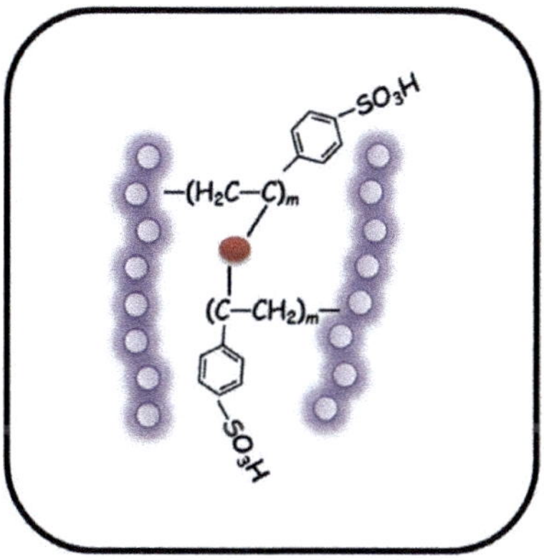

Figure 7.24 Chemical structure of radiation-grafted membrane for fuel cells.

cell is 98 mA cm^{-2} and the working life is extended to 204 h[45]; the optimized ETFE-*g*-PSSA and PFA-*g*-PSSA-like grafted films can continue to operate for more than 2000 h in low-temperature hydrogen–oxygen fuel cells or methanol fuel cells without a significant reduction in output power[46]; the best membrane with highest durability is FEP-*g*-PSSA film, which is capable of working in a 60°C hydrogen–oxygen fuel cell for more than 10,000 h after cross-linking.[43]

For the other properties such as ion-exchange capacity, proton conductivity, resistance to methanol crossover and the initial current density of the battery output power, etc., the optimized radiation synthesis of the graft film is close to or exceeds that of commercial perfluorosulfonic acid film, such as Nafion series membrane. After years of study, researchers in this field have been able to select the most appropriate sulfonation reaction conditions by changing the radiation conditions (radiation source, absorbed dose, and dose rate), grafting method, solvent selection, cross-linking agent content, and other factors to achieve the goal of optimizing the electrochemical properties of radiation-grafted membrane. However, compared with the long-term operational lifetime of Nafion membrane (60,000 h working time), the first generation of radiation-grafted film in the fuel cell operating environment suffer from rapid chemical degradation, resulting in the short running life of the fuel cell. To better understand the mechanism of the oxidative degradation of proton exchange membranes, the chain scission reactions of membranes with $OH^{\bullet}$ or $HO_2^{\bullet}$ produced by the incomplete redox reaction of H_2/O_2 on Pt-catalyzed electrode materials in fuel cells were investigated.[47] Especially $OH^{\bullet}$ free radicals, under acidic conditions, have the oxidation potential of up to 2.64 versus NHE, which is thought to be the main species attacking the polymeric membranes. In the radiation-grafted film with typical PSSA as the graft chain, the alpha hydrogen position in the styrene is most susceptible to $OH^{\bullet}$ attack. A model study shows ~22% $OH^{\bullet}$ will directly react with the styrene via hydrogen elimination or via an addition-elimination process, as shown in Fig. 7.25. As a result, PS-free radicals are formed and eventually lead to the detachment of ion-exchange groups from the polymer backbones.[47]

In contrast, owing to the protection of fluorine atoms, the sulfonic acid groups in commercially available perfluorosulfonic acid films strongly resist $HO^{\bullet}$ oxidation compared with those in radiation-grafted membranes. The reaction rate constant ($4 \times 10^8\ M^{-1}\ s^{-1}$) of $OH^{\bullet}$ with PSSA is higher than that with perfluorinated Nafion ($<10^6\ M^{-1}\ s^{-1}$) by two orders of magnitude.[47] This indicates that the PS grafts bearing immobilized sulfonic acid groups in the radiation-grafting film degrade easily, constituting the main reason for poor membrane durability. Therefore, continuous

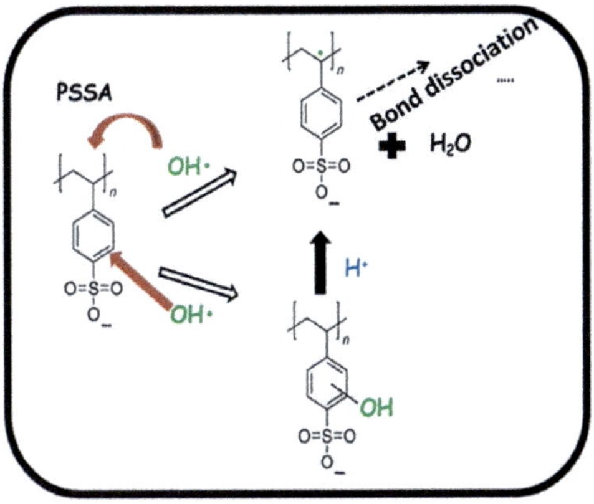

Figure 7.25 Degradation reaction of OH radicals attacking poly(styrene) chains.[47]

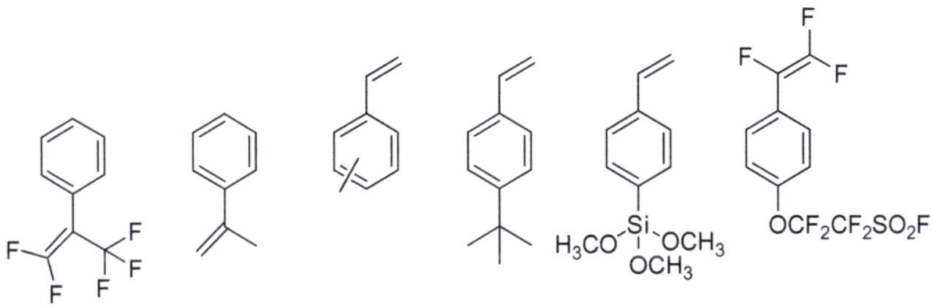

Figure 7.26 Styrene-derived monomer used in radiation-synthesized ion-exchange membrane (from left to right: α,β,β-trifluorostyrene, α-methylstyrene, *m*, *p*-methylstyrene, *tert*-butylstyrene, styryltrimethoxysilane, tetrafluorodimethyl ether sulfonyl fluoride trifluoro styrene).

efforts have been made in radiation grafting of substitute styrene monomers such as α-methylstyrene, α,β,β-trifluorostyrene, etc., or other fluorine-containing monomers to improve the chemical stability of membranes. The chemical structure of substituted styrene monomers is listed in Fig. 7.26.

Compared with styrene, α-hydrogen-substituted derivatives of styrene monomers, such as α,β,β-trifluorostyrene, and α-methylstyrene, have better chemical stability, but these monomers are not commonly available and the price is generally high. On the other hand, due to the steric hindrance effect, the kinetics of the radiation-grafting reaction of these individual monomers are very slow and the DG achieved is very low (<10%). In addition, as described above, perfluoroalkyl vinyl ether monomer may be grafted onto fluorine-containing film, followed by secondary grafting of the styrene monomer. The advantage of this grafting process is that the easily degradable PS branches are protected by perfluorinated branches, thereby increasing the chemical stability of the membrane. The rate of degradation of the grafted films using the new monomers in oxidation condition (5%–30% H_2O_2/Fe^{II}, 60–80°C) is significantly slower than that of the grafted films using styrene monomer.[47] However, to date studies on the performance and durability of such films assembled in hydrogen or methanol fuel cells are lacking.

In addition to selecting the appropriate substrate film or monomer, there are many other ways to improve the stability of the radiation-grafting film; for example, (1) adding different inorganic particles such as SiO_2 and TiO_2 to the film[48]; (2) the use of living radiation grafting or powder-based grafting, as described above. These two new grafting methods optimize the distribution of the graft chains in the membrane, so that the grafted polymers are resistant to the oxidative radicals attacking the interface as they are surrounded and well-protected by high fluoride content backbones.[2,3,12] An alternative method is the direct sulfonation of fluorine-containing base films such as PVF, PTFE, and ETFE. It is apparent that this method does not necessarily require a radiation-grafting process and it preserves the chemical stability of the fluorine-containing film to the greatest extent. However, the efficiency of the method is not sufficiently high, and proton conductivity performance needs to be further improved.[49]

7.4.4 Application of radiation-grafted membranes in alkaline fuel cells

The alkaline AEM fuel cell (ADMFC), a new generation of fuel cells developed in recent years, has many advantages, one of which is that fuel (methanol or hydrogen) in the alkaline environment has higher oxidation rate. The concept and working principle of

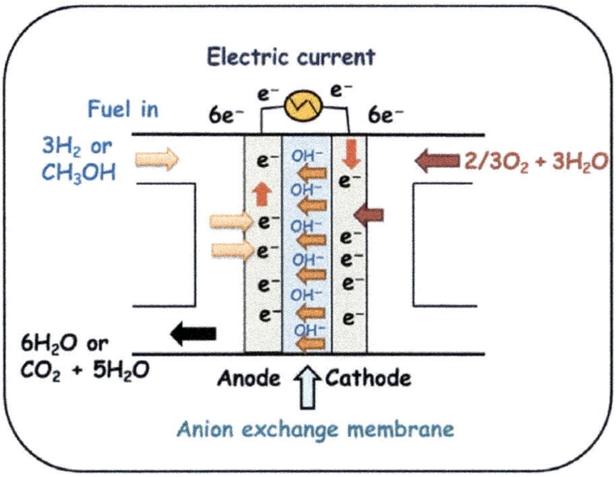

Figure 7.27 Structure and working principle of alkaline anion-exchange membrane fuel cell.

ADMFC is shown in Fig. 7.27. Unlike the membranes in PEMFC, ADMFC uses an AEM and potassium hydroxide solution as the electrolyte. The membrane is still one of the key components in ADMFC, playing the role of conducting OH-anions and separating the fuel and oxidants. The membrane's performance directly affects ADMFC battery performance and running duration.

The rapid development of ADMFC has led the way to the application of radiation synthesis of AEMs. Using chloromethylstyrene (VBC) as the grafting monomer, the AEM is prepared after amination. Its synthetic route is environmentally friendly and it is suitable for AFCs. For example, the anion-exchange capacity of the synthesized FEP-based AEM is 0.9–1.0 mmol L^{-1} and the ion conductivity is between 10 and 20 mS cm^{-1}.[50–52] After assembling the cell, the voltage of the electrode 500 mV and the battery energy density at 50°C can reach 55 mW cm^{-2}.[52] The electrochemical properties of the AEM based on ETFE are excellent. With the same ion-exchange capacity (1.03 mmol L^{-1}), the ion conductivity at 50°C is 34 mS cm^{-1} and the assembled battery has an energy density of 92 mW cm^{-2}.[53] In addition, the study found that the performance of AFCs is in relation to the thickness of the film and the compatibility of the membrane and electrode. For the same membrane material, the battery energy density (48 mW cm^{-2}) with Ag/carbon electrode is lower than that of Pt/carbon electrode (60 mW cm^{-2}).[53–55] Surprisingly, in all the VBC-based anion films, the LDPE AEM shows the most excellent energy density (823 mW cm^{-2}) with 60°C working environmental temperature.[56] However, the chemical stability of these AEMs under strong alkaline conditions is not particularly desirable, principally because OH-anions are susceptible to nucleophilic substitution reactions or Hoffman elimination reactions with amine groups.[57] Another type of application is to use *N*-vinyl imidazole as the monomer for the AEM. When this monomer is copolymerized with styrene monomer on the polymer substrate, due to the protection of the aromatic rings on the styrene, the imidazole anion-exchange groups are not susceptible to attack and therefore exhibit good electrochemical properties.[58]

The selection of chemically stable monomers, especially fluorine-containing monomers, in the process of radiation grafting will remain a trend in the future to improve the durability of fuel cell membranes. In-depth mechanistic study of the degradation of membrane in the fuel cell, especially in in situ electrochemical environments, is very helpful for designing advanced membranes and achieving longer durability. In addition, in practical applications, it is also very important to take the following factors into consideration: the interface between the grafted film and various electrode catalyst materials, the affinity of the membrane in different fuel cells or different working conditions (temperature, pressure, and humidity, etc.), electrochemical changes in mechanical properties, and so on.

After decades of continuous effort, radiation-grafting technology has now opened up a new way for the synthesis of IEMs. It has several unique advantages:

1. The flexibility of the preparation process. Based on the demands of the application, different polymer substrates and monomers are available for a variety of IEMs; the nature of the membrane can be simply changed or optimized by changing the grafting method or irradiation conditions.

2. The process is simple and well-established. Radiation-grafting technology has been developed over more than 50 years. Radiation synthesis of IEMs generally includes the following three steps: the irradiation of polymer, the introduction of the monomer solution, and the introduction of ionic groups followed by simple chemical treatment. The thickness of the polymer film material is generally between 50 and 200 μm, and an electron accelerator with energy of ~1 MeV can meet the penetration depth requirement of membrane. These three steps of radiation synthesis can be carried out separately and it is not difficult to manufacture IEMs.
3. It is efficient and environmentally friendly. Irradiation at an absorbed dose of 10–100 kGy can produce sufficient free radicals in polymer to fulfill the needs of radiation synthesis of IEM without addition of any chemical initiators.

REFERENCES

1. Strathmann H. Ion exchange membrane separation processes. In: Oyama ST, Stagg-Williams S, editors. *Membrane science and technology series*. San Diego, CA: Elsevier; 2004. p. 90–206.
2. Nasef MM, Hegazy ESA. Preparation and applications of ion exchange membranes by radiation-induced graft copolymerization of polar monomers onto non-polar films. *Prog Polym Sci* 2004;**29**(6):499–561.
3. Miwa Y, Yamamoto K, Sakaguchi M, Shimada S. Well-defined polystyrene grafted to polypropylene backbone by "living" radical polymerization with TEMPO. *Macromolecules* 2001;**34**(7):2089–94.
4. Barsbay M, Güven O, Davis TP, Barner-Kowollik C, Barner L. RAFT-mediated polymerization and grafting of sodium 4-styrenesulfonate from cellulose initiated via γ-radiation. *Polymer (Guildf)* 2009;**50**(4):973–82.
5. Barner L, Zwaneveld N, Perera S, Pham Y, Davis TP. Reversible addition–fragmentation chain-transfer graft polymerization of styrene: solid phases for organic and peptide synthesis. *J Polym Sci Polym Chem* 2002;**40**(23):4180–92.
6. Rosenberg Y, Siegmann A, Narkis M, Shkolnik S. Low-dose γ-irradiation of some fluoropolymers: effect of polymer chemical-structure. *J Appl Polym Sci* 1992;**45**(5):783–95.
7. Nasef MM, Saidi H, Dessouki AM, El-Nesr EM. Radiation-induced grafting of styrene onto poly(tetrafluoroethylene) (PTFE) films. I. Effect of grafting conditions and properties of the grafted films. *Polym Int* 2000;**49**(4):399–406.
8. Sun JZ, Zhang YF, Zhong XG. Radiation crosslinking of polytetrafluoroethylene. *Polymer (Guildf)* 1994;**35**(13):2881–3.
9. Misra B, Kumar D, Kaur I. Preparation and characterization of Teflon-FEP films grafted with acrylic acid by radiochemical method. *J Appl Polym Sci* 1995;**56**(9):1133–9.

10. Nasef M, Saidi H. Preparation of crosslinked cation exchange membrane by radiation grafting of styrene/divinylbenzene mixtures onto PFA films. *J Membr Sci* 2003;**216**(1–2):27–38.
11. Dargaville T, George G, Hill D, Wihittaker A. High energy radiation grafting of fluoropolymers. *Prog Polym Sci* 2003;**28**(9):1355–76.
12. Hietala S, Holmberg S, Karjalainen M, Nasman J, Paronen M, Serimaa R, et al. Structural investigation of radiation grafted and sulfonated poly(vinylidene fluoride), PVDF, membranes. *J Mater Chem* 1997;**7**(5):721–6.
13. Li LF, Deng B, Ji YL, Yu Y, Xie LD, Li JY, et al. A novel approach to prepare proton exchange membranes from fluoropolymer powder by pre-irradiation induced graft polymerization. *J Membr Sci* 2010;**346**(1):113–20.
14. Hasegawa S, Satou K, Narita T, Suzuki Y, Takahashi S, Morishita N, et al. Radiation-induced graft polymerization of styrene into a poly(ether ether ketone) film for preparation of polymer electrolyte membranes. *J Membr Sci* 2009;**345**(1–2):74–80.
15. Chen WW, Mesrobian R, Glines A. Graft copolymers derived by ionizing radiation. *J Polym Sci* 1957;**23**(23):903–13.
Shkolink S, Behar D. Radiation-induced grafting of sulfonates on polyethylene. *J Appl Polym Sci* 1982;**27**(6):2189–96.
16. Kim M, Saito K. Radiation-induced graft polymerization and sulfonation of glycidyl methacrylate onto porous hollow fiber membranes with different pore sizes. *Radiat Phys Chem* 2000;**57**(2):167–72.
17. Choi SH, Han JY, Jeong RJ, Lee KP. Desalination by electrodialysis with the ion-exchange membrane prepared by radiation-induced graft polymerization. *Radiat Phys Chem* 2001;**60**(4–5):503–11.
18. Qiu JY, Zhao L, Zhai ML, Ni J, Zhou H, Peng J, et al. Pre-irradiation grafting of styrene and maleic anhydride onto PVDF membrane and subsequent sulfonation for application in vanadium redox batteries. *J Power Sources* 2008;**177**(2):617–23.
19. Świtała-Żeliazkow M. Radical copolymerization of maleic acid with styrene. *Eur Polym J* 1999;**35**(1):83–8.
20. Holmberg S, Holmlund P, Nicolas R, Wilen CE, Kallio T, Sundholm G, et al. Versatile synthetic route to tailor-made proton exchange membranes for fuel cell applications by combination of radiation chemistry of polymers with nitroxide-mediated living free radical graft polymerization. *Macromolecules* 2014;**37**(26):9909–15.
21. Chen JH, Asano M, Yamaki T, Yoshida M. Chemical and radiation crosslinked polymer electrolyte membranes prepared from radiation-grafted ETFE films for DMFC applications. *J Power Sources* 2006;**158**(1):69–77.
22. Chen JH, Asano M, Yamaki T, Yoshida M. Preparation and characterization of chemically stable polymer electrolyte membranes by radiation-induced graft copolymerization of four monomers into ETFE films. *J Membr Sci* 2006;**269**(1–2):194–204.
23. Assink R, Amold JC, Hlooandsworth R. Preparation of oxidatively stable cation-exchange membranes by the elimination of tertiary hydrogens. *J Membr Sci* 1991;**56**(2):143–51.

24. Zhai ML, Chen JH, Hasegawa S, Maekawa Y. Synthesis of fluorinated polymer electrolyte membranes by radiation grafting and atom transfer radical polymerization techniques. *Polymer (Guildf)* 2009;**50**(5):1159–65.
25. Kostov G, Tumanova S. Radiation-initiated graft copolymerization of 4-vinylpyridine onto polyethylene and polytetrafluoroethylene films and anion-exchange membranes therefrom. *J Appl Polym Sci* 1997;**64**(8):1469–75.
26. Elmidaoui A, Belcadi S, Houdus Y, Cohen T, Gavach C. Perfluorinated anion-exchange membranes: preparation and preliminary tests of dialysis. *J Polym Sci Polym Chem* 1992;**30**(7):1407–12.
27. Herman H, Slade R, Varcoe J. The radiation-grafting of vinylbenzyl chloride onto poly(hexafluoropropylene-*co*-tetrafluoroethylene) films with subsequent conversion to alkaline anion-exchange membranes: optimization of the experimental conditions and characterization. *J Membr Sci* 2003;**218**(1):147–63.
28. Chakravorty B, Mukherjee R, Basu S. Synthesis of ion-exchange membranes by radiation grafting. *J Membr Sci* 1989;**41**:155–61.
29. Qiu JY, Li MY, Ni JF, Zhai M, Peng J, Xu L, et al. Preparation of ETFE-based anion exchange membrane to reduce permeability of vanadium ions in vanadium redox battery. *J Membr Sci* 2007;**297**(1–2):174–80.
30. Sollner K. On mosaic membranes. *Biochem Z* 1932;**244**:370–81.
31. Hegazy EA, El-Rehim H, Shawky H. Investigations and characterization of radiation grafted copolymers for possible practical use in waste water treatment. *Radiat Phys Chem* 2000;**57**(1):85–95.
32. Zhai ML, Qiu JY, Peng J, Xu L. Amphoteric ion exchange membranes, US, US20110318644A1[P]2011.
33. Sum E, Rychcik M, Skyllas-Kazacos M. Investigation of the V(II)/V(III) system for use in the positive half-cell of a redox battery. *J Power Sources* 1985;**16**(2):85–95.
34. Mohammadi T, Skyllas-Kazacos M. Evaluation of the chemical stability of some membranes in vanadium solution. *J Appl Electrochem* 1997;**27**(2):153–60.
35. Hu GW, Wang Y, Ma J, Peng J, Li JQ, Zhai ML. A novel amphoteric ion exchange membrane synthesized by radiation-induced grafting α-methylstyrene and *N,N*-dimethylaminoethyl methacrylate for vanadium redox flow battery application. *J Membr Sci* 2012;**407–408**:184–92.
36. Ma J, Wang SJ, Peng J, Yuan J, Yu CH, Ju XC, et al. Covalently incorporating a cationic charged layer onto Nafion membrane by radiation-induced graft copolymerization to reduce vanadium ion crossover. *Eur Polym J* 2013;**49**(7):1832–40.
37. Ma J, Wang Y, Peng J, Qiu JY, Li JQ, Zhai ML. Designing a new process to prepare amphoteric ion exchange membrane with well-distributed grafted chains for vanadium redox flow battery. *J Membr Sci* 2012;**419–420**(1):1–8.
38. Yuan J, Yu CH, Peng J, Wang Y, Ma J, Li JQ, et al. Facile synthesis of amphoteric ion exchange membrane by radiation grafting of sodium styrene sulfonate and *N,N*-dimethylaminoethyl methacrylate for vanadium redox flow battery. *J Polym Sci A* 2013;**51**(24):5194–202.

39. Hoogers G. Fuel cell components and their impact on performance. In: Hoogers G, editor. *Fuel cell technology-hand book*. New York: CRC Press; 2003. p. 1–27.
40. Nasef MM. Radiation-grafted membranes for polymer electrolyte fuel cells: current trends and future directions. *Chem Rev* 2014;**114**(24):12278–329.
41. Nasef MM, Saidi H, Nor HM. Radiation-induced graft copolymerization for preparation of cation exchange membranes: a review. *J Nucl Sci Malays* 1999;**1**:39–54.
42. Horsfall J, Lovell K. Synthesis and characterisation of sulfonic acid-containing ion exchange membranes based on hydrocarbon and fluorocarbon polymers. *Eur Polym J* 2002;**38**(8):1671–82.
43. Gubler L, Gürsel SA, Scherer GG. Radiation grafted membranes for polymer electrolyte fuel cells. *Fuel Cells* 2005;**5**(3):317–35.
44. Chen J, Maekawa Y, Asano M, Yoshida M. Chemically stable hybrid polymer electrolyte membranes prepared by radiation grafting, sulfonation and silane-crosslinking techniques. *J Polym Sci Polym Chem* 2008;**46**(16):5559–67.
45. Shen M, Roy S, Kuhlmann J, Scott K, Lovell K, Horsfall J. Grafted polymer electrolyte membrane for direct methanol fuel cells. *J Membr Sci* 2005;**251**(1–2):121–30.
46. Peck DH, Kang S, Lee BR, Jung DH. Effect of the impurities in methanol fuel on the performance of MEA for direct methanol fuel cell. In: Grube T, Stolten D, editors. *Proceedings of 18th world hydrogen energy conference, Book 1: Fuel cell basics/fuel infrastructures*, vol. 78. Essen: Forschungszentrum, Zentralbibliothek Jülich; 2010. p. 1.
47. Gubler L, Dockheer SM, Koppenol WH. Radical (HO•, H• and HOO•) formation and ionomer degradation in polymer electrolyte fuel cells. *J Electrochem Soc* 2011;**158**(7):755–69.
48. Yu H, Shi J, Zeng X, Bao M, Zhao X. A proton-exchange membrane prepared by the radiation grafting of styrene and silica into polytetrafluoroethylene films. *Radiat Phys Chem* 2009;**78**(7):497–500.
49. Paronen M, Sundholm F, Ostrovskii D, Jacobsson P, Jeschke G, Rauhala E, et al. Preparation of proton-conducting membranes by direct sulfonation. 1. Effect of radicals and radical decay on the sulfonation of poly(vinyl fluoride) films. *Chem Mater* 2003;**15**(23):4447–55.
50. Varcoe JR, Slade RCT, Yee ELH, Poynton SD, Driscoll DJ, Apperley DC. Poly(ethylene-*co*-tetrafluoroethylene)-derived radiation-grafted anion-exchange membrane with properties specifically tailored for application in metal-cation-free alkaline polymer electrolyte fuel cells. *Chem Mater* 2007;**19**(10):2686–93.
51. Slade RC, Varcoe JR. Investigations of conductivity in FEP-based radiation-grafted alkaline anion-exchange membranes. *Solid State Ionics* 2005;**176**(5–6):585–97.
52. Herman H, Slade RC, Varcoe JR. The radiation-grafting of vinylbenzyl chloride onto poly(hexafluoropropylene-*co*-tetrafluoroethylene) films with subsequent conversion to alkaline anion-exchange membranes: optimisation

of the experimental conditions and characterization. *J Membr Sci* 2003;**218**(1):147–1663.
53. Varcoe JR, Slade RC. An electron-beam-grafted ETFE alkaline anion-exchange membrane electrolyte in metal-cation-free solid-state alkaline fuel cells. *Electrochem Commun* 2006;**8**(5):839–43.
54. Varcoe JR, Slade RC, Wright GL, Chen Y. Steady-state d.c. and impedance investigations of H_2/O_2 alkaline membrane fuel cells with commercial Pt/C, Ag/C and Au/C cathodes. *J Phys Chem B* 2006;**110**(42):21041–9.
55. Poynton SD, Kizewski JP, Slade RC, Varcoe JR. Novel electrolyte membranes and non-Pt catalysts for low temperature fuel cells. *Solid State Ionics* 2010;**181**(3–4):219–22.
56. Varcoe JR, Slade RC, Yee ELH. An alkaline polymer electrochemical interface: a breakthrough in application of alkaline anion-exchange membranes in fuel cells. *Chem Commun* 2006;**3**(13):1428–9.
57. Varcoe JR, Slade RC. Prospects for alkaline anion-exchange membranes in low temperature fuel cells. *Fuel Cell* 2005;**5**(2):187–200.
58. Asazawa K, Sakamoto T, Yamaguchi S, Yamada K, Fujikawa H, Tanaka H, et al. Study of anode catalysts and fuel concentration on direct hydrazine alkaline anion-exchange membrane fuel cells. *J Electrochem Soc* 2009;**156**(4):509–12.

CHAPTER EIGHT

Radiation Preparation or Application of Graphene, Nanomaterials, Porous Polymeric Materials, and Ionic Liquids

Pei Dai[1], Shuojue Wang[1], Xinmiao Zeng[1], Maolin Zhai[1], Rong Li[2], Guozhong Wu[2], Mozhen Wang[3], Xuewu Ge[3], Yinyong Ao[1] and Jing Peng[1]

[1]College of Chemistry and Molecular Engineering, Peking University, Beijing, P.R. China
[2]Shanghai Institute of Applied Physics, Chinese Academy of Sciences, Shanghai, P.R. China
[3]Department of Polymer Science and Engineering, University of Science and Technology of China, Hefei, P.R. China

Radiation can induce cross-linking, degradation, oxidation, and reduction of different materials. Therefore, the radiation technology can be employed to prepare new functional materials. This chapter introduces the radiation technology used for the preparation of graphene, nanomaterials, porous polymeric materials, and ionic liquids (ILs).

8.1 RADIATION PREPARATION AND APPLICATION OF GRAPHENE AND ITS NANOCOMPOSITES

The discovery of graphene with a two-dimensional (2-D) structure expanded the kinds of carbon materials. Geim and Novoselovat of the University of Manchester successfully obtained graphene from graphite for the first time, and they won the 2004 Nobel Prize in physics. Graphene is a flat monolayer of carbon atoms arranged into a 2-D honeycomb lattice, and is a basic building block for carbon materials. Graphene can be wrapped up into 0-D fullerenes, rolled into 1-D nanotubes, or stacked into 3-D

Radiation Technology for Advanced Materials.
DOI: https://doi.org/10.1016/B978-0-12-814017-8.00008-1

graphite. The intrinsic breaking strength of graphene (130 GPa) is 100 times higher than that of steel with the same thickness. The thermal conductivity of graphene is up to 5300 W m^{-1} K^{-1}, which is significantly higher than that of other carbon materials such as carbon nanotubes, graphite, and diamonds. The high electron transfer rate (2.5×10^5 cm^2 V^{-1} s^{-1}), electrical conductivity (6000 S cm^{-1}), and surface area of a single graphene sheet provide graphene with excellent electrochemical performance. In addition, graphene has biocompatibility, room-temperature quantum Hall effect, and room-temperature ferromagnetism. Due to these advantages, graphene has extraordinary potential applications in nanoelectronic components and electrical energy storage devices, which may trigger an electronic revolution in the future.

However, graphene sheets with a high specific surface area tend to form irreversible agglomerates or even restack to form graphite through Van der Waals interactions, which lead to the decrease of mechanical properties, and the thermal and electronic conductivities of graphene. Furthermore, the fabrication of high-quality graphene on a large scale is difficult in industry, due to technical restrictions and the high cost. Therefore, scientists have tried to improve the properties and reduce the cost by combining metal particles or polymers with graphene to solve the above problems. Graphene-metal nanocomposites have been applied in catalysis, biosensors, super-capacitors, and fuel cells. The incorporation of graphene in the polymer matrix can increase the mechanical properties and electronic conductive ability of polymer and graphene-polymer nanocomposites could be applied in super-capacitors, solar cells and catalysis.

After the first report of isolated graphene prepared by simple mechanical cleavage of graphite crystals, many approaches have been developed for the synthesis of graphene, including epitaxial growth, chemical vapor deposition (CVD), chemical method, liquid-phase ultrasonic exfoliation of graphite powder, and chemical reduction of exfoliation. Among these methods, epitaxial growth, CVD growth, and chemical exfoliation are bottom-up fabrication methods; and mechanical cleavage, liquid-phase ultrasonic exfoliation, and chemical reduction of exfoliation are top-down methods. The graphene oxide (GO) could be reduced economically and efficiently by utilizing radiation without high temperature and pressure and poisonous reductive agents compared to traditional methods. Furthermore, radiation could be used to synthesize metal nanoparticles and polymers conveniently, as graphene-metal nanocomposites and graphene-polymer nanocomposites can be prepared under irradiation in one pot.

8.1.1 Preparation and application of radiation-reduced graphene oxide

Chen et al.[1] investigated the irradiation of GO with a 5 MeV EB at ambient temperature in air and found that the irradiated GO samples were partially reduced to graphene. However, the degree of reduction of GO is low at an absorbed dose of 500 kGy due to the weak penetrability of the EB. In contrast, γ-ray has a much greater penetration depth due to its zero charge and mass. GO can be reduced efficiently by irradiating a dispersion solution of GO in solvent.[2,3] The inset in Fig. 8.1 shows photographs of GO dispersed in ethanol/water (50 vol/vol%) before and after irradiation with an absorbed dose of 35.3 kGy, and the color of the dispersed

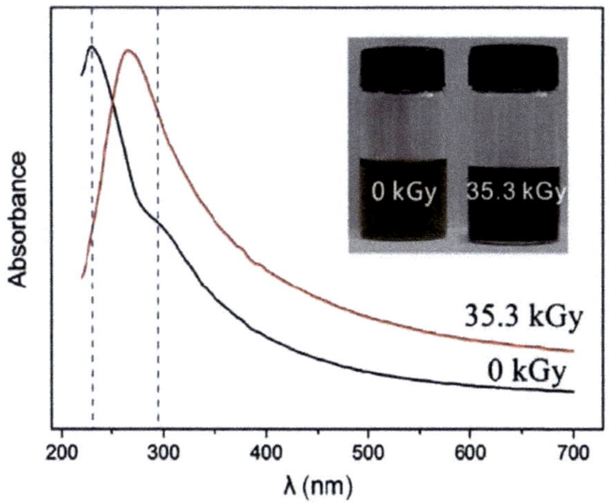

Figure 8.1 UV–vis spectra of GO dispersion before and after irradiation.[2] *GO*, graphene oxide.

GO changes from brown to black.[2] UV−vis spectra show that the shoulder peak of GO around 290 nm ($n \rightarrow \pi^{\star}$ transition of C = O bonds) disappears and the sharp absorption peak around 230 nm ($\pi \rightarrow \pi^{\star}$ transitions of aromatic C−C bonds) appears. Reduced GO (RGO) sheets with high thermal stability are obtained by γ-ray irradiation, and the thickness is around 0.9 nm, which is less than that of monolayer GO sheets ($\sim$1.2 nm). In addition, the electrical conductivity of the resultant RGO increases with increasing degree of reduction degree.[2]

The addition of reducing agent is not needed in the radiation reduction process. This is possibly attributed to the reduction of solvated electrons and other reductive free radicals by radiolysis of solvent. Fig. 8.2A indicates that GO cannot be reduced in pure

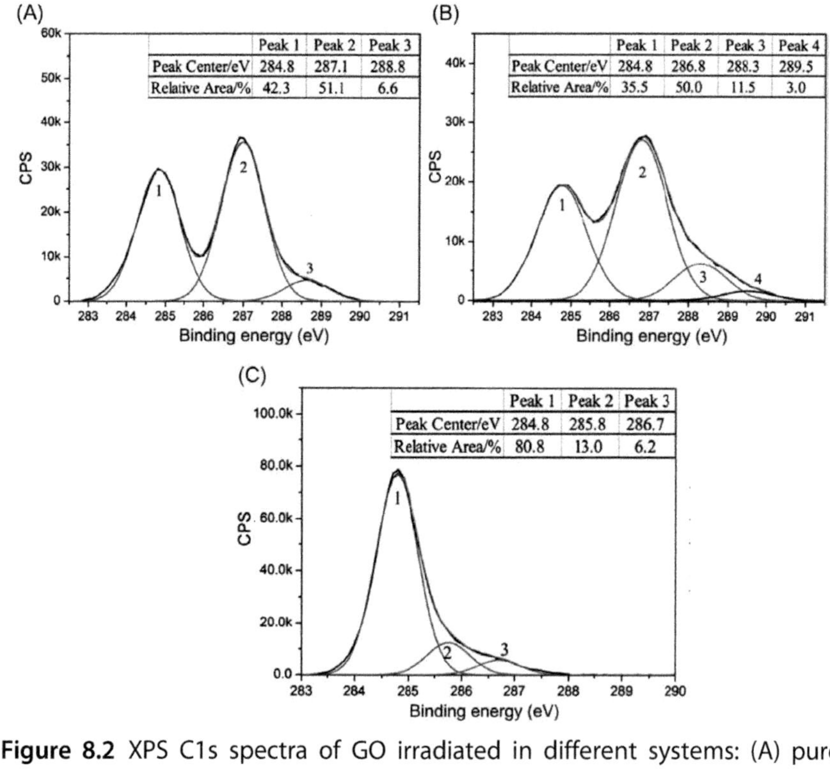

(A)	Peak 1	Peak 2	Peak 3
Peak Center/eV	284.8	287.1	288.8
Relative Area/%	42.3	51.1	6.6

(B)	Peak 1	Peak 2	Peak 3	Peak 4
Peak Center/eV	284.8	286.8	288.3	289.5
Relative Area/%	35.5	50.0	11.5	3.0

(C)	Peak 1	Peak 2	Peak 3
Peak Center/eV	284.8	285.8	286.7
Relative Area/%	80.8	13.0	6.2

Figure 8.2 XPS C1s spectra of GO irradiated in different systems: (A) pure water in N_2 atmosphere, (B) ethanol/water in air atmosphere, and (C) ethanol/water in N_2 atmosphere (absorbed dose: 35.3 kGy; dose rate: 0.882 kGy h^{-1}).[2] *XPS*, X-ray photoelectron spectroscopy; *GO*, graphene oxide.

water because of the existing oxidative •OH. O_2 is a powerful scavenger of hydrated electrons, which promotes the generation of strong oxidative products, including H_2O_2, $•O_2H$, and O_2^-. Fig. 8.2B shows that GO sheet is further oxidized when it is irradiated in ethanol/water with air. Compared with Figs. 8.2A and 8.2B, Fig.8.2C shows that removing oxygen and adding ethanol are the key steps for reduction of GO. The reactions of •OH radicals with alcohols take place mainly at the α-carbon position, and produce the reductive radicals.

The different structures of alcohol molecules affect the yield of reductive radicals. Li[2] compared the reduction level of GO by adding various alcohols, such as methanol, ethanol, isopropanol, and glycol, and found that ethanol was the best additive for this reduction system. Increasing ethanol concentration results in the increase of degree of reduction of GO. However, when the concentration of ethanol exceeds 70%, the reduction degree is decreased. This is attributed to the poor solubility of GO in ethanol compared with water.

The reduction of GO is also influenced by increasing absorbed dose. It is easy to understand that increasing the absorbed dose is helpful to produce more reductive species. Other additives can also improve the reduction of GO, such as cationic surfactants, especially with longer alkyl chains.[4] However, anionic surfactant cannot promote the reduction of GO.

Recently, well-dispersed graphene sheets have been prepared by γ-ray radiation-induced reduction of GO suspension in *N,N*-dimethyl formamide (DMF) at room temperature.[3] The dispersion behavior of RGO is shown in Fig. 8.3. RGO can be readily dispersed in isopropanol, glycol, DMF, tetrahydrofuran (THF), and *N*-methyl-2-pyrrolidone (NMP) without the addition of any surfactant or stabilizer. Micro-FTIR spectrum confirmed the existence of $N(CH_3)^{2+}$ groups on the RGO. Then it was proposed that the DMF could be hydrolyzed into $NH(CH_3)^{2+}$, and subsequent protonated $NH(CH_3)^{2+}$ groups could interact with residual COO^- groups on the edge of the as-synthesized RGO due to the electrostatic interaction. Similar to DMF dispersion, GO can be reduced

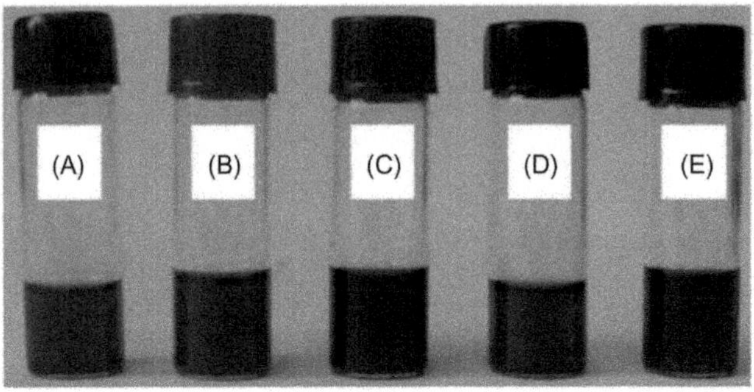

Figure 8.3 Dispersion of RGO in various solvents after standing for 2 weeks [(A) isopropanol, (B) glycol, (C) DMF, (D) THF, E: NMP].[3] *RGO*, Reduced graphene oxide; *DMF*, *N,N*-dimethyl formamide; *THF*, Tetrahydrofuran; *NMP*, *N*-methyl-2-pyrrolidone.

by the electrons generated from the radiolysis of NMP under γ-ray irradiation.[4] It is expected that the radiation-induced reduction of GO will become an economical and efficient method for a large-scale production of graphene in the future.

8.1.2 Radiation preparation and application of graphene-metal nanocomposites

Zhang et al.[5] synthesized high-quality platinum nanoparticle (PtNP)-modified graphene via one-step γ-ray radiation-induced reduction of GO and chloroplatinic acid at room temperature. A series of PtNPs/RGO composites are prepared by changing the absorbed dose, pH, and solvent composition. The carbon/oxygen (C/O) ratio is used to determine the degree of reduction of GO. As shown in Table 8.1, after γ-ray irradiation, the C/O ratio of RGO with or without PtNPs is significantly higher than that of graphite, indicating the radiation-induced reduction of GO. At pH 4.5, the amount of loaded platinum in the PtNPs/RGO composite is 14.3 wt% (PR-I), while the amount of loaded Pt in PR-V is only 1.5 wt% (PR-V) under alkaline condition. In addition, the platinum loading of PtNPs/RGO composites decreases as the absorbed dose increases, corresponding to the increase in the ratio of C/O. Therefore, the C/O ratio of RGO and the metal load of PtNPs/RGO composites show the opposite trend with different

Table 8.1 Sample abbreviations, corresponding preparation conditions and properties[5]

Name	Samples	Solvent	Does (kGy)	pH	Pt (wt%)[a]	C/O[b]
GO	Graphite oxide					2.6
RGO-I	RGO	EG	300	~4.5	0	3.9
RGO	RGO	EG	300	~12	0	4.8
PR-I	PtNPs/RGO	EG	300	~4.5	14.3	4.2
PR-II	PtNPs/RGO	EG	50	~12	1.9	4.4
PR-III	PtNPs/RGO	EG	100	~12	1.6	4.9
PR-IV	PtNPs/RGO	EG	200	~12	1.5	5.7
PR-V	PtNPs/RGO	EG	300	~12	1.5	6.6
PR-VI	PtNPs/RGO	EG-Water[c]	300	~12	2.5	6.8

GO, graphene oxide; *RGO*, reduced graphene oxide; *PtNP*, platinum nanoparticle.
[a]The values are obtained from the ICP-AES measurement.
[b]The C/O ratio is calculated according to the following equation: C/O = C (at.%)/O (at.%); the elemental atomic amount is obtained from XPS data.
[c]The EG-water volume fraction is 50 vol/vol%. EG is ethylene glycol.

pH and absorbed dose. This is because the existence of oxygen-containing functional groups attracts the metal precursors by coordination or electrostatic interaction. Therefore, the more the oxygen groups are removed, the fewer the PtNPs anchored on the RGO nanosheets. It is clear that reduction of GO is much easier and complete under strongly alkaline conditions because the reductive species can be scavenged by protons. Furthermore, under the same conditions, the C/O ratio of the PtNPs/RGO composites is higher than that of pure RGO. This suggests that the synergistic effects of metal precursors and graphene reduction may result in a higher degree of reduction of graphene.

The size and morphology of PtNPs as well as the crystalline structure of RGO nanosheets in PtNP–RGO composites were studied by high-resolution transmission electron microscope (TEM) (Fig. 8.4). As shown in Fig. 8.4A, ripples and wrinkles are formed on the sheet surface and edges of PR-V, which is an inherent feature of graphene nanosheets. The high-resolution TEM image (Fig. 8.4B) shows that the PtNPs on the graphene sheets are mostly quasi-spherical shape with few aggregates. The corresponding selected area electron diffraction pattern demonstrates the

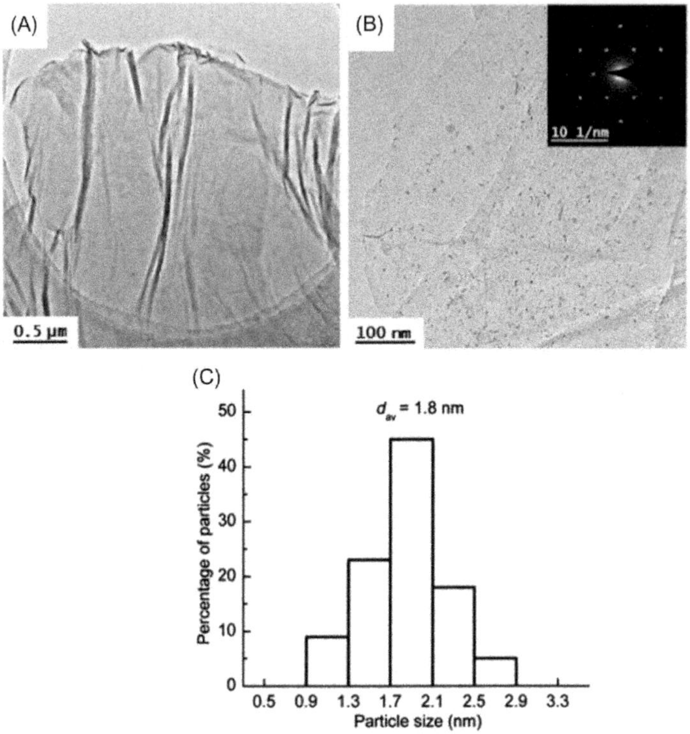

Figure 8.4 (A) PR-V low resolution, (B) high-resolution TEM photo and selected area electron diffraction pattern (illustration), (C) platinum nanoparticle particle size distribution histogram.[5]

crystalline structure of PR-V, with a regular hexagon pattern presenting an ordered graphite lattice; it further confirms the formation of RGO. The average diameter of PtNPs on the RGO sheets is 1.8 nm. The average diameter of PtNPs and the narrow size distribution within 2 nm suggest that the protection of solvents and the interaction between PtNPs and the RGO sheets successfully control the size of PtNPs at the ultrafine nanoscale.

The radiation-prepared PtNPs/RGO composite may be used as the electrode material in super-capacitors. As shown in Fig. 8.5, cyclic voltammograms (CVs) curves with a quasi-rectangular shape suggest a typical electrochemical double layer capacitive behavior of the super-capacitor for PR-V. As the scan rate is increased from 5 to 100 mV s^{-1}, both the profile and area of CV curves are retained

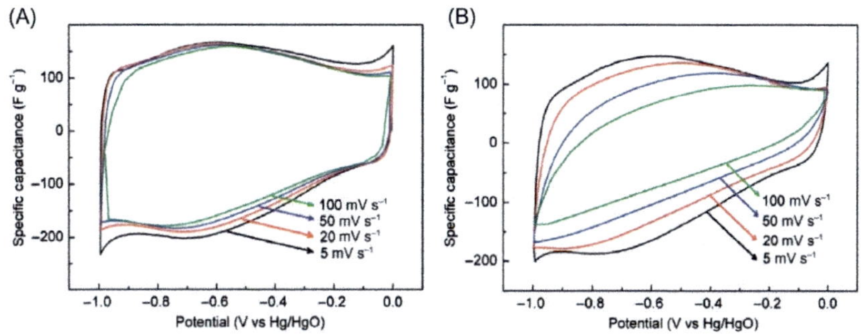

Figure 8.5 (A) PR-V composites and (B) RGO (the cyclic voltammetry curve of the super-capacitor for electrode material at different sweep speeds).[5] *RGO*, reduced graphene oxide.

very well. This is attributed to the smoothly accelerated charge/discharge process, because PtNs in the composites serve as an intercalator to provide a diffusion channel for electrolyte ions and electrons into the electrode. The PtNPs also facilitate the reduction of GO with the repair of the p-conjugation structure, which is convenient for electron transfer. The specific capacitance of PR-V reaches 144.5 F g^{-1} at 5 mV s^{-1} and reveals a high retention rate (83.5%) at 100 mV s^{-1}. Therefore, the loading of PtNPs onto RGO improves the specific capacitance and performance of the fast charging and discharging of super-capacitors.

Wang et al.[6] reported the IL-assisted radiation synthesis of graphene-silver nanoparticle (Ag-RGO NPs) composites. The C/O ratio of Ag-RGO-1^{IL} is prepared by adding IL at 160 kGy is 8.7, and the degree of reduction of graphene is improved obviously compared with that synthesized without ILs, indicating that the [EMIm]Ac plays a key role in the radiation-induced reduction of GO. This may be because the $[EMIm]^+$ can scavenge the oxidative hydroxyl radicals, thereby promoting the reduction of GO. The increase in nitrogen content for the samples prepared with [EMIm]Ac indicates the adsorption of [EMIm]Ac on GO/RGO sheets through electrostatic interaction and Π−Π interaction between GO/RGO sheets and imidazolium cations. Since the removal of oxygen-containing groups on the GO is not beneficial to the loading of silver nanoparticles (Ag NPs), the silver content decreases with the increase of the C/O ratio in the Ag-RGO composites.

8.1.3 Radiation preparation and application of graphene-polymer nanocomposites

As mentioned above, the $N(CH_3)^{2+}$ groups on RGO significantly improve the dispersion of RGO in many organic solvents. PS-RGO nanocomposites were prepared by simple solution blending, and they exhibited a low percolation threshold at 0.24 vol% of RGO.[7] The electrical conductivity of polystyrene-RGO (PS-RGO) nanocomposite is high up to 45 S m^{-1} at 2.3 vol% of RGO, which is close to the highest conductivity of PS chemically-converted graphene composites (72.2 S m^{-1} at 2.5 vol%)[7] and polycarbonate functionalized graphene nanocomposites (51.2 S m^{-1} at 2.2 vol%).[8] Compared to conventional chemical reductions, γ-ray radiation-induced reduction provides a simple approach for the preparation of highly conductive polymer nanocomposites.

Generally, the dispersion of RGO prepared by chemical reduction is poor in polar solution. Xie et al.[9] reported that GO dispersion in ethanol/water irradiated by γ-ray under N_2 atmosphere is not only reduced efficiently, but also dispersed well in ethanol or ethanol/water solutions. Next they prepared conductive grapheme-based polymer films via vacuum filtration and polymer coating using the resultant RGO. Different polymers such as polyvinyl alcohol (PVA), polydimethyl siloxane (PDMS) were used as the matrix. As shown in Fig. 8.6A, RGO-polymer composite film is flexible and optically transparent. The sheet resistance of RGO-polymer composite films decreases with increasing RGO film thickness (Fig. 8.6B). The transparencies of PVA and PDMS films are dramatically decreased after addition of RGO film. Nevertheless, the optical transmittance of RGO-polymer composite films in the visible region is still up to 45% (thickness 0.24 μm), satisfies the requirements of certain semitransparent materials. Notably, the composite films possess very strong absorption in the ultraviolet region, which makes them potential candidates as UV-shielding materials. Therefore, by

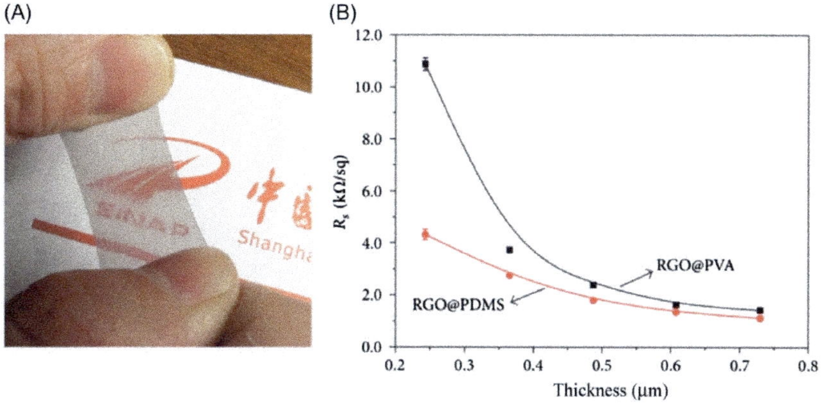

Figure 8.6 (A) Photograph of RGO@PDMS composite film, and (B) relationship of sheet resistances of composite films with thickness of RGO layer.[9] *RGO*, reduced graphene oxide; *PDMS*, polydimethyl siloxane.

changing the thickness of RGO films, the sheet resistance and optical transmittance of RGO-polymer composite films can be controlled, which offers potential for use in solar cells, electric circuits, and antistatic coatings.

Cationic surfactant cetyltrimethyl ammonium bromide (CTAB) with long alkyl chains can effectively promote the radiation-induced reduction of GO by preventing the aggregation of graphene sheets, which has been testified by X-ray photoelectron spectroscopy (XPS), X-ray diffraction, thermo gravimetric analysis, Raman spectroscopy, and Fourier transform infrared spectroscopy.[4] Furthermore, when as-prepared γ-ray reduced graphene oxide (RGOC)/polyaniline (PANI) composites are used for super-capacitor electrode materials, there is a maximum specific capacitance of 484 F g^{-1} at a current density of 0.1 A g^{-1} for the graphene produced in the presence of cationic surfactant CTAB. Fig. 8.7 indicates that the introduction of PANI provides the synergistic effect of a combination of the conducting and mechanical properties of graphene and the high pseudo-capacitance of the PANI component. In addition, the current of the composite electrode increases markedly with scanning rate, suggesting the good conductivity and capacitive behavior of RGOC/PANI composite electrodes.

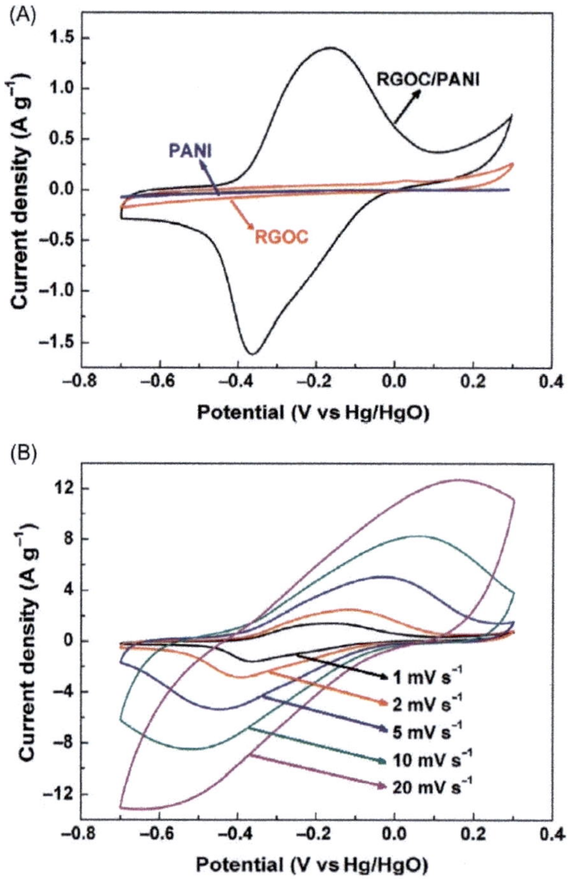

Figure 8.7 (A) CV curves of RGOC, PANI, and RGOC/PANI composite at 1 mV s^{-1}; (B) CV curves of RGOC/PANI composite at different scan rates.[4] *CV*, cyclic voltammogram; *PANI*, polyaniline.

8.1.4 Radiation preparation and application of graphene oxide-polymer nanocomposites

Due to the existence of polar groups on the surface and edges of the GO layer, GO can be dispersed well in water and some polar solvents, such as DMF, NMP, and THF. However, GO has poor dispersion in organic or polymer matrices. The dispersion of GO in water and organic solvents has been accomplished via covalent functionalization of the GO sheets with different molecules and polymers. The radiation-induced reduction of GO provides an economical and efficient method for the large-scale production of functional graphene sheets, and radiation may also be applied for the modification of GO. Zhang et al.[10] reported a simple approach to modify GO sheets with poly(vinyl acetate) (PVAc) by γ-ray radiation-induced graft polymerization. The thickness of GO-*g*-PVAc sheets is ~2.4 nm, larger than that of pristine GO (1.33 nm), and the height of the PVAc aggregates is 4.2 nm, which indicates that GO sheets are covered by a layer of grafted PVAc chains. Due to the high solubility of PVAc in organic solvents, the PVAc layer grafted onto GO sheets can form a solvation shell, which hinders the aggregation of GO sheets. As shown in Fig. 8.8A, the solutions of GO-*g*-PVAc in all these solvents are very stable, even after 2 months, indicating the excellent dispersibility and stability of GO-*g*-PVAc. Similar to PVAc, PS chains are

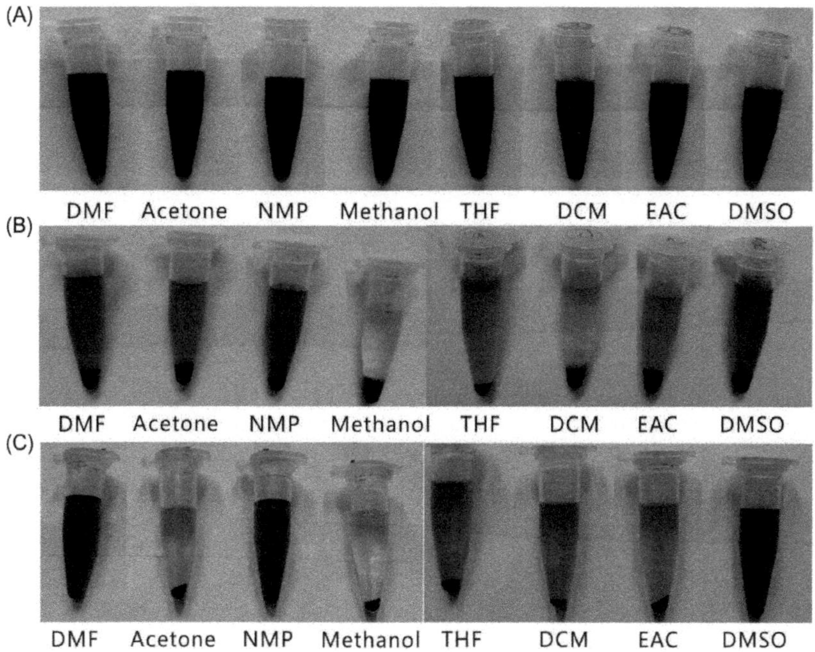

Figure 8.8 Photographs of dispersions in different organic solvents after ultra sonication: (A) GO-*g*-PVAc with degree of grafting (DG) of 28.5%, settled after 2 months; (B) GO, settled after 1 h; (C) GO/PVAc blend, settled after 1 h.[10] *GO*, graphene oxide; *PVAc*, poly(vinyl acetate).

grafted on the GO sheets and then exfoliation of GO is induced by γ-ray irradiation.[11,12]

Hydrogels are hydrophilic 3-D polymer networks that can absorb and retain a considerable amount of water with the maintenance of shape. Liu et al.[13] reported a simple and effective strategy for preparing graphene peroxide (GPO) via the radiation-induced peroxidation of GO, and they used GPO as the initiator and cross-linking agent of AAm monomer to obtain GO-*g*-polyacrylamide (GO-g-PAAm) composite hydrogels. The resultant GO-*g*-PAAm composite hydrogels exhibited excellent mechanical properties due to chemical bonding between the PAAm polymer and the GO sheets. The possible structure of GPO is shown in Fig. 8.9A. The hydroxyl, epoxy, carbonyl, and carboxyl groups can be broken under irradiation to form radicals, which can combine with oxygen to form peroxy radicals. Peroxy radicals can further react with small radicals to form peroxides. The peroxides on GPO are decomposed on heating to generate free radicals (GO•, OH•), where GO•, the radicals on graphene, can initiate the grafting polymerization of monomers onto the GPO sheets, and OH• can initiate the homopolymerization of monomers. The growing grafted polymer chains on vicinal GPO sheets can be covalently joined through mutual combination and/or be physically entangled, that is, chemically and/or physically cross-linked (Fig. 8.9B). Due to the presence of a large number of peroxides on each GPO sheet, the GPO sheets function as initiators and cross-linking centers. Scanning electron microscope (SEM) micrographs show the formation process of a GPO hydrogel in Fig. 8.9C and D. The SEM images indicate that the isolated GPO sheets are entangled by bundles of polymer chains. With the increase of reaction time, more and longer fibers are formed, and hence a 3-D network structure is formed.[13]

Nanoantibiotic materials that protect public health have a wide range of biomedical applications. The bacterial toxicity of graphene nanosheets exhibits more antibacterial activity for both Gram-positive and Gram-negative models of bacteria. Zhao et al.[14] reported that GO in suspension solution can firmly combine with cotton fabrics by adsorption or by radiation-induced cross-linking

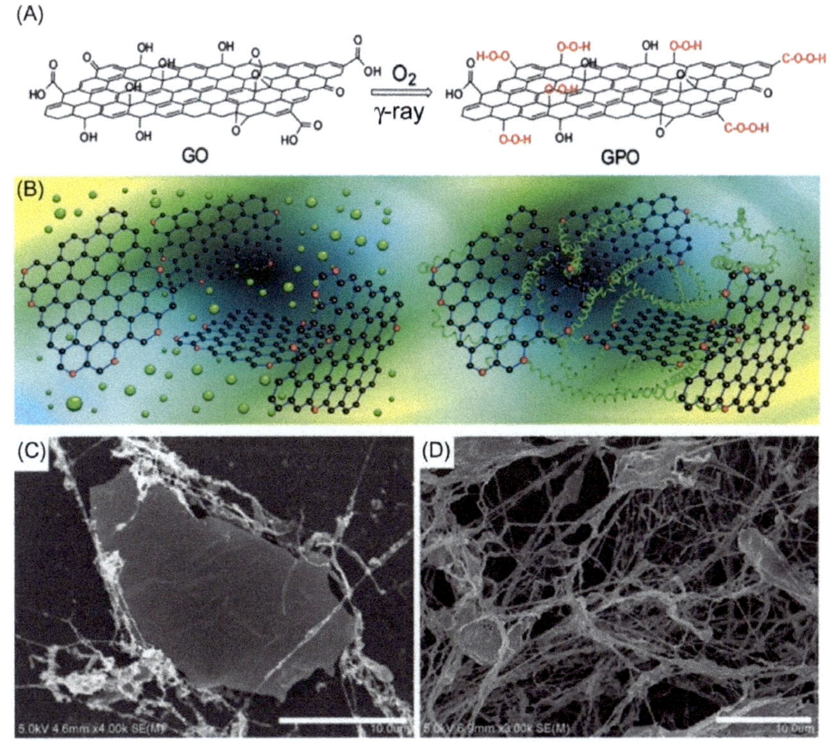

Figure 8.9 (A) Possible schematic model of GO and GPO; (B) 3-D schematic model of the formation of a GPO hydrogel, before (left side) and after (right side) polymerization (the green balls are the monomer molecules, the green spiral lines are the polymer chains, and the peroxyl groups are colored in red); (C) and (D) SEM micrographs of the formed GPO hydrogel: the reaction times are 6 min (C) and 12 min (D).[13] *GO*, graphene oxide; *GPO*, graphene peroxide; *3-D*, three-dimensional.

using triallylisocyanurate as a cross-linker. GO-modified cotton fabrics can inactivate >98% of bacteria in less than 4 h, especially the directly filtrating GO-modified cotton fabric. Cotton-GO is laundering durable, and even after being washed 100 times, its inactivation efficiency of bacteria is >90%. Moreover, GO-modified cotton fabrics are all safe after 72 h of contact with to skin. The flexible, foldable, and recyclable GO-based cotton fabric can be used for medical and food applications, and has great potential to reduce the use of antibiotics.

The surface modification of GO determines the interactions between GO and polymers, which possibly has a significant impact on the mechanical properties of polymer. Xie et al.[15] modified GO with glycidyl methacrylate (GMA) and triethylenetetramine through γ-ray irradiation. GO is also reduced simultaneously during the radiation process. The chemical bonding formed between amino groups on GO sheets and ethylene-methyl acrylate-GMA random terpolymer (ST2000) is more efficient in improving the compatibility between polyethylene terephthalate (PET) and ST2000, leading to a 2.5-fold increase in impact strength, along with a slight increase in tensile strength. However, the addition of a tiny amount of RGO sheets lacking polar groups caused the decrease in the mechanical property of the PET/ST2000 blend.

8.2 PREPARATION OF NANOMATERIALS BY RADIATION TECHNOLOGY

Nanomaterials are materials that have at least one of the dimensions within the nanometer scale range (100 nm) or are made up of such basic units which are considered the most promising materials in the 21st century. There are many kinds of nanomaterials, because of the volume effect, surface effect, quantum size effect, macroscopic quantum tunneling effect, and dielectric limit effect, resulting in a series of special properties differing from conventional materials. Nanomaterials have extensive applications in magnetic materials, optics, ceramics, semiconductors, catalysis, sensing, medicine, national defense, aviation and metallurgy. The first international conference on nanoscience and technology, held in Baltimore, United States, in July 1990, officially recognized nanomaterials science as a new branch of material science, which marked the birth of nanoscience and technology.

There are a number of methods for the preparation of nanomaterials; radiation technology is a unique approach, since solvated or hydrated electrons generated by radiolysis are employed as the reducing species. Furthermore, compared with other methods, the radiation method has many advantages: it can be operated at normal temperature and pressure or even at lower temperature; the preparation period is short with a simple process; the product has small particle size and narrow distribution and is easy to handle; the productivity is high and reprocessing is convenient. Therefore, the radiation method is promising for the fabrication of large amounts of nanoparticles under normal conditions.

8.2.1 Preparation of metal nanoparticles

8.2.1.1 Fundamentals

After 1992, the radiation method began to develop rapidly as a new method for preparing metal nanomaterials. Metal clusters of Ag, Au, Cu, Pt, Pd, Pb, Ir, Cd, and Tl were successfully prepared in aqueous solution, and the formation mechanism of metal clusters by irradiation was also explored. The basic principle of preparing nanometals by irradiation is as follows: when dilute aqueous solution containing metal ions is irradiated by high energy electrons or γ-ray, water radiolysis generates reductive species of H atoms and hydrated electrons (e_{aq}^{-}), oxidizing species (•OH free radicals), and other active species or molecules. These active species then react with the metal ions and transfer the radiant energy to the solute indirectly. Among them, standard redox potential of e_{aq}^{-} is -2.77 eV, with a strong reduction capability, which theoretically can reduce all metal ions except for alkali metals and alkaline earth metal. In order to effectively reduce the metal ions, it is usually necessary to remove the dissolved O_2 from the solution by purging with high-purity N_2 and adding free radical scavengers such as isopropanol and ethanol to remove oxidative hydroxyl radicals via H abstraction. These reductive species can gradually reduce the metal ions to metal atoms or low valence metal ions. The newly-formed metal atoms agglomerate into nuclei and grow into nanoparticles, which later precipitate out of the solution. The nanometals obtained by this method have higher crystallinity. In addition, the radiation-induced chemical reaction can be used to prepare metal oxide and sulfide nanoparticles. The radiation synthesis method can be easily combined with the sol–gel process or the polymerization process to prepare inorganic/organic hybrid nanocomposites.

The formation mechanism of metal clusters and nanoparticles by the radiation method can be summarized as follows:

$$H_2O \rightsquigarrow H\bullet, \bullet OH, e^-_{aq}, H_3O^+, H_2, H_2O_2, HO_2\bullet, H_2O^* \tag{8.1}$$

$$\bullet OH(H\bullet) + (CH_3)_2CHOH \rightarrow H_2O(H_2) + (CH_3)_2\dot{C}OH \tag{8.2}$$

$$e^-_{aq} + M_n^+ \rightarrow M_{(n-1)}^+ \tag{8.3}$$

$$(CH_3)_2\dot{C}OH + M_n^+ \rightarrow M_{(n-1)}^+ + (CH_3)_2CO + H^+ \tag{8.4}$$

$$e^-_{aq} + M^+ \rightarrow M^0 \tag{8.5}$$

$$(CH_3)_2\dot{C}OH + M^+ \rightarrow M^0 + (CH_3)_2CO + H^+ \tag{8.6}$$

$$M^+ + M^0 \rightarrow M_2^+ \tag{8.7}$$

$$M_m^{x+} + M_n^{y+} \rightarrow M_{m+n}^{(x+y)+} \tag{8.8}$$

These clusters further agglomerate to form a metal colloid.

Henglein[16–19] and Belloni[20] carried out detailed studies of the radiation reduction of metal ions and the formation of clusters using pulse radiolysis technique. Because the reaction activity of metal clusters in aqueous phase is very strong and the reaction occurs in a very short period of time, it is impossible to follow the process with traditional methods. However, the pulse radiolysis technique provides a very powerful time-resolved means to produce and observe unknown particles and measure fast-reaction kinetics. It is a powerful tool to study the rapid reaction process and short-life intermediates, and has been successfully applied in the study of free radical reaction mechanisms in homogeneous solution in the past decades. The technology in inorganic chemistry is mainly applied to the oxidation-reduction reaction in aqueous solution, which is characteristic in the study of unusual valence metal ions or metal clusters. The principle is to use high energy short pulses (picosecond–nanosecond) to irradiate the metal ion solution so as to produce a variety of active species, among which, reductive species (such as hydrated electrons) can be utilized to reduce metal ions to atoms or lower oxidized ions (unusual metal valence states) and then agglomerate into clusters. Clusters of different numbers of agglomerates, due to different surface states with different optical absorption characteristics, can be detected by fast-response absorption spectroscopy. Using pulse radiolysis to study the formation and stability of these newly dispersed phases has the following advantages: (1) the newly dispersed phase in a homogeneous medium can be quickly and uniformly generated; (2) the study of the early stage of nucleation of the dispersed phase becomes possible, and (3) the concentration of the dispersed phase can be adjusted easily by changing the metal ion concentration or the absorbed dose. To date, the formation and reaction of metal clusters in dilute aqueous solutions have been studied using pulse radiolysis techniques, such as Ag, Au, Pt, Tl, Ir, Cd, Pb, Cu, Pd, Ni, Co, Sn, In, Bi, etc. Silver is the most studied nanoparticle, and its formation mechanism is well-known.

8.2.1.2 Preparation of Ag clusters in aqueous solution

After the high energy ionization pulse irradiation of the Ag^+ aqueous solution, a reductive reaction of Ag^+ occurs after capturing the hydrated electrons:

$$Ag^+ + e_{aq}^- \rightarrow Ag^0 \quad (k = 3.5 \times 10^{10}\ \mathrm{L\ mol^{-1}\ s^{-1}}) \tag{8.9}$$

The Ag^0 has a maximum absorption peak at 360 nm. However, it quickly combines with Ag^+ to produce Ag_2^+ particles, and then Ag_4^{2+}:

$$Ag^0 + Ag^+ \rightarrow Ag_2^+ \quad (k = 5 \times 10^9\ \mathrm{L\ mol^{-1}\ s^{-1}}) \tag{8.10}$$

$$Ag_2^+ + Ag_2^+ \rightarrow Ag_4^{2+} \quad (2k = 4 \times 10^8\ \mathrm{L\ mol^{-1}\ s^{-1}}) \tag{8.11}$$

The maximum absorption peaks of Ag_2^+ and Ag_4^{2+} are 310 and 260 nm, respectively.

Janata et al.[18] found that before the formation of Ag_4^{2+}, in addition to Ag^0 and Ag_2^+, there may be another intermediate Ag_3^{2+}

(absorption peaks at 265 and 310 nm). The production of Ag_3^{2+} and dimerization occurs as follows:

$$Ag_2^+ + Ag^+ \rightarrow Ag_3^{2+} \quad (k = 2 \times 10^9\ L\ mol^{-1}\ s^{-1}) \tag{8.12}$$

$$Ag_3^{2+} + Ag_3^{2+} \rightarrow Ag_4^{2+} + 2Ag^+ \tag{8.13}$$

Ag_2^+, Ag_3^{2+}, and Ag_4^{2+} clusters are further reduced and agglomerated. In the presence of excessive silver ions, Ag_9^+ clusters may also be formed (absorption peaks at 295 and 325 nm), which can exist stably for several minutes. Its formation process may be:

$$2Ag_4^{2+} \rightarrow Ag_7^{3+} + Ag^+ \tag{8.14}$$

$$2Ag_7^{3+} \rightarrow Ag_8 + 6Ag^+ \tag{8.15}$$

$$\rightarrow Ag_9^+ + 5Ag^+ \tag{8.16}$$

When no silver ions exist (i.e., silver ions are completely reduced), Ag_2 and Ag_4 and other neutral intermediate clusters can be detected, which appear at 270 and 340 nm, respectively. However, these oligomeric clusters soon agglomerate into larger clusters and quasi-colloidal particles (Ag_q, q is more than 10 atoms, 360–370 nm) and eventually agglomerate into metal nanoparticles (Ag_m, over 380 nm). The general reaction is as follows:

$$(Ag_{n+x})^{x+} + (Ag_{m+y})^{y+} \rightarrow (Ag_{n+m+x+y})^{(x+y)+} \tag{8.17}$$

If the ions on the cluster are omitted, it can be abbreviated as

$$Ag_n + Ag_m \rightarrow Ag_{n+m} \tag{8.18}$$

It should be emphasized that the relationship between the absorption peaks of the clusters and the structures of clusters is not yet determined; in different systems, their absorption peaks are different. Of these absorption peaks, only clusters situated at 260 (275 nm under stabilizers), 295, and 325 nm remain especially stable. Researchers have found that the delocalized valence electrons of these clusters (Ag_4^{2+} and Ag_9^+) have closed shell electron configurations.[18] Therefore, they are named magic number clusters, for instance, the Ag_4^{2+} with two valence electrons and Ag_9^+ with eight valence electrons. The term "magic number" refers to the phenomenon that properties of clusters with a certain number of atoms suddenly increase or decrease compared to the total trends when some clusters' properties vary with the number of atoms in the cluster.

The development of silver clusters in aqueous solution is as follows: the silver ions are first reduced to silver atoms (Ag^0, 360 nm) before they recombine into oligomeric clusters ($Ag_4^{2+} + Ag_3^{2+}$, 265 and 310 nm), and then develop into magic number clusters (Ag_4^{2+}, 260 nm; Ag_9^+, 295 and 325 nm), quasi-colloidal particles (Ag_q, q means a dozen atoms, 360 and 370 nm), and finally agglomerate into metal nanoparticles (Ag_m, above 380 nm).

The reactivity of metal clusters is related to its electrochemical potential, which is closely related to the cluster size (namely the agglomeration number n). Therefore, it is important to study the relationship between the electrode potential of the cluster and its agglomeration number n. For a solution containing Ag_n cluster and Ag^+ ions, its typical electrode equilibrium can be summarized as

$$Ag_n \Leftrightarrow Ag_{n-1} + Ag^+ + e^- \tag{8.19}$$

Using the electrode potential of the silver cluster as a function of the agglomeration number n, their relationship is shown in Fig. 8.10[21]:

The figure above shows that the potential of the silver electrode becomes more positive as the agglomeration n increases. When n increases, that is, $n \rightarrow \infty$, the silver electrode comes close to the potential of silver electrode $+0.799$ V. When the n value is too small, the electrode potential strongly depends on the n value. When $n = 1$, that is, as a free silver atom, the standard oxidation-reduction potential is -1.8 V; when $n = 2$ or $n = 3$, the standard oxidation-reduction potentials are -0.1 and -0.9 V, respectively, with obvious odd and even oscillation characteristics. Since the thermodynamic properties of larger silver clusters are unclear, cluster potential values greater than three cannot be expressed more specifically.

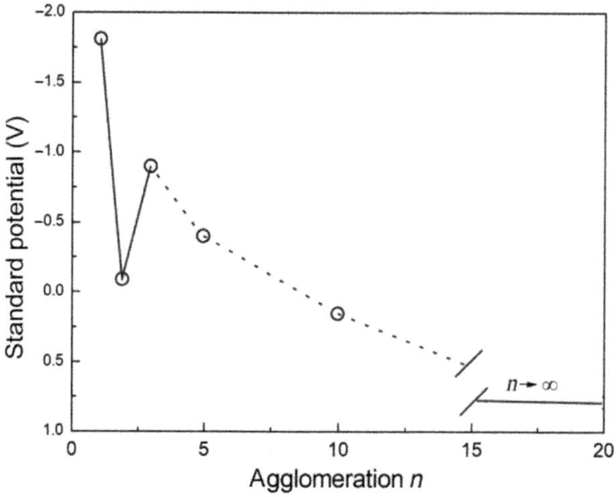

Figure 8.10 Relationship between silver cluster's standard oxidation-reduction potential $E(Ag^+/Ag_n)$ and agglomeration n.[21]

There have been many reports on the preparation of silver nanoparticles by radiation methods. The use of stabilizers is also important to obtain stable silver nanoparticle colloids. Long et al. controlled the size of silver nanoparticles prepared upon irradiation by adding chito-oligosaccharide to the aqueous solution as a stabilizer. The size of the prepared silver nanoparticles can be controlled between 5 and 15 nm and can be evenly dispersed in sodium chloride solutions.[22] Subsequently, Liu et al., used sodium alginate as a stabilizer to prepare silver nanoparticles. The silver nanoparticles prepared by irradiation in sodium alginate solution were stable after storage at room temperature for several hours and 6 months, respectively. As shown in Fig. 8.11, there was no significant difference in absorbency and appearance between them. These results show that the prepared silver nanoparticles have excellent stability, and the particle sizes are controllable.[23]

8.2.1.3 Preparation of Cu clusters in aqueous solution

Ershov et al.[24–28] investigated the formation mechanism of copper colloid in aqueous solution using pulsed radiolysis techniques. In the presence of sodium formate, e_{aq}^- and CO_2^- free radicals are produced upon irradiation (absorption peak at 240 nm); they can

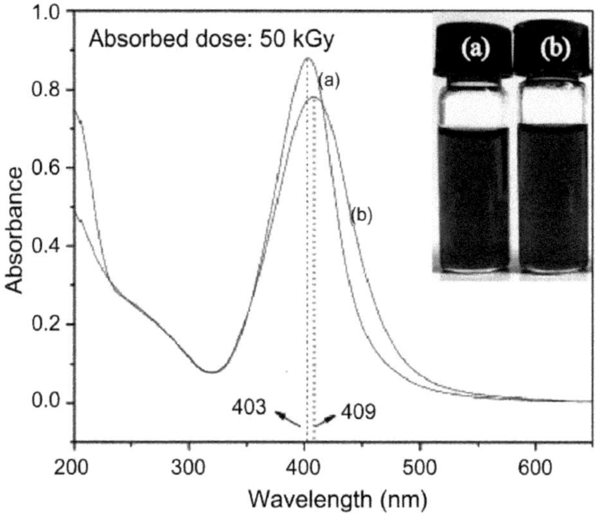

Figure 8.11 Ultraviolet absorption spectrum and appearance of sodium alginate solutions of silver nanoparticles (with initial concentration of 10 mmol L^{-1}) for different periods of time [(A) several hours; (B) 6 months].[23]

reduce Cu^{2+} into Cu^+ (weak absorption at 265 nm). Cu^+ can be further reduced into Cu^0 by e^-_{aq} and can also react with CO_2^- to generate $CuCO_2$ (when pH = 7.3, absorption peak is at 480 nm; at lower pH, the absorption peak is shifted to 385 nm). Since $E^\circ_{(Cu+/Cu^0)}$is − 2.7 V, which is lower than the standard potential of CO_2^- (−2.0 V), the following reaction is unable to occur:

$$CuCO_2 \rightarrow Cu^0 + CO_2 \tag{8.20}$$

Decay of $CuCO_2$ is attributed to the following two reactions:

$$CuCO_2 + Cu^{2+} \rightarrow 2Cu^+ + CO_2 \tag{8.21}$$

$$CuCO_2 + Cu^+ \rightarrow Cu_2^+ + CO_2 \tag{8.22}$$

The absorption peak of Cu_2^+ occurs at 350 nm. The intermediates $CuCO_2$ and Cu_2^+ are formed in 100 μs, and then they nucleate in a period of microseconds to seconds, and finally form into large agglomerations. The formation of copper colloid under the reduction of CO_2^- free radicals is as follows:

$$Cu^{2+} \rightarrow CuCO_2 \rightarrow Cu_2^+ \rightarrow \text{Nucleation} \rightarrow \text{Copper Colloid} \tag{8.23}$$

Studies of the radiation reduction process of $CuSO_4$ aqueous solution in the presence of surfactant (PVA, PAA) have been reported by Khatouri's group.[29] Under the protection of PVA with a pH of 2.6, the peak shape of the absorption spectrum does not change with the increase of absorbed dose. There is a weak absorption at 570 nm, where the color of solution changes from colorless to light red.

Other metal ions are also employed to produce clusters under ionizing radiation. These metal clusters continue to grow into colloidal particles, and eventually aggregate into nanoparticles precipitating from the solution. Generally, in order to prevent the nanoparticles from growing too large, a certain amount of surfactant such as PVA and sodium dodecyl benzene sulfonate (SDBS) is used as a stabilizer. Surfactants can modify the surface of nanoparticles to diminish the surface energy to increase its stability. Zhu et al. investigated the influence of surfactants on the preparation of silver nanoparticles. It was found that appropriate concentration of surfactants prevented further aggregation of silver nanoparticles. When SDBS is absent, the particle size of the silver nanoparticle is 47 nm; when the SDBS concentration is 0.1 mol L^{-1}, the particle size is reduced to $\sim$10 nm. However, the continuing increase of SDBS concentration has little impact on the particle size. IL can also be used as a stabilizer for metal nanoparticles. As shown in Fig. 8.12, gold nanoparticles can be produced in IL and its mixture with water under irradiation, and the particle size of nanoparticles can vary from 10 to 50 nm.[30]

8.2.1.4 Preparation of nanoalloy, -metal oxide, and -metal sulfide materials

Since the oxidation and reduction potentials of precious metal ions are high, they are easy to reduce using ionizing radiation. The manipulations are not complicated, and the preparation of many types of precious metal nanoparticles has been achieved. In contrast, the redox potentials of active metal ions are lower. This make them easily oxidized in the production period. Consequently, they are difficult to prepare using conventional methods.

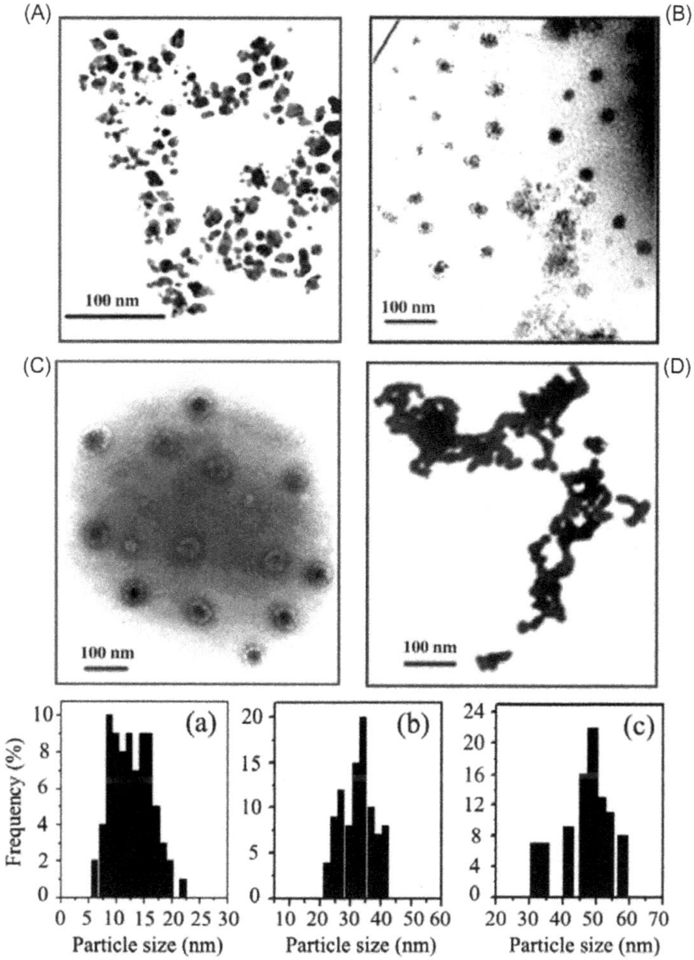

Figure 8.12 Transmission electron microscopy (TEM) images and particle size distribution of gold nanoparticles prepared using different systems: (A) ionic liquids under quaternization, (B) ionic liquids and iso-Propyl alcohol (IPA) under quaternization, (C) ionic liquids and water under quaternization, and (D) water.[30]

Although the radiation method for the preparation of nanometer alloy powders starts relatively late, many successful preparations of alloys colloid have laid the foundations for the synthesis of nanometer alloys. Research shows that adjusting the concentration of the metallic salts, and controlling the metal ions of the restored potentiometer as well as the response rate can enable many metal ions to restore simultaneously and pile up orderly on crystal nucleus, in order to prepare nanoalloy particles, such as Ag–Cu, Au–Cu, Cu–Pd, etc. Wang et al.[31] and Zhang et al.[32] loaded Pt atom (see Fig. 8.13) and Pt/Ru nanoparticles respectively on multiple-wall carbon nanotubes through γ-ray irradiation reduction. The modified carbon nanotubes were then used for the preparation of a fuel battery separator and showed good electrical catalytic performance.

The radiation method can also be used to prepare oxide nanopowders. Mn_2O_3 and MnO_2 powders can be prepared using the radiation reduction method by controlling the pH value of the

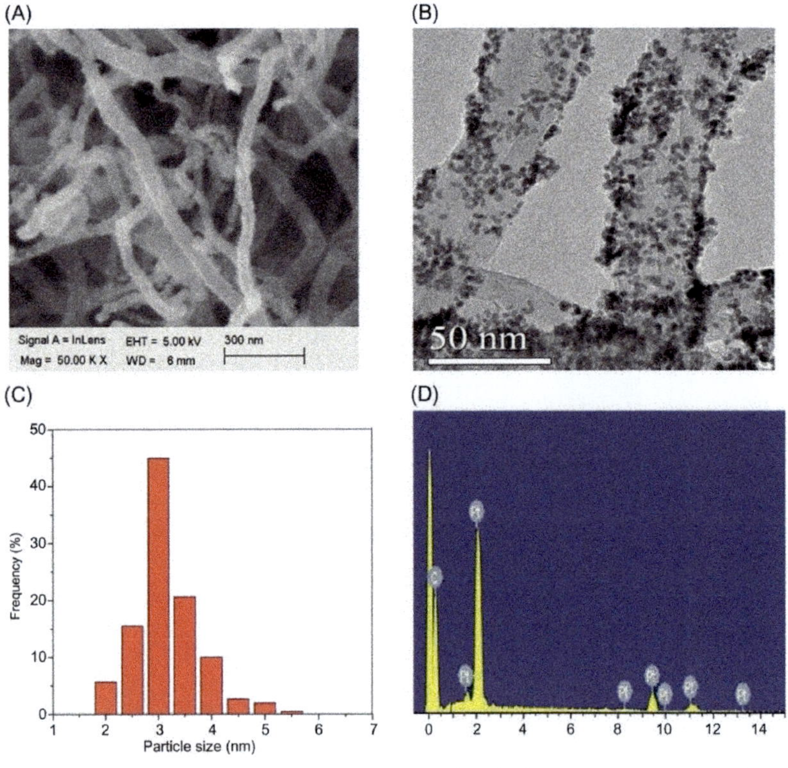

Figure 8.13 Multiwalled carbon nanotubes loaded with Pt nanoparticles: (A) scanning electron microscope image, (B) transmission electron microscope, (C) particle size distribution map, and (D) energy dispersive X-ray spectroscopy image.[31]

solution, the OH radical scavenger, and the concentration of $KMnO_4$. The Mn_2O_3 nanocrystalline was firstly prepared, and the average size of the MnO_2 nanocrystalline can be as small as 6 nm. However, the preparation conditions of single-phase powder are very strict in order to prevent the formation of two-phase coexistent products.

Amorphous oxide nanopowders can also be synthesized using radiation method. When a salt solution with high concentration is irradiated, the particle concentration of the obtained reduced product is correspondingly large. When these particles are clustered, their diffusion rate to the nuclear surface is higher than growth rate of the nucleus. The growth rate of particles on the nuclear surface may be too slow for orderly arrangement in the lattice, followed by the formation of amorphous products. Because they have the characteristics of both amorphous materials and nanomaterials, nanoamorphous materials have very important theoretical and practical values in catalysis, microwave absorption, and corrosion resistance. At present, amorphous oxide powders such as nano-MoO_2 and Fe_2O_3 are prepared by irradiation, and the average particle size is $\sim$10 nm. It is generally considered that this is a product of the reoxidation of highly active nanometal particles. The nanoamorphous powder prepared by radiation method is small in particle size and its distribution range is narrow.

Nanometal sulfide materials can also be produced by radiation. Hayes et al. first reported the synthesis of CdS by γ-ray irradiation. However, only CdS colloidal solution was formed and no CdS nanoparticles were obtained, since the solution concentration was very low (10^{-4} mol L^{-1}).[33] Yin et al.[34] successfully prepared evenly distributed CdS (size: $\sim$2.3 nm) and ZnS (size: $\sim$3.6 nm) nanoparticles under the condition of high concentration by the radiation method. Sodium thiosulfate was used as the sulfur source. In addition, the effects of absorbed dose, oxidizing free radical scavengers, surfactants and reactant concentration on the final products were investigated. Furthermore, Yin et al,. also produced lead sulphide (PdS) nano powder using two approaches, and made a preliminary exploration of the preparation of amorphous sulfide using radiation.

Qiao produced CdS and ZnS nanoparticles in an aqueous system, and discussed the impact of different sulfur sources on the product phase.[35] Meanwhile, CS_2 was used as the sulfur source to successfully produce PbS and CuS nanocrystalline in a nonwater system.

8.2.2 Preparation of nanocomposite materials

The radiation method plays an important role in the preparation of nanocomposites of polymer carriers. Normally, these materials are produced by mixing preprepared nanopowders with high polymers, or by first attracting metal ions to the polymers, followed by reducing them with reductants. However, the particles and polymers are formed separately and nanoparticles are distributed unevenly. For the latter method, particle content is low because this is completely dependent on the attraction of ions to high polymers. Nevertheless, the radiation method provides a simple and effective approach to producing nanocomposites of polymer carriers. This technique requires only simple reaction equipment and moderate conditions, and can produce inorganic crystallines and polymers at one time. With this method, organic monomers and metal salts can be mixed together in aqueous phase or microemulsion at molecule level. When the mixture is irradiated by γ-ray, the polymerization of monomers and the reduction of ions occur simultaneously. This leads to the even distribution of particles in the dispersion phase. Because the polymerization rate of monomer is faster than that of ion reduction, the system viscosity increases, and then limits the further polymerization of nanoparticles, thus leading to small particles with the size of several nanometers in the dispersion phase. When the γ-ray irradiation is combined with the microemulsion method, there are advantages in the preparation of nanomaterials in particular. Generally, when preparing nanomaterials with microemulsion, it mainly relies on the collision and recombination of microemulsion droplets with exchange materials, thus producing nanoparticles. Nevertheless, due to the probability distribution of such collisions, the nanoparticles produced are usually of unsatisfactory dispersibility and the particle size is also unevenly distributed.

However, when adopting the radiation emulsion method, the reaction occurs directly in each "pool," no longer relying on the collision of microemulsion droplets, which thus can improve the homogeneity of the product particles, and adjust the microemulsion parameters, that is, it is possible to adjust the "pool" size and control the particle size of the product in a certain range. In addition, the shape of microemulsion varies in accordance with the ratio of water, oil, and emulsifier. As water droplets can be present in the shape of balls, rods or flakes, we can use, theoretically, water droplets in microemulsion to prepare a variety of morphologies of nanomaterials. It is difficult to use ordinary chemical methods to initiate the reaction without affecting the microphase structure of the microemulsion; whereas, the radiation emulsion method uses radiation instead of additional reactants to initiate the chemical reaction, which is directly effecting the preparation of nanometer materials with specific microphase structures.

8.3 PREPARATION OF POROUS POLYMERIC MATERIALS USING RADIATION TECHNIQUE

High energy γ-ray irradiation can initiate the polymerization of monomers directly at room or even lower temperature without the use of chemical initiators, which brings some unique advantages to irradiated systems, including a steady and controllable polymerization process in the whole system and no initiator residues. As a consequence, the γ-ray radiation initiation technique can be combined with the various phase separation processes in emulsion systems to produce high-performance porous polymeric materials.

8.3.1 Porous polymeric microspheres prepared via radiation-initiated seeded emulsion polymerization

Porous polymeric materials represent the fusion of typical characteristics of porous and polymer materials, including high specific surface area, low density, and good processability. They have been extensively used in the catalysis and separation fields, and have attractive potential applications in the areas of space, biomedicine, chemical engineering, and new energies. For example, porous polymer can be used to achieve the highly-efficient separation and purification of giant biomolecules, viruses, and some components of cells, since they can be easily modified with different ligands. Some chemically-modified porous polymeric materials have the capability to adsorb and separate radioactive nuclear wastes so as to protect the environment from being polluted. Therefore, as one of the numerous new materials developed in the 20th century, porous polymeric materials have attracted the strong interest of researchers and have become one of the interdisciplinary hotspots. A wide range of different types of porous polymeric materials have been fabricated, such as hollow polymer fibers, porous polymer films, porous polymer microspheres, and three-dimensionally porous polymer framework materials. The unique functional properties and the application areas of these porous polymeric materials depend on their topologies, pore sizes and size distributions, polymer matrix properties, and surficial chemical compositions. International Union of Pure and Applied Chemistry (IUPAC) suggested classifying pores into three types according to their size, that is, macropores (pores with width exceeding ~50 nm), mesopores (pores of widths between 2 and 50 nm), and micropores (pores with widths not exceeding ~2 nm). These three types of pores have their own filling mechanisms based on different absorption thermodynamics and kinetics. The key to improving the functionality of porous polymeric materials and expanding their applications is how to achieve multiple regulations on the pore structure and functionalized modification on the pore surfaces. This also stimulates the further development of the fabrication methods for new porous polymers.

In recent decades, new fabrication methods for porous polymeric materials have been proposed and can be categorized according to the pore formation mechanisms, including direct template synthesis, the self-assembly of block copolymers, and direct polymerization method. The former two methods need appropriate templates and block copolymers with well-designed molecular chain structures, which limit the preparation of complex porous structures. In the direct polymerization method, the pore structures can be constructed before or during the polymerization process of monomers based on phase separation principles in multiphase systems. Due to the compositional diversity of multiphase systems, researchers can design and control a variety of phase separation processes to obtain the expected unique pore structures. For this reason, the direct polymerization method has always been popular for material scientists since it provides a much broader design space for novel porous polymer materials.

Emulsions are the most familiar multiphase systems, and emulsion polymerization is a well-known mature polymer synthesis technique. The preparation of specially-constructed porous polymer materials via an emulsion polymerization process in a delicate monomer emulsion system is now an important subject in the theoretical and practical research field of porous materials. This opens the way for the radiation technique to serve in the development of advanced porous polymeric materials. When high energy particles (high energy EB or γ-ray) travel through the continuous phase (water, monomer, etc.), active free radicals and ions are produced, even at temperatures below freezing point, which initiate the polymerization of monomers directly without any help of extra chemical initiators. In this section, we introduce the progress of the fabrication of porous polymer microsphere materials taking advantage of γ-ray radiation-initiated seeded emulsion polymerization.

Before discussing the application of γ-ray irradiation in the seed emulsion system, it is important to consider the common mechanisms in the formation of porous polymer microspheres through seed swelling polymerization: monodispersed polymer microspheres, generally prepared by dispersion polymerization or surfactant-free emulsion polymerization, are first dispersed homogeneously in the continuous phase (usually water or an aqueous solution) as seed microspheres. Then monomer or monomer dissolved in an inert solvent (containing initiators) is added to the seed dispersion, which penetrates into and swells the seed microspheres due to their strong affinity. The seed microspheres grow gradually until swelling equilibrium is reached. At this moment, the polymerization reaction of monomers is initiated. The newly-formed polymer is separated from the inert solvent. Finally, porous polymer microspheres are obtained after the inert solvent is removed. In this method, the inert solvent acts as the porogen. Because the polymerization reaction happens in the swollen seed microspheres, the size of the porous microspheres obtained can be controlled by a well-designed swelling process. However, for conventional thermally-initiated polymerization, the swelling equilibrium of the seed microspheres will be broken with the increasing temperature. The phase separation of the newly-formed polymers is inevitably affected by the temperature. This causes difficulty with the morphological control of the porous microspheres and increases the possibility of deviation from the expected morphology. On the other hand, the morphology of the obtained porous microspheres also depends on the selection of the porogen, that is, the hydrophobic solvent of monomer. The inert solvent should meet the following requirements. First of all, it should dissolve little in water (strongly hydrophobic), but is a good solvent for both monomer and seed microspheres. Second, it should be a poor solvent for the newly-formed polymer (to ensure the phase separation between the solvent and polymer during the polymerization process). Lastly, it should be eliminated easily (to form the pores). Based on the above requirements, commonly-used inert solvents are hydrophobic and volatile hydrocarbons such as toluene, nonpolar alkane, and some mixed organic solvents, which are generally so highly toxic that their application is limited in industry due to environment protection concerns. Therefore, if the polymerization reaction can be initiated by γ-ray irradiation at room temperature and the toxic organic porogen can be replaced by water (i.e., the continuous phase), it will certainly be beneficial for the morphological control of the porous product and make the whole preparation more green and environmentally-friendly. This idea has been explored for the past 10 years by Ge and Wang's group at the University of Science and Technology of China.

8.3.1.1 Cage-like hollow polymer microspheres prepared by γ-ray initiated seeded emulsion polymerization

If water needs to spontaneously penetrate swollen polymer seed microspheres during the swelling process, it is evident that there should be strong hydrophilic groups in the seed microspheres so that water diffuses into the seed microspheres driven by the strong interactions between the water molecules and hydrophilic groups (e.g., hydrogen bond interactions). The introduction of strong hydrophilic sulfonate groups ($-SO_3H$) is a good way to improve the hydrophilicity of nonpolar materials because it can be easily achieved by a simple sulfonation reaction between the hydrophobic polymer chain (e.g., PS) and the concentrated sulfuric acid.[35,36] Therefore, the monodispersed PS seed microspheres shown in Fig. 8.14A and B were first obtained by the dispersion polymerization of styrene in water with the initiator azodiisobutyronitrile and the stabilizer poly(4-vinyl pyrrolidone). The size of the seed microspheres can be controlled in the range of 800 nm and several μm. The PS seed microspheres were sulfonated by concentrated sulfuric acid for a certain time to obtain the sulfonated PS (SPS) microspheres shown in Fig. 8.14C and D. The degree of sulfonation

Figure 8.14 SEM and TEM images of PS (A, B) and SPS (C, D) microspheres.[37] *SPS*, sulfonated PS.

(DS) greatly depends on sulfonation time, temperature, and the weight ratio of PS seed microspheres to concentrated sulfuric acid. XPS and element analysis can be used to analyze quantitatively the DS of the surface and the SPS microspheres.[37]

SPS, acting as the seed microspheres, can be homogeneously and stably dispersed in water without any help of stabilizer due to the strong hydrophilic sulfonate groups, and even at the early stage of the addition of the monomer methyl methylacrylate (MMA), they play a role as particle stabilizers for MMA droplets to form a typical Pickering emulsion system, as shown in Fig. 8.15A[36]. However, the initial Pickering emulsion is unstable. Along with the diffusion of MMA into the SPS microspheres, the amount of MMA monomer in the big droplets gradually decreases with the standing time, while the volume of the SPS seed microspheres expands correspondingly due to monomer swelling. As a consequence, the initial Pickering emulsion droplets ultimately disintegrate into myriads of individual MMA-swollen SPS microspheres, as displayed in Fig. 8.15B.

It should be noted that MMA has the ability to dissolve SPS microspheres. This means the mobility of SPS molecular chains in seed microspheres is improved with the swelling of MMA monomers, which opens the channel for water molecules to permeate into SPS seed microspheres driven by the strong hydrophilicity of sulfonate groups. At the same time, the amphiphilic SPS molecular

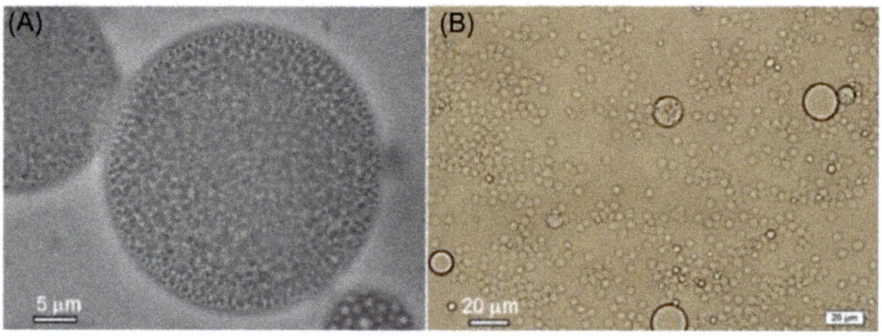

Figure 8.15 Optical images of emulsion at different stages after the addition of MMA monomer into the aqueous dispersion of SPS microspheres under stirring [(A) 1 h; (B) 24 h].[36] *MMA*, monomer methyl methylacrylate; *SPS*, sulfonated PS.

chains have enough mobility to adjust their conformation. They act as surfactants to stabilize the permeated water phase, forming the inside water phase in the swollen SPS seed microspheres. This in turn promotes the diffusion of water to the inner part of the swollen seed microsphere until the osmotic pressure balance is reached. The above monomer swelling and water diffusion processes cause the large volume expansion of the SPS seed microspheres. Fig. 8.16A shows the morphologies of the SPS seed microspheres with an original average diameter of 1.6 μm after being swollen by MMA monomers for 48 h after free-standing at room temperature, followed by freeze-drying. It is clear that the diameters of the seed microspheres have risen to ~5 μm, and most importantly, a porous skeleton structure of the swollen seed microspheres is already present. The collapsed structure also indicates the SPS seed microspheres are in a swollen state with a certain degree of mobility before freeze-drying. This is direct evidence of the proposed dynamic monomer swelling-water osmosis process in the seed microspheres, which has already been proved by many other similar systems including those with nonpolymerizable organic solvent in place of MMA monomer.[37,38]

When the MMA-swollen SPS microspheres are irradiated by ^{60}Co γ-ray under certain conditions at room temperature, MMA

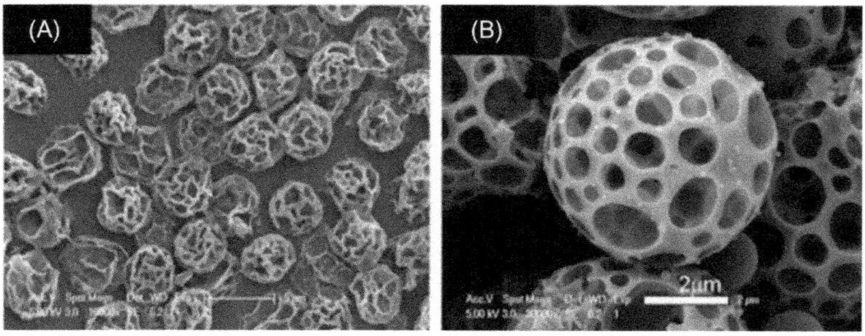

Figure 8.16 SEM images of SPS seed microspheres after being swollen by MMA monomers for 48 h after free-standing at room temperature, followed by different treatment processes: (A) direct freeze-drying; and (B) exposure in the ^{60}Co γ-ray radiation field (dose rate: 80 Gy min^{-1}, total absorbed dose: 9.6 kGy) followed by vacuum drying for 24 h.[38] *SPS*, sulfonated PS; *MMA*, monomer methyl methylacrylate.

monomers are initiated to polymerize in situ into solid poly(methyl methylacrylate) (PMMA), resulting in the formation of the solid SPS/PMMA complex skeleton of the microspheres. Therefore, the porous structure of the swollen seed microspheres remains perfect after γ-ray radiation-induced in situ polymerization, and cage-like hollow microspheres appear after the evaporation of the inside water phase, as displayed in Fig. 8.16B. The sizes of the cage-like hollow microspheres and the pores nearly correspond to those before polymerization (Fig. 8.16A). This preparation route fully reflects the advantage of the radiation initiation technique for morphological control in complex multiple-phase systems compared with thermal initiation. In the latter case, monomer swelling and water diffusion behavior vary constantly, due to the enhanced movement of molecules with the increase of temperature, resulting in the loss of the morphological control of the microspheres after polymerization.

8.3.1.2 Cross-linked porous polymer microspheres prepared by γ-ray initiated seeded emulsion polymerization

Based on the above discussion of the pore formation mechanism of dynamic monomer swelling-water osmosis in seeded SPS microspheres, it can be inferred that the pore structures in the seed microspheres can be subtly adjusted by all the factors, which can affect the monomer swelling and water osmosis processes. These factors involve the structural and compositional parameters of SPS seed microspheres (size, molecular weight, DS, cross-linking degree, etc.), the polarity and dose of monomer, and swelling conditions (temperature, agitation, etc.). Using the radiation-initiated polymerization technique, all these subtle morphological regulation can be captured and finally we can obtain porous polymer microspheres with diverse pore structures and whole morphologies, even specific functionalities. A particulate example is if cross-linked SPS microspheres are used as the seed, porous polymer microspheres with new morphologies totally different from the cage-like hollow structure can be fabricated.

The kinetics of the swelling and the polymerization of monomers in cross-linked polymer microspheres are more complicated than those in non-cross-linking seed microspheres.[36–40] In general, the main morphology of monomer-swollen non-cross-linking seed microspheres is a core–shell structure, that is, a monomer-swollen seed polymer core encapsulated with a shell comprised mainly of monomer. The polymerization of the monomers is initiated from the monomer shell and proceeds homogeneously in the all seed microspheres, resulting in isotropic spherical core–shell structured complex microspheres,[39,40] like the perfectly spherical cage-like hollow microspheres shown in Fig. 8.17B prepared from non-cross-linking SPS microspheres. However, according to thermodynamic analysis, the phase separation process in cross-linked polymer microspheres is determined by three factors, namely, the monomer-polymer mixing force arising from the absorption of monomer, the cross-linking network elastic-retractile force associated with the change in the configuration of the polymer network upon swelling, and the microsphere-water interfacial tension force caused by the increase of the surface area on swelling.[41] Microspheres with a variety of nonspherical morphologies such as ellipses and egg-like sheets, symmetrical and asymmetrical

Figure 8.17 SEM images of multihollow microspheres prepared by swelling CSPS seed microspheres by styrene after free-standing at room temperature followed by exposure to ^{60}Co γ-ray irradiation: (A) walnut-like porous microspheres (diameter: 3 μm, dosage of DVB: 0.7%, dose rate: 35 Gy min^{-1}; absorbed dose: 45 kGy); (B) plum-like porous microspheres (diameter: 3 μm, dosage of DVB: 2%, dose rate: 35 Gy min^{-1}; absorbed dose: 45 kGy); (C) walnut-like porous microspheres (diameter: 5 μm; dosage of DVB: 2%; dose rate: 50 Gy min^{-1}; absorbed dose: 45 kGy).[41] *DVB*, divinyl benzene; *CSPS*, sulfonated cross-linked PS.

bi-hemispheres, ice-cream cones, and popcorn-like shapes, have been created based on various dynamic processes of the phase separation achieved through different combinations of seed microspheres and monomers at a condition of chemical initiation.[42] Similarly, when sulfonated cross-linked PS (CSPS) microspheres (using divinyl benzene (DVB) as the cross-linking agent) are selected as the seed microspheres, and swollen with styrene monomers, followed by exposure under γ-ray irradiation, porous latex particles with a series of special morphologies, including spherical multihollow, plum-like, and walnut-like, as shown in Fig. 8.17, are produced, rather than spherical cage-like hollow microspheres with smooth surfaces.[43–45]

The reason for the creation of these special morphologies is the phase separation of the newly-formed PS phase at the interface between the cross-linked CSPS microspheres and the outside continuous water phase, or the inside water phase formed accompanied by the swelling of CSPS seed microspheres, as shown in Fig. 8.18.[46] The polymer-water interfacial tension is in inverse proportion to the interfacial curvature radius, indicating that the new polymer phase is readily created on the interface with a smaller curvature radius.[47] It has been reported that the new polymer phase is hard to form on the outer surface of solid cross-linked seed microspheres with a diameter of above 2 μm after being swollen at room temperature.[41] In the case of Fig. 8.17, the diameters of all CSPS seed microspheres are above 2 μm. Therefore, the newly-formed PS phase occurs basically at the interfaces between the inside water phase and the seed microspheres, causing the diffusion of the swelling monomers into the inner part of the seed microspheres. As a result, the volume of the inner water phase decreases, and the whole swollen seed microsphere shrinks, leading to the formation of grooved surface morphology. The inset TEM images in Fig. 8.17A and C clearly show a multiporous structure in the inner part of the walnut-like microspheres. The increase of cross-linking degree in the seed microspheres produces much higher cross-linking network elastic-retractile force and promotes the phase separation of the newly-formed polymer phase, causing increased

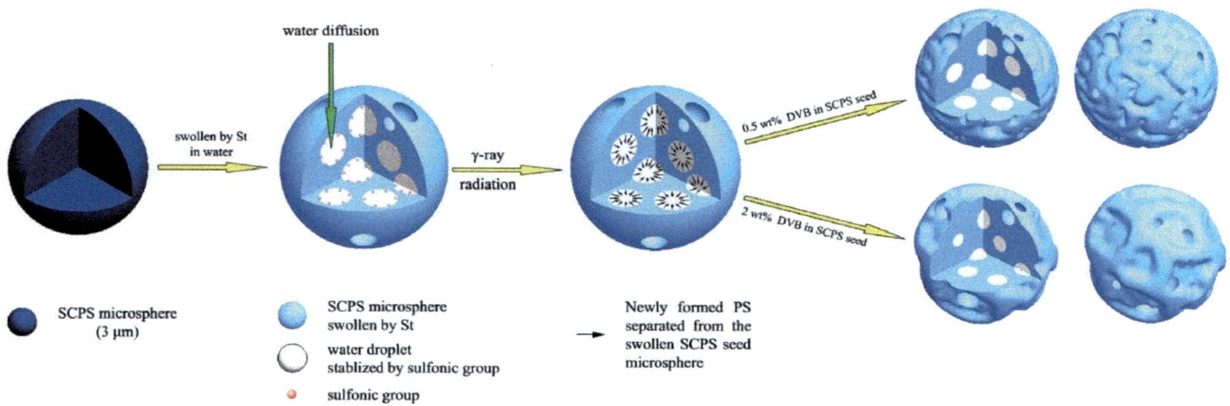

Figure 8.18 Formation mechanism of walnut-like and plum-like multihollow microspheres prepared from the CSPS seed microspheres with a size above 2 μm via ^{60}Co γ-ray radiation seed emulsion polymerization.[46] *CSPS*, sulfonated cross-linked PS.

shrinkage of the seed microspheres and the formation of plum-like microspheres, as shown in Fig. 8.17B and C.

On the other hand, when the diameter of the seed CSPS microspheres is lower than 500 nm, the new polymer phase can be produced at the interfaces between the seed microspheres and both the inside and outside water phases, as illustrated in Fig. 8.19. The growth of the new polymer phase in the two opposite directions may balance the stress in the microspheres so that spherical multiporous microspheres are obtained.

The actual morphologies of microspheres prepared from the CSPS seed microspheres with a size of 170 nm via ^{60}Co γ-ray radiation seed emulsion polymerization are exhibited in Fig. 8.20A and B, in which the sulfonation time of the CSPS seed microspheres are 5 and 8 h respectively. It can be seen that spherical microspheres with inner porous structures and regular surfaces are obtained. The sulfonation degree of seed microspheres affects the diffusion of styrene and water, resulting in different volumes of inner water phase and ultimate pore sizes of the microspheres produced.[44] It should be noted that compared with the non-cross-linking seed microspheres (Fig. 8.16A), CSPS seed microspheres can maintain their shapes even after being swollen by styrene for 15 h, as shown in Fig. 8.20C and D. The results indicate a similar pore structure in the swollen seed microspheres before and after the polymerization of styrene initiated by γ-ray irradiation. Once again, we can see the advantage of the radiation-initiated polymerization technique in fixing the morphologies formed before polymerization, which not only helps us to further study and verify the multiphase separation theories, but also makes the fabrication of porous microspheres designable and replicable, which are of great significance for the industrial production of fine-structured porous polymer microspheres.

8.3.2 Interconnected porous polymer materials prepared via radiation-initiated polymerization in high internal phase emulsions

Polymer matrices with interconnected pores have attracted extensive interest due to their wide practical application in

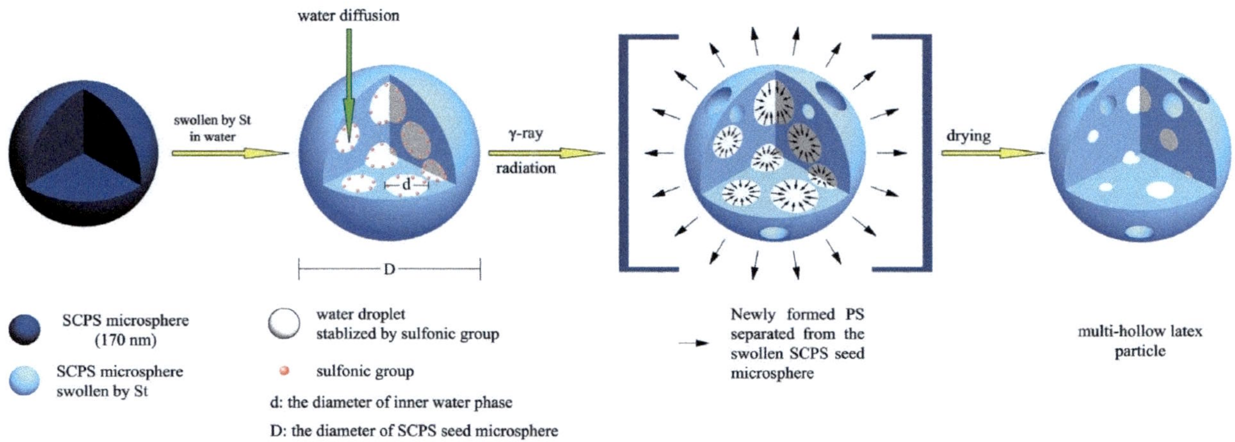

Figure 8.19 Formation mechanism of spherical multihollow microspheres prepared from CSPS seed microspheres with a size of 170 nm via ^{60}Co γ-ray radiation seed emulsion polymerization.[44] *CSPS*, sulfonated cross-linked PS.

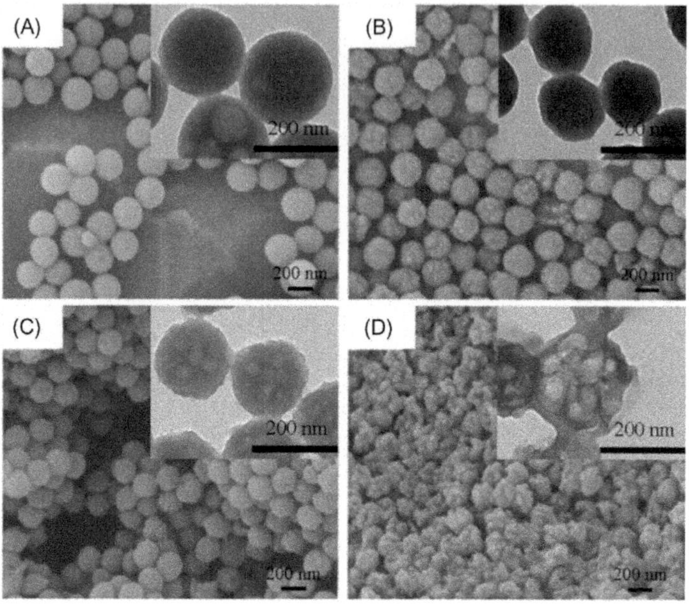

Figure 8.20 SEM and TEM (inset) images of CSPS seed microspheres with a size of 170 nm and different sulfonation times after being swollen by styrene followed by different treatments: (A) sulfonation for 5 h, irradiated by ^{60}Co γ-ray after being swollen; (B) sulfonation for 8 h, irradiated by ^{60}Co γ-ray after being swollen; (C) sulfonation for 5 h, directly freeze-drying after being swollen; and (D) sulfonation for 8 h, directly freeze-drying after being swollen (dose rate:35 Gy min^{-1}, absorbed dose: 45 kGy).[44] *CSPS*, sulfonated cross-linked PS.

chromatographic separation, catalysis, sensors, tissue engineering, and environment science. Many methods have been developed for the preparation of interconnected porous polymer materials, including the supercritical fluid method, the self-assembled colloidal crystal template method, and the sacrificed polymer template method. The high internal phase emulsion (HIPE) template method is one of the most effective ways for the synthesis of macro porous polymer materials (pore diameter exceeds 50 nm) because of its high design flexibility and the ability to precisely regulate the size and size distribution of the pores and channels. HIPE is defined by Dissent as an emulsion in which the total volume fraction of the droplet phases (namely, internal phase) exceeds 74.05%,[48] which is equivalent to the maximum space occupied by monodispersed incompressible spherical particles in a unit cube. HIPE with a continuous oil (monomer) phase and dispersed water droplets, that is, water-in-oil (W/O) emulsion, is mostly used as the template for interconnected porous polymer materials, since once the monomers in HIPE are initiated to polymerize, the whole emulsion system rapidly transforms into a polymer monolith with interconnected macropores, which gives the product the name of PolyHIPE. Recently, double (or multiple) emulsions formed by dispersing W/O HIPE in water have been explored for use as the template for porous polymer microspheres with interconnected macropores. Some hydrophilic interconnected porous polymer materials can be synthesized based on the oil-in-water HIPE template. Naturally, the stability of HIPE plays an essential role in the formation of the inner pore structure and the surface morphology of polyHIPE, making a stable HIPE template the key to successful synthesis. Many factors, including the properties and concentration of the surfactant, the composition of the emulsion, the temperature, and the mixing technique, can affect the stability of HIPE. It is well-known that thermal motion promotes the collision of dispersed droplets and accelerates the coalescence rate of droplets. As a consequence, the emulsion stability decreases remarkably with the increase of temperature. The same is true for HIPE with styrene or MMA as the monomer phase.[49] To date, most PolyHIPEs have been prepared using chemical initiator at a relatively high temperature. In this case, a large amount of surfactant is often introduced to maintain the stability of HIPE, which causes extra waste of surfactant and increases the difficulty of product purification. Liu's group at the University of Science and Technology of China has made encouraging progress in the preparation of functional interconnected porous polymer materials at room temperature using radiation-induced polymerization in HIPE.[49–53] For a typical W/O HIPE template composed of styrene (St), DVB, water, and the surfactant Span 80 (sorbitan monooleate, SP80), only 1.4 wt% of SP80 is needed to obtain macroporous PolyHIPE (P(St-DVB)) via ^{60}Co γ-ray radiation-initiated polymerization. However, the corresponding dosage is at least up to 5 wt% using a chemical initiator.[49] For a W/O HIPE template composed of MMA, DVB, water, and the surfactant Span 80 (SP80), PolyHIPE (P(MMA-DVB)) can be easily fabricated at room temperature under ^{60}Co γ-ray irradiation, but cannot be formed using the chemical initiation method from the same HIPE template.[49] Therefore, the radiation initiation technique is more effective for the preparation of PolyHIPEs and other novel porous materials based on HIPE systems.

An interesting example is the one-step synthesis of MnO_2-nanoparticle-loaded PS composite foam by γ-ray radiation-initiated polymerization of HIPE. First, a mixed surfactant composed of SP80 and 12-acryl-oxy-9-octadecenoic acid (AOA, produced in-house), neopentyl glycol diacrylate (cross-linking agent), and isopropanol (radical scavenger) are dissolved in St monomer. Next, the aqueous solution of $KMnO_4$ is dropped slowly into the above monomer phase. After that, the system is agitated vigorously until a homogenous and viscous HIPE forms. The HIPE is irradiated by ^{60}Co γ-ray to initiate the polymerization of St and the reduction of $KMnO_4$ into MnO_2 nanoparticles. Finally, PS foam loaded with MnO_2 nanoparticles with a size of 10–12 nm is one-step synthesized, as shown in Fig. 8.21. If the single surfactant AOA is used in this HIPE system, the pore size of the obtained PS foam reduces with the increase of the dosage of $KMnO_4$. The load capacity of MnO_2 nanoparticles in the PS foam can reach up to 15.4%. Moreover, the loaded MnO_2 nanoparticles have excellent catalytic performance, similar to free MnO_2 nanoparticles.[50]

The magnetic Fe_3O_4 nanoparticles can be immobilized on the surfactant AOA through the esterification reaction between the –OH groups on the surface of Fe_3O_4 nanoparticles and the carboxyl groups on AOA molecules. After the Fe_3O_4 nanoparticles are hyhrophobically modified to a certain extent, they are homogeneously dispersed in the oil phase (St-DVB), acting as Pickering emulsifier to stabilize the water droplets in HIPE, as shown in Fig. 8.22.[51] The polymerization of St in HIPE is initiated at room temperature by ^{60}Co γ-ray irradiation, leading to the formation of

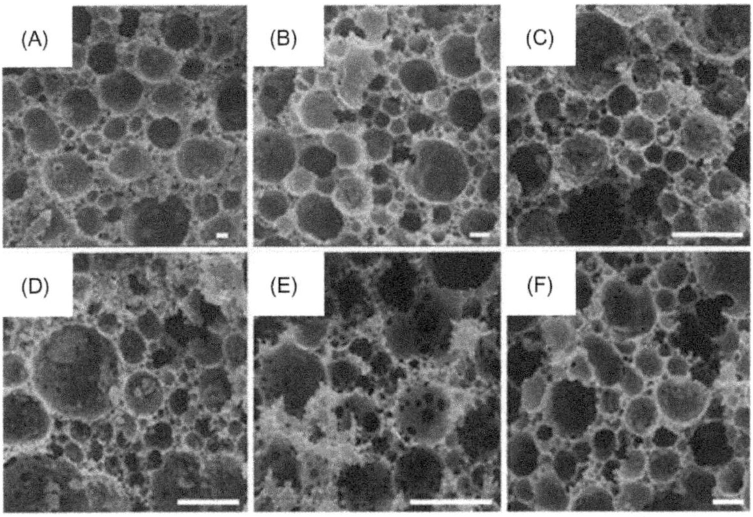

Figure 8.21 SEM images of PS foam loaded with MnO_2 nanoparticles prepared at different concentration of $KMnO_4$: (A) $C_{KMnO_4} = 0.16\ mol\ L^{-1}$; (B) $C_{KMnO_4} = 0.26\ mol\ L^{-1}$; (C) $C_{KMnO_4} = 39\ mol\ L^{-1}$; (D) $C_{KMnO_4} = 0.48\ mol\ L^{-1}$; (E) $C_{KMnO_4} = 0.16\ mol\ L^{-1}$; (F) $C_{KMnO_4} = 0.26\ mol\ L^{-1}$ (the fraction of aqueous phase weight in HIPE: 80%; scale bar: 10 μm).[50] *HIPE*, high internal phase emulsion.

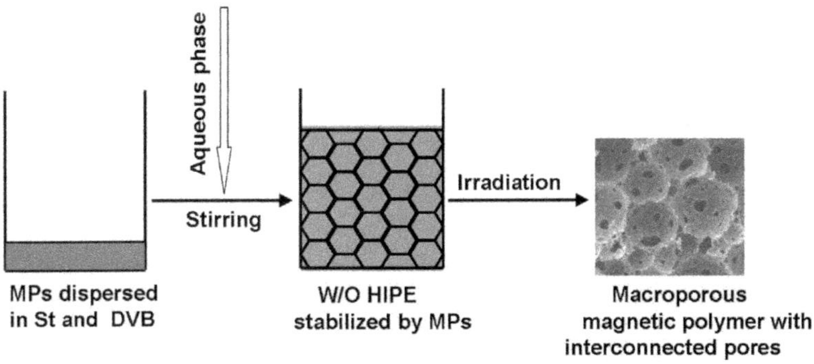

Figure 8.22 Formation of W/O HIPE stabilized by AOA-modified Fe_3O_4 nanoparticles and the morphology of PolyHIPE prepared by γ-ray radiation-initiated polymerization.[51] *W/O*, water-in-oil; *HIPE*, high internal phase emulsion; *AOA*, 12-acryl-oxy-9-octadecenoic acid.

interconnected porous PolyHIPE materials with uniform pore size distribution. At the same time, the Young's modulus of the prepared PolyHIPE is also enhanced to a certain degree, and can reach up to 69.7 MPa. The compressive strength of the PolyHIPE also rises to 5.29 MPa. Above all, PolyHIPE exhibits outstanding magnetic properties so that they can be easily separated in the external magnetic field, indicating potential applications as carriers for adsorption and separation.

The active epoxy groups are extensively used to hydrophilically modify nonpolar porous polymer materials and enhance the reactivity for further functional modifications.[54] A common method for the introduction of epoxy groups is the block (or graft) copolymerization of GMA. However, it is difficult to maintain the stable status of HIPE with a monomer phase containing GMA and DVB at high temperature because of the strong hydrophilicity of GMA. An interconnected porous PolyHIPE based on (P(GMA-DVB)) can still be fabricated when the system is irradiated by ^{60}Co γ-ray irradiation at room temperature. Two emulsifiers, polyglycerin oleate (PGEO) and SP80, are used to prepare W/O HIPEs containing GMA-DVB. After ^{60}Co γ-ray irradiation, both HIPEs transform into interconnecting porous PolyHIPEs, as shown in Fig. 8.23.[52] The pore size of the PolyHIPE with SP80 as the

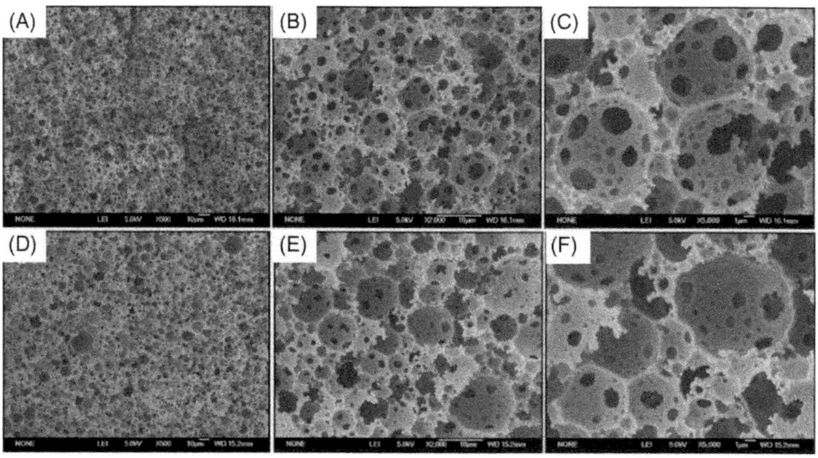

Figure 8.23 SEM images at different magnifications of PolyHIPEs prepared via γ-ray radiation-initiated polymerization of W/O (GMA-DVB/H_2O) HIPEs stabilized by different emulsifiers: (A)–(C) polyglycerin oleate; (D)–(F) Span 80.[52] *W/O*, water-in-oil; *HIPE*, high internal phase emulsion; *DVB*, divinyl benzene.

emulsifier is larger than that with PGEO, which may relate to the stronger ability of PGEO to inhibit droplet coalescence than SP80. These epoxy-modified PolyHIPEs have good adsorption ability for phenol in cigarette smoke.[54]

The polymerization of HIPE results in the solidification of the emulsion system. Therefore, the macroscopic shape of the derived porous PolyHIPEs is in accord with the geometric shape of the reactor, basically being blocky or columnar monoliths, which conveniently fill the chromatographic columns. Particulate porous PolyHIPEs materials with regular shapes have been seldom reported. Some attempts to fabricate interconnecting porous PolyHIPE microspheres have been carried out recently through the precipitation polymerization of the HIPEs dispersed in water. An example is illustrated in Fig. 8.24.[53] First, a certain amount of water is added drop-wise and slowly to an oil phase composed of St, ethylene glycol dimethacrylate, and the ammonium salt of AOA, to

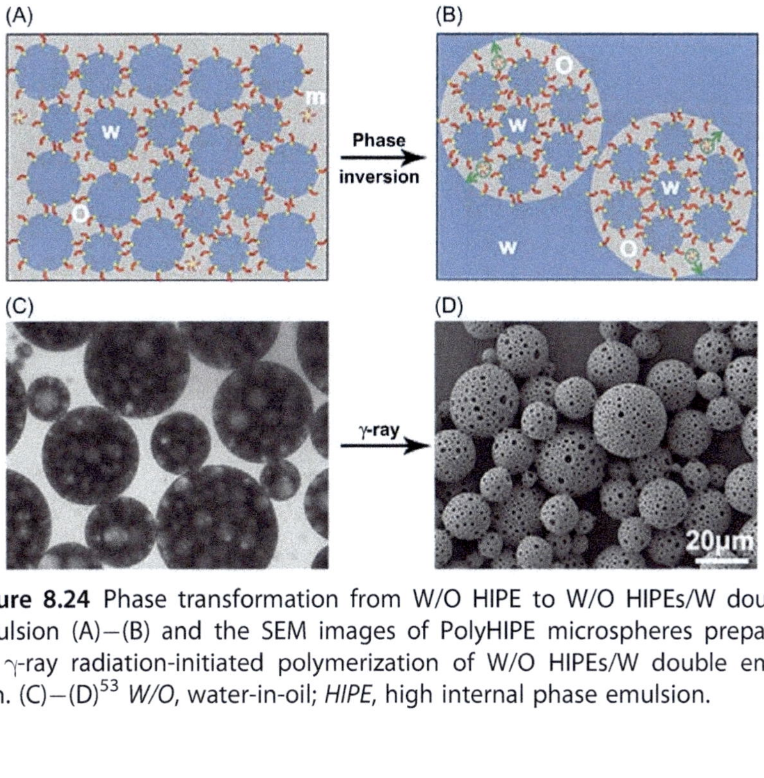

Figure 8.24 Phase transformation from W/O HIPE to W/O HIPEs/W double emulsion (A)–(B) and the SEM images of PolyHIPE microspheres prepared via γ-ray radiation-initiated polymerization of W/O HIPEs/W double emulsion. (C)–(D)[53] *W/O*, water-in-oil; *HIPE*, high internal phase emulsion.

form a W/O HIPE. Then, a large amount of water is continuously added to the HIPE until a double emulsion with W/O HIPEs emulsion as the droplet phase and water as the continuous phase is formed. The W/O HIPEs/W emulsion is irradiated by ^{60}Co γ-ray irradiation, and interconnected porous PolyHIPE microspheres with a size of 10–20 μm are obtained. The phase transformation process from W/O HIPE to W/O HIPEs/W double emulsion in this synthesis route is greatly influenced by a number of factors, such as the concentration of AOA and the salt (NaCl), the polarity of the monomer, the pH of the emulsion, the method of adding water, and the agitation mode. It is noted that St can be replaced by acrylic monomers such as MMA and *n*-butyl acrylate to prepare the interconnected macroporous polymer microspheres.[53]

Based on the above research progress in the fabrication of novel porous polymeric materials, the advantages of the radiation-initiated polymerization technique are obvious and sometimes irreplaceable in the synthesis routes based on the diverse phase separation mechanisms in multiphase systems. It can not only produce high-purity products, but more significantly, can avoid the complication of the thermodynamic and kinetic processes in systems caused by the increase of the temperature, making the principle of synthesis clearer and the synthesis routes more designable and repeatable. From this point of view, the perfect combination of the phase separation principles and the radiation chemical effect in multiphase systems may lead to the development of an infinite number of novelly and uniquely structured polymeric and polymeric-inorganic composite materials, which means the radiation technique has great potential to serve for the creation of the advanced materials.

8.4 RADIATION EFFECTS OF IONIC LIQUIDS

Room-temperature ionic liquids (RTILs) with melting points at or below ambient temperature are liquid organic salts composed entirely of anions and cations. Compared with traditional volatile molecule solvents, RTILs have many unique properties due to the strong interaction between ions, such as excellent conductivity, low vapor pressure, nonflammability, wide electrochemical window, and physicochemical stability. With the rapid development of green chemistry, the emergence of RTILs has injected new vitality into the traditional extraction and separation technologies. Due to their attractive properties, such as excellent thermal and chemical stability, negligible vapor pressure, designable properties, and favorable solvating properties for a range of polar and nonpolar compounds, RTILs combined with traditional extracting agents are considered as promising systems in the field of extraction and separation. Since highly-efficient extraction of radioactive metal ions from aqueous solution can be achieved by using RTILs in combination with extracting agents rather than conventional organic solvents, RTILs are promising solvents instead of conventional organic solvents in the reprocessing of spent nuclear fuel. However, during the practical application in the treatment of spent nuclear fuel, RTILs and extracting agents are exposed to high energy. Therefore, studies of the radiation effect of RTILs and their extraction for metal ions are of great importance in the assessment of the feasibility of their application in the reprocessing of liquid nuclear waste.

8.4.1 Radiation effect on imidazolium ionic liquids

8.4.1.1 Influence of radiation on physical properties of imidazolium ionic liquids

The radiation effect of imidazolium ILs was studied under γ-irradiation. The density, surface tension, and refraction index of the ILs changed slightly by an absorbed dose of $\sim$500 kGy. Hagiwara et al.[55] found that the viscosity of ILs was mainly controlled by Van der Waals' force and hydrogen bonding in the system. Berthon et al.[56] pointed out that 1-butyl-3-methylimidazolium chloride ([C_4mim]Cl) produced strong polar substances after γ-irradiation, which increased the hydrogen bonding in the system, resulting in a significant increase in the viscosity of the system. Kou et al.[57] used acetone and DMF as molecule probes to study the variation of absorption peaks corresponding to carbonyl groups of molecule probes with different polar solvents. As the polarity of solvent increased, the absorption peak corresponding to carbonyl groups moved toward low wave numbers. Yuan et al.[58] confirmed that some strong polar substances were produced in irradiated $[C_4mim][PF_6]$ by micro Fourier transform infrared spectrometer (micro-FTIR) spectroscopy, and speculated that the formation of strong polar substances contributed to the increase of viscosity in the system.

8.4.1.2 Radiation-induced color changes of imidazolium ionic liquids

During the radiation process, colorless and transparent ILs show color changes. A new absorption band at 230–400 nm corresponding to colored radiolytic products was found in irradiated samples and the absorption band enhanced with increased absorbed dose.[59] The UV–vis spectra of ILs with different cations after irradiation were investigated. The absorption band at 200–400 nm of irradiated ILs with ammonium cations was much weaker than that of imidazolium ILs, indicating that the change of UV–vis spectra was mainly related to the radiolysis of cations.[59] Furthermore, the effect of anions of IL on the radiation-induced color was investigated. Due to the electron density of $[PF_6]^-$ being lower than that of $[NTf_2]^-$, the imidazolium cation of $[C_4mim][PF_6]$ absorbed higher radiation energy than that of 1-butyl-3-methylimidazolium bis(trifluoromethylsulfonyl)imide ($[C_4mim][NTf_2]$), leading to more colored radiolytic products and higher absorption in UV–vis range.[59] In order to further study the reasons for the radiation-induced darkening of ILs, Yuan et al.[60] studied the decolorization of irradiated ILs treated with nitric acid and $KMnO_4$. After treatment, the intensity of the peak at 290 nm decreased obviously and the decoloration of irradiated ILs occurred, suggesting that the peak at 290 nm was assigned to the colored species and the species could be oxidized (Fig. 8.25). Treatment with $KMnO_4$ is a homogeneous decolorization method, but some problems exist such as heavy metal pollution and the recovery difficulty of ILs.[60] Heterogeneous

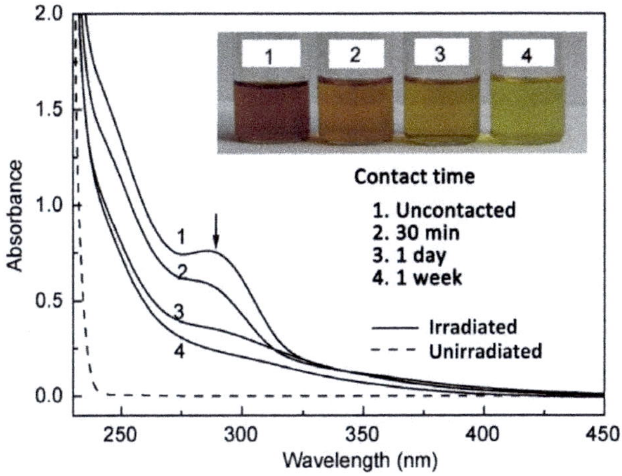

Figure 8.25 UV–vis spectra of irradiated $[C_4mim][NTf_2]$ after decoloration.[60]

decolorization can be achieved by using nitric acid. The decolorization of irradiated $[C_4mim][NTf_2]$ and irradiated 1-butyl-3-1-methylpyrrolidinium bis[(trifluoromethyl)sulfonyl]imide ([BMPyrr] $[NTf_2]$) was achieved easily by nitric acid washing.[60] The irradiated ILs was also treated using ozone with strong oxidizing property. Compared to oxidation by $KMnO_4$ and nitric acid, the ozone bleaching method is more simple and effective. This method provides an effective way of IL recycling in practical applications.

Furthermore, the color of irradiated ILs can be attenuated after dilution by acetonitrile.[60] The change of the aggregation state of ILs was confirmed by fluorescence spectroscopy, leading to the formation of super-conjugated structure and color change. By changing the structure of imidazolium cations, it was found that the coupling reaction or the complex reaction of the cationic radical could be inhibited when the C(2)-H on the imidazolium ring was substituted by methyl, resulting in reduced amounts of cationic dimer, and less color change than that of $[C_4mim][NTf_2]$. Cationic dimer in irradiated ILs was also confirmed by high-resolution electrospray ionization mass spectrometry (ESI-MS). Therefore, the color change of imidazolium ILs during radiolysis is mainly related to the structure of imidazolium cations. Dehydrogenation reaction of C(2)-H on the imidazolium ring happened during irradiation, leading to the formation of radiolytic products containing double bonds. The formation of imidazolium dimer and aggregates of ILs, which are mainly colored substances, was the main reason for the radiation-induced darkening of ILs. The absence of any significant change in the various spectroscopic results was an indication that the radiolytic species are formed in very small quantities. Since less than 1% of $[C_4mim][NTf_2]$ underwent radiolysis after irradiation at 550 kGy; this change cannot be detected by micro-FTIR and nuclear magnetic resonance (NMR).[60]

8.4.1.3 Identification of radiolytic products of imidazolium ionic liquids

The first study on the radiation effect of ILs was carried out by Allen et al. in 2002.[61] The radiation effects on several hydrophilic imidazolium ILs were evaluated and they noted that less than 1% of ILs underwent radiolysis when exposed to an absorbed dose of 400 kGy. The radiation chemical yields of methyltributylammonium cation ($[N_{1444}]^+$) and bis[(trifluoromethyl)sulfonyl]imide anion ($[NTf_2]^-$) of $[N_{1444}][NTf_2]$ were determined to be -0.38 and $-0.25\ \mu mol\ J^{-1}$ (the negative before values means the loss of matter), respectively. Therefore, $[N_{1444}][NTf_2]$ had high radiation stability.[62] Wang et al. studied the effects of γ-radiation on $[C_4mim][PF_6]$[59] and 1-butyl-3-methylimidazolium tetrafluoroborate ($[C_4mim][BF_4]$).[63] They found that the radiation chemical yields of F^- ($G(F^-)$) of several kinds of ILs are related mainly to the bond strength and anion structure of F–X (X represents atoms directly connected with F atoms). The radiation stability of $[C_4mim]Cl$ could be improved by adding $FeCl_3$, which indicated that the solvated electrons could be captured by $[FeCl_4]^-$ to reduce the damage to ILs.[64] In addition, the radiation stability of $[C_4mim]Cl$ had a direct relationship with the added amount of $FeCl_3$.[65]

Zhai et al.[64,65] investigated the radiolytic products of $[C_4mim][PF_6]$ and $[C_4mim][NTf_2]$. In order to separate water-soluble radiolytic products from the organic phase, the irradiated ILs were washed with D_2O, and then the aqueous phase was analyzed using various spectroscopic methods. AS D_2O-washing is an efficient

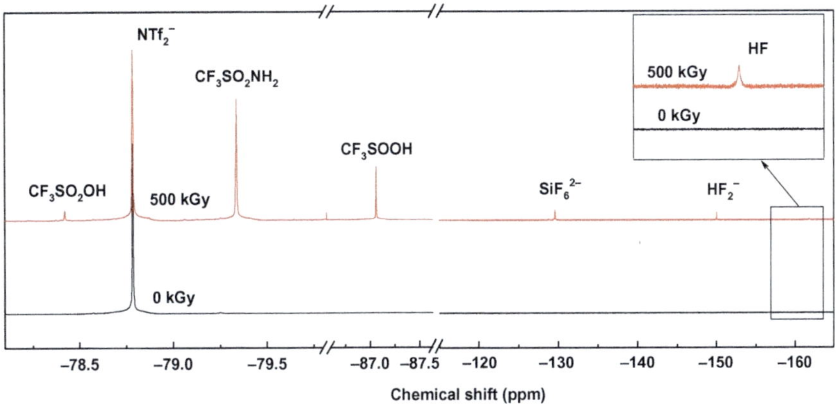

Figure 8.26 ^{19}F NMR spectra of [C$_4$mim][NTf$_2$] before and after irradiation.[66] *NMR*, nuclear magnetic resonance.

way to avoid the influence of H_2O in micro-FTIR and NMR analysis, the water-soluble radiolytic products of ILs can be detected by normal spectral analysis methods. The trace water-soluble acidic radiolytic products of [C$_4$mim][NTf$_2$] were identified as CF_3SO_2OH, CF_3SOOH, $CF_3SO_2NH_2$, HF, and H_2SO_3 (Fig. 8.26).[66] In addition, the water-soluble radiolytic products of [C$_4$mim][PF$_6$] were also analyzed by the D_2O-washing method. The main radiolytic products (HF and HOP(O)F_2) of [C$_4$mim][PF$_6$] were identified and their amount was quantified by ^{19}F NMR.[67] Compared with the radiation chemical yields of acidic radiolytic products of [C$_4$mim][NTf$_2$], the radiation stability of [C$_4$mim][PF$_6$] was better than that of [C$_4$mim][NTf$_2$] and was clearly influenced by the chemical structure of anion. Wu et al.[68] also found that the radiation chemical yield of $[PF_6]^-$ was less than that of [NTf$_2$]. It can be deduced that $[PF_6]^-$ has a simple chemical structure and lower P-F bond energy than that of $[NTf_2]^-$.

8.4.1.4 α-Radiation effect on imidazolium ionic liquids

The separation of transplutonium elements, especially for minor actinides–lanthanides (MA–Ln), is a huge challenge during the advanced fuel cycle, because these elements have very similar chemical properties. Moreover, most long-lived MA elements produce α-ray. Therefore, extracting agents and solvents are used under high level α-ray radiation. In spent nuclear fuel reprocessing, the penetrating capacity of α-ray is very weak and almost absorbed by the extraction system. Therefore, a study of α-radiation effects on the extraction system is extremely important. Simulation of α-radiation effects can be achieved through irradiation by using He ion beams. This has some advantages, such as the lack of radioactive operation; and the samples can be further studied and characterized easily.

Ao et al.[69] studied the α-radiation effects on ILs simulated by He^+ beams from a heavy ion accelerator, and designed the equipment (Fig. 8.27). The radiolytic products of ILs irradiated by He^+ beams were analyzed and studied. Compared with γ-irradiated $[C_4mim][NTf_2]$, the absorbance of the γ-irradiated samples at 285 nm was much higher than that of He^+ beam irradiated samples (Fig. 8.28). The reason is that when ILs are irradiated under He^+ beams, a large number of cationic radicals are generated in a small volume. The recombination of cationic radicals reduces the possibility of damage to other IL molecules, avoiding further radiolysis of ILs. Similarly, He^+ irradiated $[C_4mim][NTf_2]$ was analyzed by the D_2O-washing method, and the water-soluble radiolytic products of irradiated samples were studied using various spectroscopic methods. Compared with the results of $[C_4mim][NTf_2]$ under γ-ray, it was found that the radiolysis of $[C_4mim][NTf_2]$ under He^+ beam was similar to that under γ-irradiation. The main water-soluble radiolytic products were assigned to CF_3SOOH, $CF_3SO_2NH_2$, and HF.[69] Compared with the amount of radiolytic

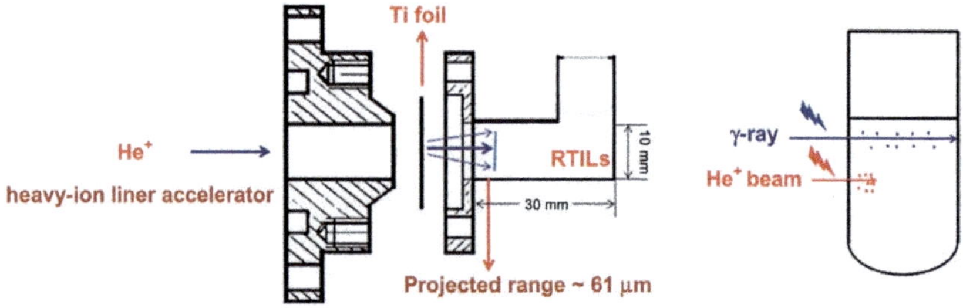

Figure 8.27 The equipment for simulation of α-radiation by He^+ beam.[69]

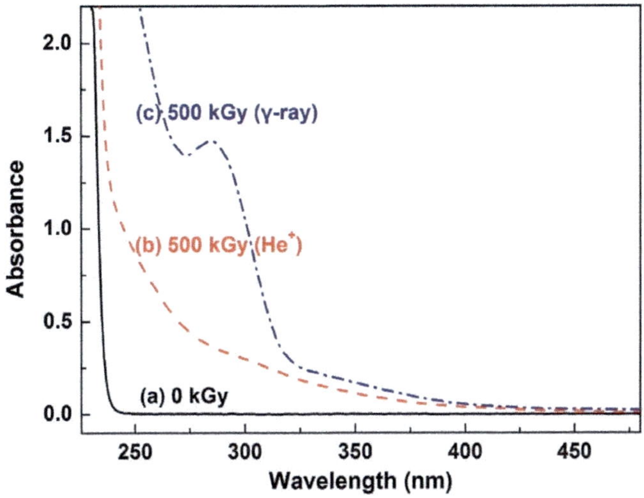

Figure 8.28 Influence of different radiation ray on radiolysis of ionic liquids.[69]

products of [C_4mim][NTf_2] under γ-irradiation, under the He^+ beam was less, which is attributed to the recombination of [C_4mim][NTf_2] cationic radicals during He^+ beam radiation.

8.4.2 Radiation effect on imidazolium ionic liquid extraction systems

The removal of ^{90}Sr from nitric acid slurries prior to further waste disposal can greatly reduce the cost of subsequent treatment and disposal of spent nuclear fuel. Crown ethers with high efficiency and selectivity in Sr^{2+} binding have been considered for the extraction of isotopes from spent nuclear fuel. In nuclear applications, the extraction system is exposed to the strong ionizing radiation conditions generated by radionuclides. The results reported in the literature show that the radiation chemical yields of 0.44 M 18-crown-6 (18C6) in benzene are $-0.13-0.11\ \mu mol\ J^{-1}$ within an absorbed dose range of 100–500 kGy.[70] For 0.01 M 18C6 and 0.01 M dicyclohexano-18-crown-6 (DCH18C6) in a water solution system, which were exposed to an absorbed dose of 250 kGy, the radiation chemical yields of crown ethers in water solution were 0.06 and $0.09\ \mu mol\ J^{-1}$, respectively.[71] In order to evaluate the possibility of ILs as solvents for the separation of radioactives, the radiation effect on the Sr^{2+} partitioning of crown ether/ILs system was studied by Zhai et al.[64,65] When IL was irradiated, the extraction ability of DCH18C6 in combination with irradiated IL decreased with the increase of absorbed dose. It was proposed that the radiolysis of DCH18C6 happened during the irradiation of DCH18C6/ILs. Therefore, the structure of crow ethers affects the radiation stability of the extraction system. Three kinds of extraction systems based on $[C_4mim][NTf_2]$ and crown ethers (DCH18C6, 4′,4″(5″)-di-*tert*-butyldicyclohexano-18-crown-6 (DtBuCH18C6), and benzo-18-crown-6 (B18C6)) were designed for the assessment of radiation effects. The concentration of crown ethers in $[C_4mim][NTf_2]$ after irradiation was quantified by ultra-performance liquid chromatography and quadrupole time-of-flight mass spectrometry (Fig. 8.29). The yields for radiolytic destruction of DCH18C6 and DtBuCH18C6 in $[C_4mim][NTf_2]$were calculated as $\sim -0.06\ \mu mol\ J^{-1}$ and that of B18C6 in $[C_4mim][NTf_2]$ was $\sim -0.09\ \mu mol\ J^{-1}$. The radiation stability of the crown ether/ $[C_4mim][NTf_2]$ system is listed as follows: DCH18C6/$[C_4mim]$ $[NTf_2]$ > DtBuCH18C6/$[C_4mim][NTf_2]$ > B18C6/$[C_4mim]$

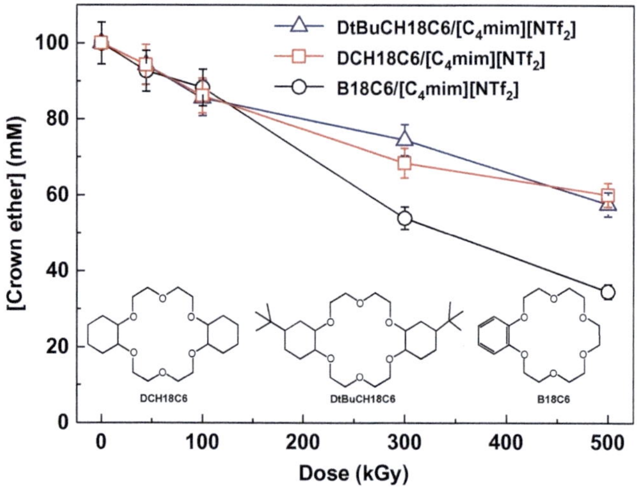

Figure 8.29 Concentration of crown ethers in extraction systems after irradiation.[72]

[NTf_2].[72] B18C6, containing a single benzene structure, resulted in the crown ether ring of B18C6 being easily attacked by free radicals during irradiation.

When extracting agents are present in ILs, active radical species are most likely to react with the extractant to form a new substance. By using high-resolution ESI-MS, the radiolytic products of the crown ether/[C_4mim][NTf_2] system can be identified.[72] Methyl, trifluoromethyl, butyl, and [C_4mim]$^+$ radicals are generated from [C_4mim][NTf_2] under γ-irradiation.[56,62] The radiolytic products of crown ether/RTILs systems are mainly substituted crown ethers formed between crown ethers and active radicals such as methyl, trifluoromethyl, butyl, and [C_4mim]$^+$ radicals, which are formed by the irradiation of crown ether/[C_4mim][NTf_2] (Fig. 8.30). For the purpose of ascertaining the chemical structure of the alkyl substituted radiolytic products of crown ether/[C_4mim][NTf_2] systems, the energy difference between different substituted radiolytic products was calculated using density functional theory. It was proposed that the alkyl substituted products of crown ethers were mainly on the cavities of crown ethers. Similarly, the influence

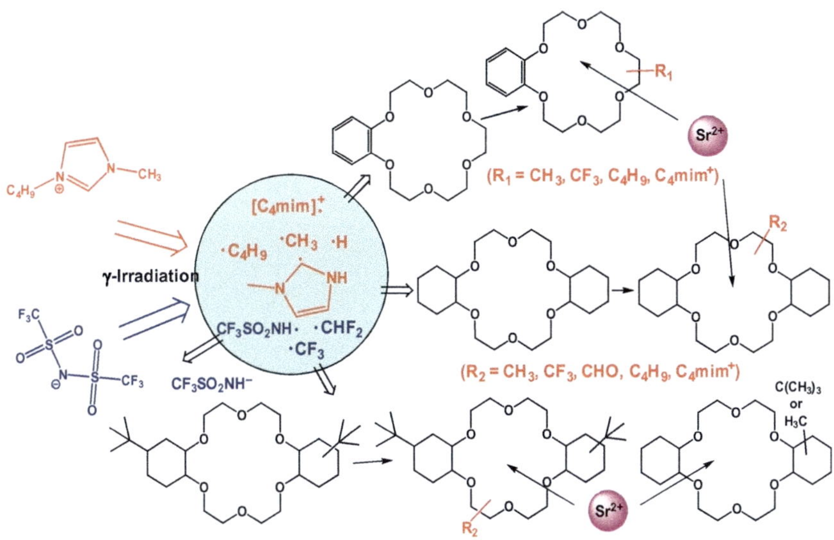

Figure 8.30 Proposed simplified radiolysis pathways of crown ether/[C_4mim][NTf_2].[72]

of the radiation effect on a 2,6-di(5,6-diisoButyl-1,2,4-triazin-3-yl) pyridine (isobutyl-BTP)/1-ethyl-3-methylimidazolium bis(trifluoromethylsulfonyl)imide ($[C_2mim][NTf_2]$) extraction system was assessed. The radiolytic products of isobutyl-BTP/$[C_2mim][NTf_2]$ system were confirmed as the substituted compounds formed between isobutyl-BTP and radicals of $[C_2mim][NTf_2]$, especially imidazolium cation radicals.[73] Related research shows that the radiolytic products of extraction systems are assigned dominantly to substituted extracting agents formed between extracting agents and active radicals of ILs.[73]

8.4.3 Influence of radiation on the extraction ability of ionic liquid extraction systems

The extractability of Sr^{2+} using crown ether in combination with irradiated ILs ($[C_4mim][PF_6]$ and $[C_4mim][NTf_2]$) was investigated.[74] The extraction efficiencies for Sr^{2+} of extraction system decreased obviously with increased absorbed dose. Ao et al.[74] studied the influence of hydrogen ions (nitric acid) on the extraction of Sr^{2+} from aqueous solutions by un-irradiated $[C_4mim][NTf_2]$ in combination with DCH18C6. A significant decline of distribution ratios (D_{Sr}) was observed with an increased concentration of hydrogen ions. The competition between H^+ and Sr^{2+} to interact with crown ether was the dominant reason for the inhibition of Sr^{2+} extraction (Fig. 8.31). The coordination ability of DCH18C6 decreases when it interacts with H_3O^+, and the cation exchange

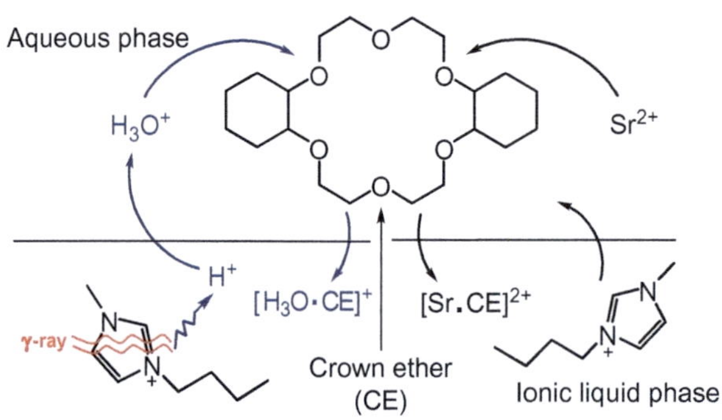

Figure 8.31 Influence of γ-radiation on DCH18C6/$[C_4mim][PF_6]$ system.[74]

mechanism is inhibited by hydrogen ions. Due to the understanding of the radiolysis of ILs, it is known that some acidic products can be produced during t irradiation, and this affects the properties of extraction system. However, since most of these acidic products are water soluble, water washing can be used to remove them. Therefore, the irradiated [C_4mim][NTf_2] was washed five times with deionized water before Sr^{2+} extraction to ensure the acidic radiolytic products were totally removed, and a neutral upper aqueous phase was obtained. It was found that the Sr^{2+} partitioning in irradiated [C_4mim][NTf_2] can be largely recovered by removing acidic radiolytic products through water-washing treatment. Subsequently, an extended X-ray absorption fine structure (EXAFS) spectrum was employed to study the influence of irradiated RTILs on DCH18C6-Sr coordination structure. The Sr K-edge EXAFS of the Sr−DCH18C6 complexes in un-irradiated and 550 kGy irradiated [C_4mim][NTf_2] suggest no obvious change in the coordination structure of Sr−DCH18C6 complexes. The results clearly indicate that the irradiation of [C_4mim][NTf_2] does not change the interaction between Sr and crown ether. In order to assess the influence of radiation on the extractability of crown ether/[C_4mim][NTf_2] systems, the extractability of Sr^{2+} using irradiated crown ether/[C_4mim][NTf_2] solution was investigated further.[74] The Sr^{2+} partitioning of DCH18C6/[C_4mim][NTf_2] system decreased with the increase of absorbed dose. The extraction ability of DCH18C6/[C_4mim][NTf_2], DtBuCH18C6/[C_4mim][NTf_2] and B18C6/[C_4mim][NTf_2] system for Sr^{2+} decreased by 14.4%, 18.2%, and 30.2% at 500 kGy, respectively. DCH18C6/[C_4mim][NTf_2] and DtBuCH18C6/[C_4mim][NTf_2] still showed higher radiation stability and good extractability for Sr^{2+} after irradiation than that of the B18C6/[C_4mim][NTf_2] system. This result is consistent with the analysis of the change of the content of crown ethers in [C_4mim][NTf_2] after irradiation.[72] Although the crown ether structure was changed and their extraction performance was partially reduced, DCH18C6/[C_4mim][NTf_2] and DtBuCH18C6/[C_4mim][NTf_2] systems still retained higher extraction ability even at 500 kGy. These results show that

DCH18C6/[C_4mim][NTf_2] and DtBuCH18C6/[C_4mim][NTf_2] systems have high radiation stability and can be used potentially in the reprocessing of spent nuclear fuel.

N-donor heteropolycyclic ligands have been widely investigated and proved to be effective in MA–Ln separation. 2,6-bis(1,2,4-triazin-3-yl)-pyridines (BTPs) was considered as one of representative compounds of N-donor heteropolycyclic ligands. These ligands still showed good extraction performance ($D_{Am} > 100$, $S_{Am/Eu} > 100$) in strong acid conditions (1 mol L^{-1} HNO_3).[75] Accordingly, the Dy^{3+} extraction ability of BTPs/[C_4mim][NTf_2] was studied by Ao et al.[69] They found that CA-BTP was an extremely sensitive extractant for acidity, due to the protonation of extractant and the inhibition of the cation exchange mechanism by hydrogen ions. In irradiated [C_4mim][NTf_2], Dy^{3+} partitioning evidently reduced at 50 kGy, but an abnormal increase of Dy^{3+} partitioning was observed at 100 kGy,[74] which was mainly attributed to the precipitation formed between Dy^{3+} and radiolytic products (F^- and SO_3^{2-}) (Fig. 8.32). Compared to γ-irradiated sample, the influence of extraction of Dy^{3+} by He^+ beam irradiation was much less than that of γ-ray irradiation.

Similarly, the precipitation between Eu^{3+} and irradiated 2,9-bis(5,5,8,8-tetramethyl-5,6,7,8-tetrahydro-benzo[1,2,4]triazin-3-yl)-1,10-phenanthroline (BTPhen)/1-alkyl-3-methylimidazolium bis(trifluoromethylsulfonyl)imide ([$C_n$mim][NTf_2]) extraction

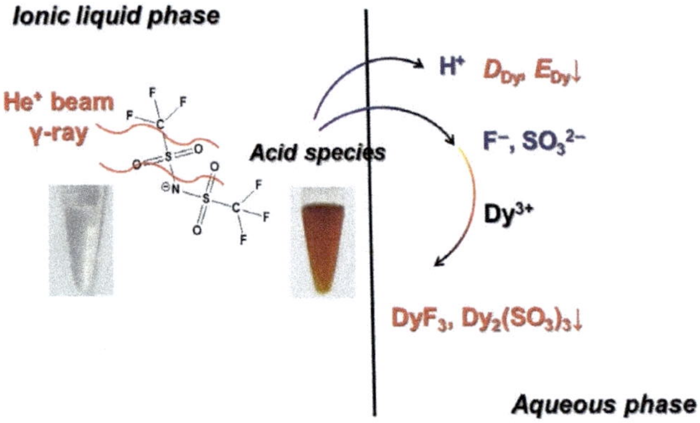

Figure 8.32 Influence of irradiated [C_4mim][NTf_2] on Dy^{3+}extraction.[74]

system was identified as EuF_3 and $Eu_2(SO_3)_3$. Although precipitation between Eu^{3+} and radiolytic products of $[C_2mim][NTf_2]$ leads to an increment in the removal of Eu^{3+} for irradiated BTPhen/$[C_2mim][NTf_2]$, the influence of radiolytic products of $[C_2mim][NTf_2]$ can be easily avoided by enhancing the acidity during extraction or washing the irradiated samples with water before extraction. Consequently, γ-radiation has no obvious effect on the extraction performance of BTPhen/$[C_2mim][NTf_2]$ system in the presence of 1 M nitric acid. Therefore, the BTPhen/$[C_2mim][NTf_2]$ system shows excellent extraction capacity and high radiation stability.[73]

8.4.4 Radiation synthesis of poly(ionic liquids) gels for metal ion removal

Poly(ionic liquids) (PILs) are a special kind of polyelecrolyte, which contain structural units of ILs. Although most of ILs exists in the liquid state at room temperature, PILs are usually in solid state. To date, there has not been a clear and universally acknowledged definition for PILs, and there is not a sharp boundary between PILs and common polyelectrolytes. Usually, the polymers from ILs monomers can be regarded as PILs (Fig. 8.33). Conventional polyelectrolytes are water-soluble and produce positive and negative charges when ionized in water. The anions of typical PILs are

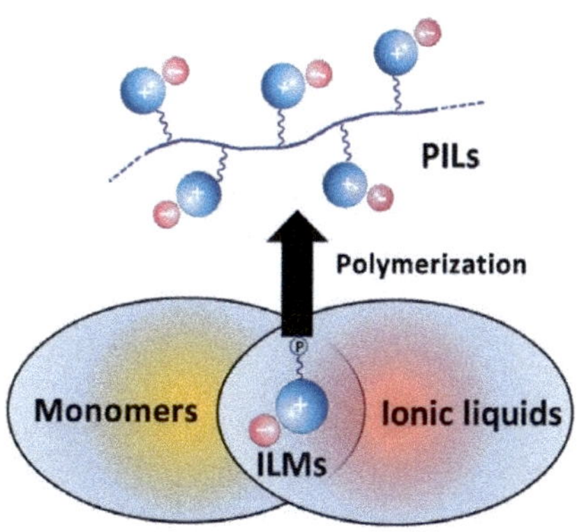

Figure 8.33 Relationship between ILs and PILs (polycation type).[76] *IL*, ionic liquid; *PIL*, poly(ionic liquid).

hydrophobic anions, including $[PF_6]^-$, $[NTf_2]^-$, and $[N(C_2F_5SO_2)_2]^-$. The cations of typical PILs are imidazolium, pyridinium, pyrrolidinium, ammonium, and phosphonium. PILs can be classified into three kinds depending on the charge on their main chains, namely polycations, polyanions, and polyzwitterions. Polycations contain cationic structural units, and polyanions contain anionic structural units. Polyzwitterions contain both cationic and anionic structural units. Moreover, random, alternative and block copolymers containing structural units of ILs also belong to PILs.

Along with the development of ILs, PILs are attracting more and more attention. Recently, many PILs have been reported for application in various fields, and these PILs are mainly synthesized by radical polymerization and cross-linking induced by chemical initiators or ultraviolet radiation. The resultant PILs gels can be applied as catalysts, and 1-vinyl-3-methylimidazolium tetrachloroferrate ($[VBIm][FeCl_4]$) microgel can catalyze the Friedel–Crafts reaction. The $[VBIm][FeCl_4]$ microgel catalyst can be easily separated from the reaction system by centrifugation or external magnetic field. It can bear ten catalytic cycles with the yield keeping above 99%.[77] PILs gels can also be used to absorb organic solvents. The gel from dodecyl methacrylate and vinylimidazolium ILs monomer has better absorption behavior of organic solvents than that of bulk poly(dodecyl methacrylate) (PDDMA) gel, and may be used for the treatment of oil spill oils.[78]

Wang et al.[79] found that γ-ray can initiate the polymerization and cross-linking of 1-allyl-3-vinylimidazolium chloride (AVImCl) ILs (Fig. 8.34). Compared with other methods to prepare PILs,[77] the radiation method is very effective and can be applied at room

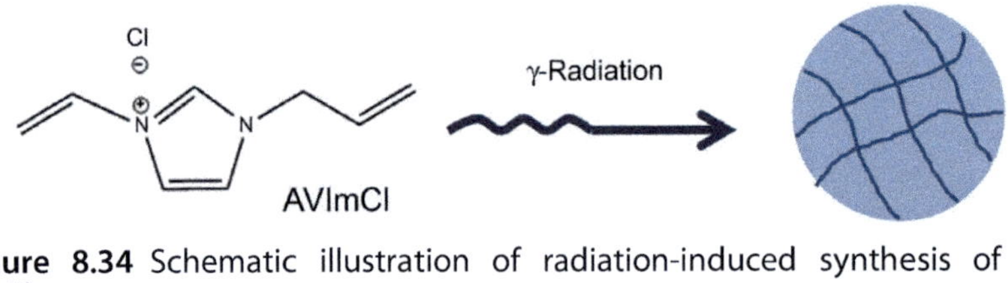

Figure 8.34 Schematic illustration of radiation-induced synthesis of PIL gel.[79] *PIL*, poly(ionic liquid).

temperature, and there is no need for a chemical initiator. The radiation-synthesized PILs gel shows excellent adsorption ability for Cr(VI), being able to remove 90% of the Cr(VI) in a solution in 90 min, and the adsorption equilibrium is achieved at 240 min. It can remove almost 100% of the Cr(VI) in a solution at pH 2–6, and the adsorption capacity is 283 mg g^{-1}. Liquid waste containing Cr(VI) often has many coexisting ions, for example, Na^+, Ca^{2+}, Mg^{2+}, Cu^{2+}, Ni^{2+}, Zn^{2+}, Cl^-, NO_3^-, which may affect the adsorption of Cr(VI). At high concentrations of other ions, the PILs gel can still adsorb Cr(VI) efficiently, with a removal of 88% Cr(VI). Ion-exchange adsorbents for Cr(VI) can be renewed by NaCl or NaOH solution. Then, the PILs gel was desorbed by NaCl or NaOH solution, and the desorption efficiencies were ~59% and ~76%, respectively. Furthermore, NaCl and NaOH mixed solution may provide higher desorption efficiency. Therefore, PILs gel has good reusability.[79]

Han et al.[80] developed the radiation synthesis of PILs gel and applied the resulting gel for Re(VII) adsorption (Fig. 8.35). Instead of bulk ILs monomer, an aqueous solution containing 1-vinyl-3-ethylimidazolium bromide ILs monomer and cross-linker (3,3′-divinyl-1,1′(1,6-hexanediyl)) was used as the precursor for γ-ray

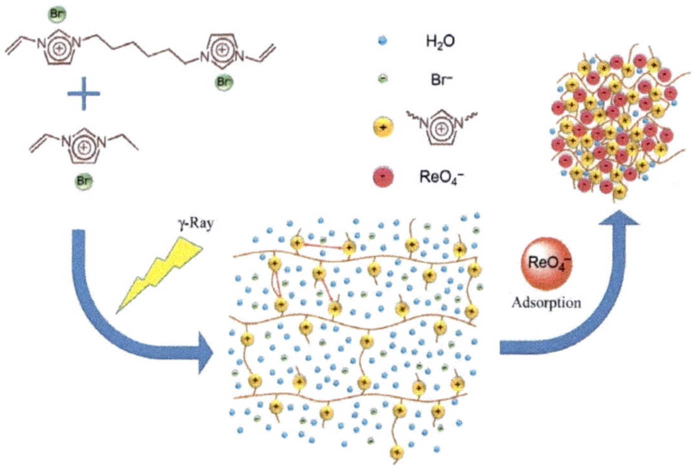

Figure 8.35 Schematic illustration of PIL gel for Re(VII) adsorption.[80] *PIL, poly(ionic liquid).*

radiation-induced polymerization and cross-linking. A gel fraction of ~95% was achieved with an absorbed dose of 40 kGy. Compared with radiation-induced polymerization and cross-linking of AVImCl, this system has higher gel fraction under a lower absorbed dose, indicating that the radiation method is more effective for this system. The resultant PILs gel can adsorb perrhenate (ReO_4^-) with a high adsorption capacity of $\sim 8.6 \times 10^2$ mg g^{-1} (Re/gel). The adsorption mechanism is ion-exchange, confirmed by IR and XPS.

REFERENCES

1. Chen L, Xua Z, Li J, Min C, Liu L, Song X, et al. Reduction and disorder in graphene oxide induced by electron-beam irradiation. *Mater Lett* 2011;**65**(8):1229–30.
2. Zhang BW, Li L, Wang Z, Xie S, Zhang Y, Shen Y, et al. Radiation induced reduction: an effective and clean route to synthesize functionalized graphene. *J Mater Chem* 2012;**22**(16):7775–81.
3. Zhang YW, Ma HL, Zhang Q, Peng J, Li J, Zhai M, et al. Facile synthesis of well-dispersed graphene by gamma-ray induced reduction of graphene oxide. *J Mater Chem* 2012;**22**(26):13064–9.
4. Cai X, Zhang Q, Wang S, Peng J, Zhang Y, Ma H, et al. Surfactant-assisted synthesis of reduced graphene oxide/polyaniline composites by gamma irradiation for supercapacitors. *J Mater Sci* 2014;**49**(16):5667–75.
5. Zhang Q, Zhang Y, Gao Z, Ma HL, Wang S, Peng J, et al. A facile synthesis of platinum nanoparticle decorated graphene by one-step γ-ray induced reduction for high rate supercapacitors. *J Mater Chem C* 2013;**1**(2):321–8.
6. Wang S, Zhang Y, Ma HL, Zhang Q, Xu W, Peng J, et al. Ionic-liquid-assisted facile synthesis of silver nanoparticle-reduced graphene oxide hybrids by gamma irradiation. *Carbon* 2013;**55**(0):245–52.
7. Pham VH, Cuong TV, Dang TT, Hur SH, Kong BS, Kim EJ, et al. Superior conductive polystyrene—chemically converted graphene nanocomposite. *J Mater Chem* 2011;**21**(30):11312–16.
8. Yoonessi M, Gaier JR. Highly conductive multifunctional graphene polycarbonate nanocomposites. *ACS Nano* 2010;**4**(12):7211–20.
9. Xie SY, Zhang B, Wang C, Wang Z, Li L, Li J. Building up graphene-based conductive polymer composite thin films using reduced graphene oxide prepared by gamma-ray irradiation. *Sci World J* 2013;**2013**(11). 954324(1-7).
10. Zhang BW, Zhang Y, Peng C, Yu M, Li L, Deng B, et al. Preparation of polymer decorated graphene oxide by gamma-ray induced graft polymerization. *Nanoscale* 2012;**4**(5):1742–8.
11. Chen L, Xu Z, Li J, Li Y, Shan M, Wang C, et al. A facile strategy to prepare functionalized graphene via intercalation, grafting and self-exfoliation of graphite oxide. *J Mater Chem* 2012;**22**(27):13460–3.

12. Chen L, Xu Z, Li J, Zhou B, Shan M, Li Y, et al. Modifying graphite oxide nanostructures in various media by high-energy irradiation. *RSC Adv* 2014;**4**(2):1025–31.
13. Liu J, Chen C, He C, Zhao J, Yang X, Wang H. Synthesis of graphene peroxide and its application in fabricating super extensible and highly resilient nanocomposite hydrogels. *ACS Nano* 2012;**6**(9):8194–202.
14. Zhao J, Deng B, Lv M, Li J, Zhang Y, Jiang H, et al. Graphene oxide-based antibacterial cotton fabrics. *Adv Healthcare Mater* 2013;**2**(9):1259–66.
15. Xie L, Duan G, Wang W, Wang M, Wu Q, Zhou X, et al. Effect of γ-ray-radiation-modified graphene oxide on the integrated mechanical properties of PET blends. *Ind Eng Chem Res* 2016;**55**(29):8123–32.
16 Butler J, Henglein A. Elementary reactions of the reduction of Tl^{+} in aqueous solution. *Radiat Phys Chem* 1980;**15**:603–12.
17. Henglein A. Small-particle research: physicochemical properties of extremely small colloidal metal and semiconductor particles. *Chem Rev* 1989;**89**(8):1861–73.
18. Mosseri S, Henglein A, Janata E. Reduction of dicyanoaurate (I) in aqueous solution: formation of nonmetallic clusters and colloidal gold. *J Phys Chem* 1989;**93**(18):6791–5.
19. Henglein A, Ershov BG, Malow M. Absorption-spectrum and some chemical-reactions of colloidal platinumin aqueous-solution. *J Phys Chem* 1995;**99**(38):14129–36.
20. Delcourt MO, Keghouche N, Belloni J. Radiation reduced platinum sols: characterization and catalytic efficiency. *Nouv J Chim* 1983;**7**(2):131–6.
21. Henglein A. Remarks on the electrochemical potential of small silver clusters in aqueous solution. *Ber Bunsen Ges Phys Chem* 2010;**94**(5):600–3.
22. Long DW, Wu GZ, Chen SM. Preparation of oligochitosan stabilized silver nanoparticles by gamma irradiation. *Radiat Phys Chem* 2007;**76**(7):1126–31.
23. Liu YS, Chen SM, Zhong L, Wu G. Preparation of high-stable silver nanoparticle dispersion by using sodium alginate as a stabilizer under gamma radiation. *Radiat Phys Chem* 2009;**78**(4):251–5.
24. Ershov BG, Henglein A. Optical spectrum and some chemical properties of colloidal thallium in aqueous solution. *J Phys Chem* 1993;**97**(13):3434–6.
25. Mills G, Henglein A. Radiation chemical formation of colloidal iridium and mechanism of catalysed hydrogen formation by radicals. *Radiat Phys Chem* 1985;**26**(4):385–90.
26. Henglein A, Gutierrez M, Janata E, Ershov BG. Absorption spectrum and chemical reactions of colloidal cadmium in aqueous solution. *J Phys Chem* 1992;**96**(11):4598–602.
27. Gingerich KA, Cocke DL, Miller F. Thermodynamic investigation of the lead molecules Pb2, Pb3, and Pb4 by mass spectrometry. *J Chem Phys* 1976;**64**(10):4027–33.
28. Henglein A, Janata E, Fojtik A. Reduction of Pb^{2+} in aqueous solution: early steps and colloid formation, and the atom→metal transition. *J Phys Chem* 1992;**96**(12):4734–6.
29. Khatouri J, Mostafavi M, Amblard J, Belloni J. Radiation induced copper aggregates and oligomers. *Chem Phys Lett* 1992;**191**(3–4):351–6.

30. Chen SM, Liu YD, Wu GZ. Stabilized and size-tunable gold nanoparticles formed in a quaternary ammonium-based room-temperature ionic liquid under γ-irradiation. *Nanotechnology* 2005;**16**(10):2360–4.
31. Wang HD, Sun XQ, Ye Y, Qiu S. Radiation induced synthesis of Pt nanoparticles supported on carbon nanotubes. *J Power Sources* 2006;**161**(2):839–42.
32. Zhang XY, Ye Y, Wang HD, Yao SD. Deposition of platinum–ruthenium nano-particles on multi-walled carbon nano-tubes studied by gamma-irradiation. *Radiat Phys Chem* 2010;**79**(10):1058–62.
33. Hayes D, Swayambunathan V, Micic OI, Nenadovic MT, Dan M. Radiolytic production and properties of ultrasmall CdS particles. *J Phys Chem* 1989;**93**(11):4603–8.
34. Yin YD. *Radiation synthesis of nanomaterials [D]*. Anhui: University of Science and Technology of China; 1998.
35. Qiao ZP. *γ-ray radiation preparation of nanocrystalline sulfide compounds and their polymer composites [D]*. Anhui: University of Science and Technology of China; 1998.
36. Souzy R, Ameduri B. Functional fluoropolymers for fuel cell membranes. *Prog Polym Sci* 2005;**30**(6):644–87.
37. Van der Bruggen B. Chemical modification of polyethersulfone nanofiltration membranes: a review. *J Appl Polym Sci* 2009;**114**(1):630–42.
38. Yuan Q. *The mechanism on the formation of multihollow polymer spheres through sulfonated polystyrene particles [D]*. Anhui: University of Science and Technology of China; 2009. in Chinese.
39. Stubbs JM, Durant YG, Sundberg DC. Polymer phase separation in composite latex particles. 1. Considerations for the nucleation and growth mechanism. *CR Chim* 2003;**6**(11):1217–32.
40. Guyut A, Landfester K, Schork FJ, Wang C. Hybrid polymer latexes. *Prog Polym Sci* 2007;**32**(12):1439–61.
41. Sheu HR, El-Aasser MS, Vanderhoff JW. Phase-separation in polystyrene latex interpenetrating polymer networks. *J Polym Sci Polym Chem* 1990;**28**(3):629–51.
42. Sheu HR, El-Aasser MS, Vanderhoff JW. Uniform nonspherical latex-particles as model interpenetrating polymer networks. *J Polym Sci Polym Chem* 1990;**28**(3):653–67.
43. Ge XP, Wang MZ, Wang H, Yuan Q, Ge XW, Liu HR, et al. Novel walnut-like multihollow polymer particles: synthesis and morphology control. *Langmuir* 2010;**26**(3):1635–41.
44. Li BX, Xu YF, Wang MZ, Ge XW. Morphological control of multihollow polymer latex particles through a controlled phase separation in the seeded emulsion polymerization. *Langmuir* 2013;**29**(48):14787–94.
45. Li BX, Wang MZ, Ge XW. Synthesis of plum-like and walnut-like polymer microspheres by gamma-ray irradiation induced seed-swelling polymerization. *J Radiat Res Radiat Process* 2014;**32**(1):43–9 (in Chinese).
46. Kim JW, Larsen RJ, Weitz DA. Synthesis of nonspherical colloidal particles with anisotropic properties. *J Am Chem Soc* 2006;**128**(44):14374–7.

47. Morton M, Kaizerman S, Altier MW. Swelling of latex particles. *J Colloid Sci* 1954;**9**(4):300–12.
48. Lissant KJ, editor. *Emulsions and emulsion technology.* New York: Marcel Dekker; 1974.
49. Mao DL, Li TT, Liu HR, Li ZC, Shao H, Li M. Preparation of macroporous polyHIPE foams via radiation-induced polymerization at room temperature. *Colloid Polym Sci* 2013;**291**(7):1649–56.
50. Wang SF, Zhang ZC, Liu HR, Zhang W, Qian Z, Wang BB. One-step synthesis of manganese dioxide/polystyrene nanocomposite foams via high internal phase emulsion and study of their catalytic activity. *Colloid Polym Sci* 2010;**288**(9):1031–9.
51. Li TT, Liu HR, Zeng L, Yang S, Li ZC, Zhang JD, et al. Macroporous magnetic poly(styrene–divinylbenzene) nanocomposites prepared via magnetite nanoparticles-stabilized high internal phase emulsions. *J Mater Chem* 2011;**21**(34):12865–72.
52. Yang S, Zeng L, Wang YP, Sun XH, Sun PJ, Liu HM, et al. Facile approach to glycidyl methacrylate-based polyHIPE monoliths with high epoxy-group content. *Colloid Polym Sci* 2014;**292**(10):2563–70.
53. Li ZC, Liu HR, Zeng L, Liu HW, Yang S, Wang YM. Preparation of high internal water-phase double emulsions stabilized by a single anionic surfactant for fabricating interconnecting porous polymer microspheres. *Langmuir* 2014;**30**(41):12154–63.
54. Kim M, Kiyohara S, Konishi S, Tsuneda S, Saito K, Sugo T. Ring-opening reaction of poly-GMA chain grafted onto a porous membrane. *J Membr Sci* 1996;**117**(1-2):33–8.
55. Hagiwara R, Ito Y. Room temperature ionic liquids of alkylimidazolium cations and fluoroanions. *J Fluorine Chem* 2000;**105**(2):221–7.
56. Berthon L, Nikitenko SI, Bisel I, Berthon C, Faucon M, Saucerotte B, et al. Influence of gamma irradiation on hydrophobic room-temperature ionic liquids [BuMeIm]PF_6 and [BuMeIm](CF_3SO_2)(2)N. *Dalton Trans* 2006;**250** (21):2526–34.
57. Tao GH, Zou M, Wang XH, Chen ZY, Evans DG, Kou Y. Comparison of polarities of room-temperature ionic liquids using FT-IR spectroscopic probes. *Aust J Chem* 2005;**58**(5):327–31.
58. Yuan LY, Peng J, Xu L, Zhai ML, Li JQ, Wei GS. Influence of gamma-radiation on the ionic liquid C(4)mim PF(6) during extraction of strontium ions. *Dalton Trans* 2008;**252**(45):6358–60.
59. Wang SJ, Liu JZ, Yuan LY, Cui ZP, Peng J, Li JQ, et al. Towards understanding the color change of 1-butyl-3-methylimidazolium bis(trifluoromethylsulfonyl)imide during gamma irradiation: an experimental and theoretical study. *Phys Chem Chem Phys* 2014;**16**(35):18729–35.
60. Yuan LY, Peng J, Xu L, Zhai ML, Li JQ, Wei GS. Radiation-induced darkening of ionic liquid [C(4)mim][NTf2] and its decoloration. *Radiat Phys Chem* 2009;**78**(12):1133–6.
61. Allen D, Baston G, Bradley AE, Gorman T, Haile A, Hamblett I, et al. An investigation of the radiochemical stability of ionic liquids. *Green Chem* 2002;**4**(2):152–8.

62. Bosse E, Berthon L, Zorz N, Monget J, Berthon C, Bisel I, et al. Stability of [MeBu3N][Tf2N] under gamma irradiation. *Dalton Trans* 2008;**8** (7):924–31.
63. Wang SJ, Ao YY, Zhou HY, Yuan LY, Peng J, Zhai ML. Research progress in radiation effects on ionic liquids. *Acta Phys Chim Sin* 2014;**30** (9):1597–604.
64. Yuan LY, Peng J, Xu L, Zhai ML, Li JQ, Wei GS. Radiation effects on hydrophobic ionic liquid C(4)mim NTf(2) during extraction of strontium ions. *J Phys Chem B* 2009;**113**(26):8948–52.
65. Yuan LY, Xu C, Peng J, Xu L, Zhai ML, Li JQ, et al. Identification of the radiolytic product of hydrophobic ionic liquid [C4mim][NTf2] during removal of Sr^{2+} from aqueous solution. *Dalton Trans* 2009;**38**(38):7873–5.
66. Ao YY, Chen J, Wang Y, Chen HB, Li JQ, Zhai ML. Radiation effect of carboxyl-functionalized task-specific ionic liquids on UO_2^{2+} removal: experimental study with DFT validation. *J Phys Chem B* 2017;**121** (8):1893–9.
67. Ao YY, Xu M, Peng J, Li JQ, Zhai ML, Wu GZ. Identification of radiolytic products of [C4mim][PF6] under γ-irradiation. *Nucl Sci Tech* 2015;**26** (3):47–50.
68. Huang W, Chen SM, Liu YS, Fu HY, Wu GZ. Fluoride ion yield and absorption spectral analysis of irradiated imidazolium-based room-temperature ionic liquids. *Radiat Phys Chem* 2011;**80**(4):573–7.
69. Ao YY, Zhou HY, Yuan WJ, Wang SJ, Peng J, Zhai ML, et al. α-Radiolysis of ionic liquid irradiated with helium ion beam and the influence of radiolytic products on Dy3 + extraction. *Dalton Trans* 2014;**43**(14):5580–5.
70. Mincher BJ, Modolo G, Mezyk SP. Review article: the effects of radiation chemistry on solvent extraction: 2. A review of fission-product extraction. *Solvent Extr Ion Exch* 2009;**27**(3):331–53.
71. Seliverstov A, Ershov B. Radiation-chemical transformations of dicyclohexyl-18-crown-6 in aqueous solutions. *High Energy Chem (Engl Transl); (United States)* 1988;**22**(3):173–6.
72. Ao YY, Yuan WJ, Yu TL, Peng J, Li JQ, Zhai ML, et al. Radiolysis of crown ether-ionic liquid systems: identification of radiolytic products and their effect on the removal of Sr^{2+} from nitric acid. *Phys Chem Chem Phys* 2015;**17**(5):3457–62.
73. Yuan WJ, Ao YY, Zhao L, Zhai ML, Peng J, Li JQ, et al. Influence of radiation effect on extractability of an isobutyl-BTP/ionic liquid system: quantitative analysis and identification of radiolytic products. *RSC Adv* 2014;**4** (93):51330–3.
74. Ao YY, Peng J, Yuan LY, Cui ZP, Li C, Li JQ, et al. Identification of radiolytic products of [C4mim][NTf2] and their effects on the Sr^{2+} extraction. *Dalton Trans* 2013;**42**(12):4299–305.
75. Drew MGB, Foreman MRSJ, Hill C, Hudson MJ, Madic C. 6,6′-bis-(5,6-diethyl-[1,2,4]triazin-3-yl)-2,2′-bipyridyl the first example of a new class of quadridentate heterocyclic extraction reagents for the separation of americium(III) and europium(III). *Inorg Chem Commun* 2005;**8**(3):239–41.

76. Yuan J, Mecerreyes D, Antonietti M. Poly(ionic liquid)s: an update. *Prog Polym Sci* 2013;**38**(7):1009–36.
77. Dobbelin M, Jovanovski V, Llarena I, Claros Marfil LJ, Cabanero G, Rodriguez J, et al. Synthesis of paramagnetic polymers using ionic liquid chemistry. *Polym Chem* 2011;**2**(6):1275–8.
78. Pourjavadi A, Doulabi M, Hosseini SH. Novel polyelectrolyte gels as absorbent polymers for nonpolar organic solvents based on polymerizable ionic liquids. *Polymer* 2012;**53**(25):5737–42.
79. Wang SJ, Ma HL, Peng J, Zhang YW, Chen J, Wang LL, et al. Facile synthesis of a novel polymeric ionic liquid gel and its excellent performance for hexavalent chromium removal. *Dalton Trans* 2015;**44**(16):7618–25.
80. Han D, Li XX, Peng J, Xu L, Li JQ, Li HB, et al. A new imidazolium-based polymeric ionic liquid gel with high adsorption capacity for perrhenate. *RSC Adv* 2016;**6**(73):69052–9.

INDEX

Note: Page numbers followed by "*f*" and "*t*" refer to figures and tables, respectively.

D

N

O

P

Q

R

T

V

W

X

Y

CPI Antony Rowe
Chippenham, UK
2019-01-28 22:40